Desalination Engineering

Desalination Engineering

Editor: Taylor Watson

MURPHY & MOORE
www.murphy-moorepublishing.com

Published by Murphy & Moore Publishing
1 Rockefeller Plaza,
New York City, NY 10020, USA
www.murphy-moorepublishing.com

Desalination Engineering
Edited by Taylor Watson

International Standard Book Number: 978-1-63987-155-1 (Hardback)

Cataloging-in-Publication Data

Desalination engineering / edited by Taylor Watson.
p. cm.
Includes bibliographical references and index.
ISBN 978-1-63987-155-1
1. Saline water conversion. 2. Saline water conversion plants. 3. Water-supply engineering. I. Watson, Taylor.
TD479 .D47 2021
628.167--dc23

Contents

Preface

The process of removing salt and other minerals from the target substance is called desalination. The discipline of desalination engineering deals with the removal of salts and minerals from seawater to produce water suitable for consumption by humans. Solar distillation, vacuum distillation, forward osmosis, membrane distillation and multi stage flash distillation are some of the techniques used for the desalination process. Most of the commercial units use the reverse osmosis method for purification of sea water. Some of the major challenges faced in desalination engineering are high energy consumption, brine disposal, membrane performance, scaling and protection of marine life. This book attempts to understand the multiple branches that fall under the discipline of desalination engineering and how such concepts have practical applications. The various studies that are constantly contributing towards advancing technologies and evolution of this field are examined in detail. This book is appropriate for students seeking detailed information in this area as well as for experts.

This book is a comprehensive compilation of works of different researchers from varied parts of the world. It includes valuable experiences of the researchers with the sole objective of providing the readers (learners) with a proper knowledge of the concerned field. This book will be beneficial in evoking inspiration and enhancing the knowledge of the interested readers.

In the end, I would like to extend my heartiest thanks to the authors who worked with great determination on their chapters. I also appreciate the publisher's support in the course of the book. I would also like to deeply acknowledge my family who stood by me as a source of inspiration during the project.

Editor

1

Phase Equilibria and Phase Separation of the Aqueous Solution System Containing Lithium Ions

Long Li, Yafei Guo and Tianlong Deng

Abstract

Brines including seawater, concentrated seawater after desalinization, salt lake, oil/gas water, and well bitter are widely distributed around the world. In order to promote the comprehensive utilization and effective protection of the valuable chemical resources existing in brines such as freshwater, lithium, sodium, potassium, and magnesium salts, the systematic foundation and application foundation research including phase equilibria and thermodynamic properties for the salt-water electrolyte solution are essential, especially for solid lithium salts and their aqueous solution systems.

Keywords: thermodynamics, phase equilibria, aqueous solution, lithium salts

1. Introduction

1.1. Lithium resources situation

Lithium is the lightest alkali metal, which plays a growing role in numerous processes such as rechargeable batteries, thermonuclear fusion, medical drugs, lubricant greases, ceramic, glasses, dyes, adhesives, and electrode welding [1–19]. Lithium is a critical energy material and a strategic resource for the twenty-first century. Consequently, the market demands for lithium resources are increasing around the world [20, 21].

Lithium naturally occurs in compound forms because of its high reactivity. Economic concentrations of lithium are found in brines, minerals, and clays in various parts of the world. Brines and high-grade lithium ores are the present sources for all commercial lithium production. The global lithium reserve is estimated at 14.0 megatons [22], which is 74.5 megatons of lithium carbonate equivalent. Lithium reserves are mainly distributed in South America, China, and Australia [23].

Generally, lithium is obtained from two major resources: the lithium mineral ores including spodumene and petalite ores and the containing-lithium brine resources including seawater, underground water and salt lake brine [24–27]. Currently, the former is well exploited while the latter is being developed by industries with relatively low efficiency. Nonetheless, more than 60% of the total lithium amount exists in sea water and brines [24]. Therefore, a great potential exists for obtaining lithium from aqueous sources, if an efficient lithium recovery technology can be developed. Separation and extraction of lithium from either sea water or brine is carried out on a semi-industrial scale and industrial scale in the USA from salt lakes [28], in Japan from thermal water [29, 30], in Israel from the Dead Sea [31], and in China from underground brines and salt lakes [22].

In addition, the most important physical and chemical processes occurring in brines are evaporation, concentration, crystallization, precipitation, dissolution, and phase transformation. It is obvious that phase equilibria and thermodynamics can explain above phenomenon and even guide those processes effectively. Therefore, it is particularly meaningful to engage the research on phase equilibria and thermodynamics properties of lithium-containing aqueous solution systems for describing the geochemical evolution of containing lithium brines and exploiting valuable lithium resources.

2. Phase equilibria of lithium-containing salt-water systems

Brines, including seawater, concentrated seawater after desalinization, salt lake, oil/gas water, and well bitter are all complex multi-component salt-water systems, whose study and application is mainly in reference to the solubility of salts in the water and the solid-liquid equilibrium rule. Hence, solid-liquid phase equilibria form the basis for salt-water systems, which in turn are used in the chemical industry for the separation of lithium [32].

2.1. Stable phase equilibria of lithium-containing salt-water systems

Early in the 1960s, Soviet scholars had conducted research on stable phase equilibria of lithium-containing salt-water systems [33–35]. The solubilities of the systems (Li^+, Na^+, Mg^{2+}//Cl^- – H_2O), (Li^+, Na^+, Mg^{2+}//SO_4^{2-} – H_2O), (Li^+, Na^+, K^+//SO_4^{2-} – H_2O) in the temperature range from 288.15 to 373.15 K were determined. Three types of double salts containing lithium, $2Li_2SO_4{\cdot}Na_2SO_4{\cdot}K_2SO_4$, $Li_2SO_4{\cdot}3Na_2SO_4{\cdot}12H_2O$, and $Li_2SO_4{\cdot}K_2SO_4$ were found for the first time and the physicochemical properties of these were measured. All these studies could provide the solubility data to extract lithium resources in sea water and other brines.

Researchers in China also studied stable phase equilibria of complex system (Li^+, Na^+, K^+, Mg^{2+}//Cl^-, CO_3^{2-}, SO_4^{2-}, borate–H_2O) and its subsystems. Some research results are shown in **Table 1**. In addition, depending on the aquatic chemical types, the lithium-containing salt-water systems can be divided into lithium-containing chloride system, sulfate system, carbonate system, and borate system.

In order to apply the hydride salting-out effect to the separation of lithium and magnesium, phase equilibria of the quaternary system H^+, Li^+, Mg^{2+}//Cl^- – H_2O and its subsystems at

Category	Lithium-containing systems	Temperature (K)	Reference
LiCl	H^+, Li^+, Mg^{2+}//Cl^-–H_2O	273.15	[36]
	Li^+, Na^+, K^+, Mg^{2+}//Cl^-–H_2O	298.15	[37]
	Li^+, Mg^{2+}//Cl^-–H_2O	298.15	[38]
	Li^+, NH^{4+}//Cl^-–H_2O	273.15,298.15, 323.15	[39]
	Li^+, Rb^+, Mg^{2+}//Cl^-–H_2O	323.15	[40]
	Li^+, Na^+, K^+, Sr^{2+}//Cl^-–H_2O	298.15	[41]
	Li^+, Mg^{2+}//Cl^-–H_2O	288.15	[42]
Li_2SO_4	Li^+, K^+//SO_4^{2-}–H_2O Li^+, Mg^{2+}//SO_4^{2-}–H_2O	298.15	[43]
	Li^+, K^+, Mg^{2+}//SO_4^{2-}–H_2O	298.15	[44]
	Li^+, K^+, Mg^{2+}//Cl^-, SO_4^{2-}–H_2O	298.15	[45]
	Li^+, Na^+, K^+, Mg^{2+}//SO_4^{2-}–H_2O	298.15	[46]
	Li^+, K^+//SO_4^{2-}–H_2O	288.15	[47]
	Li^+//Cl^-, SO_4^{2-}–H_2O	308.15	[48]
	Li^+, Mg^{2+}//SO_4^{2-}–H_2O	288.15	[42]
Li_2CO_3	Li^+, Na^+, K^+//CO_3^{2-}–H_2O	298.15	[49]
	Li^+, K^+//Cl^-, CO_3^{2-}–H_2O	298.15	[50]
	Li^+, Na^+//Cl^-, CO_3^{2-}–H_2O	298.15	[51]
	Li^+, K^+//Cl^-, CO_3^{2-}–H_2O	298.15	[52]
	Li^+, Na^+, K^+//CO_3^{2-}–H_2O	288.15	[53]
$Li_2B_4O_7$	Li^+//Cl^-, SO_4^{2-}, $B_4O_7^{2-}$–H_2O	298.15	[54]
	Li^+, Mg^{2+}//SO_4^{2-}, $B_4O_7^{2-}$–H_2O Li^+//CO_3^{2-}, $B_4O_7^{2-}$–H_2O	298.15	[55]
	Li^+, K^+//CO_3^{2-}, $B_4O_7^{2-}$–H_2O	298.15	[56]
	Li^+, Na^+//CO_3^{2-}, $B_4O_7^{2-}$–H_2O Li^+, K^+//CO_3^{2-}, $B_4O_7^{2-}$–H_2O	298.15	[57]
	Li^+, Na^+, K^+//Cl^-, $B_4O_7^{2-}$–H_2O	298.15	[58]
		288.15	[59]
		298.15	[60]
	Li^+, Na^+, K^+//CO_3^{2-}, $B_4O_7^{2-}$–H_2O	288.15	[61]
$LiBO_2$	Li^+//SO_4^{2-}, BO_2^-–H_2O	288.15, 298.15	[62]
	Li^+//Cl^-, BO_2^-–H_2O	288.15, 298.15	[63]
	Li^+//Cl^-, BO_2^-–H_2O	308.15	[64]
	Li^+//Cl^-, SO_4^{2-}, BO_2^-–H_2O	298.15	[65]
	Li^+//SO42, CO_3^{2-}, BO_2^-–H_2O	288.15, 298.15, 308.15	[66]
	Li^+//Cl^-, BO_2^-–H_2O	323.15	[67]
	Li^+//SO_4^{2-}, BO_2^-–H_2O		

Table 1. Stable phase equilibria of lithium-containing salt-water systems.

273.15, 293.15, and 313.15 K had been researched [36, 68]. Phase distribution of the quaternary system was confirmed and the salting-out effect was investigated preliminarily, which provides the physical chemistry foundation for lithium-preparation technique in brines.

Furthermore, the phase equilibria and phase diagram of various salt-water systems containing lithium had been reported heavily. However, the experimental temperature was almost focused on 298.15 K. With the maturity of research techniques and the development of instruments, phase equilibria of multi-component systems at multiple temperatures should be the research focus point in the future. It is worth mentioning that the structures of borate are complicated and have diversified aggregation forms because of the changes of pH, boron contents, types of coexisting ions, and the concentration conditions in the brines. Most studies on the borate-type salt lake brine are mainly for lithium tetraborate. Our group has made excellent progress on phase equilibria of salt-water system containing different species of lithium borates [62–67].

2.2. Metastable phase equilibria of lithium-containing salt-water systems

In the process of seawater and salt lake brine evaporation, the metastable phenomenon is ubiquitous. Because of the conditions of temperature, wind speed, and humidity in the natural environment, brine systems are in a metastable state. Early in the eighteenth century, Van't Hoff had already found that some phase regions disappeared and some enlarged in the stable phase diagram when the salt-water system was in the process of simulating evaporation [32]. So it is useful for the metastable phase diagram to extract the products which cannot be obtained in the stable phase diagram [69].

Some of metastable phase equilibria of lithium-containing salt-water systems are shown in **Table 2**. A lot of lithium-containing systems of chloride, sulfate, carbonate, and borate were researched. The concentration of lithium salt becomes higher in the sulfate-type salt lake brine in the final evaporation period. In our group, metastable phase equilibria of the

Category	Lithium-containing systems	Temperature (K)	Reference
LiCl	Li^+,Na^+,Mg^{2+}//Cl^-–H_2O	308.15	[70]
	Li^+,Na^+,Ca^{2+}//Cl^-–H_2O	288.15	[71]
	Li^+,K^+//Cl^-–H_2O	298.15	[72]
	Li^+,K^+//Cl^-–H_2O	323.15	[73]
	Li^+,K^+,Rb^+//Cl^-–H_2O	298.15	[74]
	Li^+,K^+,Rb^+//Cl^-–H_2O	323.15	[75]
Li_2SO_4	Li^+,Mg^{2+}//Cl^-,SO_4^{2-}–H_2O	298.15	[76]
	Li^+,Na^+//Cl^-,SO_4^{2-}–H_2O	273.15	[77]
	Li^+,Na^+,Mg^{2+}//SO_4^{2-}–H_2O	263.15	[78]
	Li^+,Mg^{2+}//SO_4^{2-}–H_2O	323.15	[79]
	Li^+,Mg2//Cl^-,SO_4^{2-}–H_2O	323.15	[80]

Category	Lithium-containing systems	Temperature (K)	Reference
	$Li^+,Mg^2//Cl^-,SO_4^{2-}–H_2O$	308.15	[81]
	$Li^+,K^+//Cl^-,SO_4^{2-}–H_2O$	308.15	[82]
	$Li^+,Mg^2//Cl^-,SO_4^{2-}–H_2O$	273.15	[83]
	$Li^+,Na^+//SO_4^{2-}–H_2O$	288.15, 308.15	[84]
	$Li^+,Na^+//SO_4^{2-}–H_2O$	308.15, 348.15	[85]
	$Li^+,Na^+,K^+//Cl^-,SO_4^{2-}–H_2O$	308.15	[86]
	$Li^+//Cl^-,SO_4^{2-}–H_2O$	308.15	[48]
	$Li^+,K^+//SO_4^{2-}–H_2O$	288.15, 323.15	[87]
	$Li^+,K^+//SO_4^{2-}–H_2O$	308.15	[88]
Li_2CO_3	$Li^+,Na^+//SO_4^{2-},CO_3^{2-}–H_2O$	288.15	[89]
	$Li^+,K^+//Cl^-,CO_3^{2-}–H_2O$	298.15	[90]
$Li_2B_4O_7$	$Li^+,Na^+,K^+//Cl^-,B_4O_7^{2-}–H_2O$	298.15	[60]
	$Li^+,Na^+//SO_4^{2-},B_4O_7^{2-}–H_2O$	288.15	[91]
	$Li^+,Na^+,K^+//CO_3^{2-},B_4O_7^{2-}–H_2O$	288.15	[92]
	$Li^+,K^+//Cl^-,SO_4^{2-},B_4O_7^{2-}–H_2O$	288.15	[93]
	$Li^+,K^+//SO_4^{2-},B_4O_7^{2-}–H_2O$	288.15	[94]
	$Li^+//Cl^-,CO_3^{2-},B_4O_7^{2-}–H_2O$	298.15	[95]
	$Li^+,K^+//CO_3^{2-},B_4O_7^{2-}–H_2O$	273.15	[96]
	$Li^+,K^+//CO_3^{2-},SO_4^{2-},B_4O_7^{2-}–H_2O$	273.15	[97]
	$Li^+,K^+//CO_3^{2-},B_4O_7^{2-}–H_2O$	288.15	[98]

Table 2. Metastable phase equilibria of lithium-containing salt-water systems.

lithium-containing sulfate system (Li^+, Na^+, K^+, $Mg^{2+}//Cl^-$, SO_4^{2-} – H_2O) and its subsystems at different temperature were determined, which has great help for industrial production and the comprehensive utilization of lithium-containing sulfate system salt lakes.

3. Thermodynamics of lithium salts and their aqueous solution systems

In the long-term production activities and scientific practice, it is troublesome for some production process of the new technology, new processes, or new product development, because there is no data and phase diagram of the relevant salt-water system. In addition, experimental determination on solubilities of multi-component systems is a complex work and it is virtually impossible for researchers to investigate all the salt-water systems. However, it is well-known that phase diagram is a geometry description of phase relation in the system under the condition of thermodynamic equilibrium. In theory, the phase diagram should be able to be gained based on the principles of thermodynamics [99].

As to the classical electrolyte theory, Debye-Hückel theory is only suitable for the dilution solution with a concentration below 0.1 m (molality) and it is unusable to solve the thermodynamic behaviors and to predictive the dissolution equilibria for the complex salt lake brine systems [100]. Pitzer theory [101], which was developed on the basis of Debye-Hückel ion-interaction theory, characterizing thermodynamics properties of electrolyte solution with brief and terse form is widely used either in geochemical behaviors of natural waters and mineral deposits or in the predictions of solubility of salt-water systems. A series of calculated expressions for the activity coefficient and osmotic coefficient of any electrolytes in multi-component systems were proposed by Pitzer [102]. After measuring the thermodynamic parameters, such as osmotic coefficient, activity coefficient, heat of dissolution, heat of dilution, heat of mixing, and specific heat, it is easy to calculate and fit the relative model-parameter theoretically and solubility on the basis of Pitzer and its extended ion-interaction model to promote the development of theory and practice, such as the new field of calculating phase diagram and its application [32].

At present, the domestic and foreign research methods of electrolyte solution of thermodynamic properties are mainly isopiestic method, electromotive force method, calorimetric method, conductivity method, hygrometry, density method, and so on. The isopiestic method and electromotive force method are the most common experimental method to be widely used in measuring the thermodynamic properties such as permeability and activity coefficient. They complement each other. The basic property of matter which is the change of energy can be measured directly by the calorimetric method. In recent years, the calorimetric method is widely implemented in the research of solution thermodynamic properties [103].

In 1992, Yao et al. [104] measured the osmotic and activity coefficients of aqueous mixtures of LiCl and $MgCl_2$ in the range of low concentration to near crystallization limits by the isopiestic method. The Pitzer single-salt parameters and the mixed parameters were calculated by the osmotic and activity coefficients, which were applicative for Pitzer's equation. The predicted solubilities for the system studied using Pitzer's approach were shown to be in reasonable agreement with experimental results from references. The osmotic coefficients of aqueous mixtures of Li_2SO_4 and $MgSO_4$ had been reported from 1.4 to 13.5 $mol{\cdot}kg^{-1}$ at 298 K using the isopiestic method by Zhang et al [105]. In the ranges of 0.2–8.7 and 0.6–12.7 $mol{\cdot}kg^{-1}$ the osmotic coefficients of Li_2SO_4 and $MgSO_4$ were also reported, respectively. The predicted solubilities for this system using Pitzer's approach showed good agreement with experimental results. Yang et al. [106] measured isopiestic molalities and water activities for the $Li_2B_4O_7$–LiCl–H_2O system at 298.15 K using an improved isopiestic apparatus, and the two types of osmotic coefficients were calculated and compared. Pitzer's primary model with minor modifications, in combination with the chemical equilibria, was used to represent the experimental data for the complex $Li_2B_4O_7$–LiCl–H_2O system.

Based on the principle of isopiestic method, the osmotic and activity coefficients can be obtained. However, it might be difficult to get these coefficients with the strict thermodynamics calculation in the mixed electrolyte solutions. The activity coefficients of KCl and LiCl in KCl–LiCl aqueous mixtures have been studied at 298.15 K in ionic strength range of 0.1–4.0 $mol{\cdot}kg^{-1}$ by Li et al. [107]. The data were fitted to Pitzer's equation using the regression method, Pitzer parameters were obtained. The activity confidents of aqueous LiCl in the system LiCl–$MgCl_2$–H_2O were determined at 298.15 K in the total ionic strength range from 0.05 to 6.0 mol·kg^{-1} with the electromotive force method using a lithium-selective electrode and Ag/AgCl electrode by Wang et al. [108]. The activity coefficients of the experiment were compared with which calculated by the Pitzer equation with the known parameters of experimental osmotic coefficients.

The process of chemical reaction, dissolution, dilution, and mixing are often accompanied with heat changes. Reaction heat data, such as dissolution heat, dilution heat, mixing heat, and special heat, were determined critically by the calorimetry technology, which can work out thermodynamic enthalpy, entropy, Gibbs free energy, and thermodynamic equilibrium constant. Hence, the calorimetric method became the research highlights to measure the thermodynamic properties (dissolution heat, dilution heat, mixing heat, and capacity heat).

There are some thermodynamic parameters measured by the calorimetric method. In 1965, Wu [109] and Wood [110] researched heats of mixing of a variety of aqueous containing lithium solutions of the same ionic strength at 298.15 K. Some of them were in the different concentration. The concentration dependence of the heats of mixing indicated that like-charged ion pairs were important contributors to the heat of mixing. The enthalpies of dilution of lithium in the range 0.1–1.0 m had been measured at 303.15 K with a microcalorimeter by Leung et al. [111]. The relative apparent enthalpies of these solutions had been determined with the aid of an extended form of Debye-Hückel limiting law. Enthalpies of solution of Li_2SO_4 and $Li_2SO_4{\cdot}H_2O$ in water at 298.15 K were investigated [112]. The molar enthalpies of solution extrapolated to infinite dilution at 298.15 K were $\Delta_{sol}H_m{}^{\infty}$ (Li_2SO_4) = −(30502 ± 170) J·mol^{-1} and $\Delta_{sol}H_m{}^{\infty}$ ($Li_2SO_4{\cdot}H_2O$) = −(17899 ± 152) J·mol^{-1}. The value for the monohydrate lithium sulfate had been calculated by assuming that the excess water in the sample was present as an aqueous saturated solution. There were presented preliminary specific heat capacities of lithium sulfate solution.

The standard molar enthalpy of the formation of some lithium borates were determined [113–116]. The results were $\Delta_fH_m{}^{\theta}$ ($LiBO_2{\cdot}2H_2O$) = −(1627.46 ± 0.90) kJ·mol^{-1}, $\Delta_fH_m{}^{\theta}$ ($LiBO_2{\cdot}8H_2O$) = −(3397.00 ± 0.94) kJ·mol^{-1}, $\Delta_fH_m{}^{\theta}$($LiB_5O_8{\cdot}5H_2O$) = −(5130.25 ± 4.05) kJ·mol^{-1}, $\Delta_fH_m{}^{\theta}$ ($Li_2B_4O_7{\cdot}3H_2O$) = −(4290.86 ± 3.31) kJ·mol^{-1}, $\Delta_fH_m{}^{\theta}$($Li_3B_5O_8(OH)_2$ (I)) = −(4724.1 ± 4.2) kJ·mol^{-1}, $\Delta_fH_m{}^{\theta}$($Li_3B_5O_8(OH)_2$ (II)) = −(4723.8 ± 4.2) kJ·mol^{-1}, and $\Delta_fH_m{}^{\theta}$($Li_4[B_5O_{13}(OH)_2]{\cdot}3H_2O$) = − (7953.8 ± 6.6) kJ·mol^{-1}.

Meanwhile, the thermodynamic properties in solution system contained lithium borates had been widely researched. Zhang et al. [117] determined the molar heat capacities of the aqueous $Li_2B_4O_7$ solution at a concentration of 0.0187 mol·kg^{-1} in the temperature range from 80 to 355 K by a precision automated adiabatic calorimetry. The enthalpies of dilution for the aqueous $Li_2B_4O_7$ solutions from 0.0212 to 2.1530 mol·kg^{-1} at 298.15 K have been measured [118]. The relative apparent molar enthalpies and relative partial molar enthalpies of the solvent and solute for the aqueous $Li_2B_4O_7$ system were also calculated. The thermodynamic properties of the binary aqueous system $Li_2B_4O_7$–H_2O were represented with the extended Pitzer ion-interaction model. And the enthalpies of dilution, $\Delta_{dil}H_m$, have been also measured for the LiCl–$Li_2B_4O_7$–H_2O system at T = 298.15 K [119]. A suitable microcalorimetric method was used to obtain the better data of the enthalpies of dilution for the ternary system LiCl–$Li_2B_4O_7$–H_2O at a low concentration. The relative apparent molar enthalpies have been determined and the relationships between apparent molar enthalpies and ionic strength at different molar fractions of $Li_2B_4O_7$ were obtained. Li [120] measured the heats of dilution and heat capacities of eutectic point solution system $Li_2B_4O_7$–Li_2SO_4–LiCl–H_2O and subsystems $Li_2B_4O_7$–Li_2SO_4–H_2O and $Li_2B_4O_7$–LiCl–H_2O to cover the ionic strength range from 19 to 0.1 at 298.15 K. The data of the heat of dilution were extrapolated to infinite dilution by use of the Debye-Hückel limiting law to obtain relative apparent molar enthalpies.

In our group, the heat capacities of aqueous solution systems ($Li_2B_4O_7$–H_2O) $m = 0.00415$–0.4208 $mol \cdot kg^{-1}$·at $T = 298.15$, 308.15 and 323.15 K were determined experimentally using the Setaram BT 2.15 microcalorimeter [121]. On the basis of experimental data, the apparent molar heat capacities at different concentrations and temperatures were calculated, and the relationship equations between apparent molar heat capacity and solution concentration of lithium tetraborate at 298.15, 308.15 and 323.15 K were obtained. On the other hand, the Pitzer single salt parameters of lithium tetraborate at different temperatures were fitted on the basis of the Pitzer ion-interaction theory of the electrolytes on the apparent molar heat capacity.

So far, the thermodynamic parameters of lithium salts are still scarce. The Pitzer single salt parameters and the mixing ion-interaction parameters at different temperatures have not been established yet. So, more works on the thermodynamics parameters of lithium salts and their aqueous solution systems at multi-temperatures are essential.

4. Conclusion

With the gradually increasing demands of lithium salt resources as the lithium energy battery, to exploit the lithium-containing brine resources including seawater, concentrated seawater after desalinization, salted lake, oil/gas field water, and well bitter is essential. Therefore, studies on phase equilibria and phase separation of the lithium-containing brine systems are significant to guide the comprehensive utilization of those lithium-containing brine resources around the world. In this chapter, the following three main aspects were discussed: Firstly, the stable and metastable phase equilibria of lithium-containing salt-water multi-systems including the different types of chloride, sulfate, carbonate, and borate brines at different temperatures were summarized. Secondly, a series of valuable thermodynamic properties of standard molar enthalpy of formation for solid lithium salts such as $LiCl \cdot H_2O$, $Li_2SO_4 \cdot H_2O$, $LiBO_2 \cdot 2H_2O$, $LiBO_2 \cdot 8H_2O$, $Li_2B_4O_7 \cdot 3H_2O$, $Li_3B_5O_8(OH)_2$, and $Li_4[B_5O_{13}(OH)_2] \cdot 3H_2O$, and the thermodynamic properties (dissolution heat, dilution heat, mixing heat, and capacity heat) for their relatively aqueous solutions were obtained combined by the isothermal dissolution equilibrium method, isopiestic method, and adiabatic calorimetry, and so on. Thirdly, on the basis of classical Debye-Hückel electrolyte theory, the extended modern electrolyte model, developed by Pitzer KS and his coworkers to express the activity coefficient and osmotic coefficient of any electrolytes in multi-component lithium-containing systems, was successfully used to obtain a series parameters such as the model parameter fitting, thermodynamic property calculation, and the solubility prediction.

Acknowledgements

Financial supports from the National Natural Science of China (21276194, 21306136, U1407113, U1607123), the Chinese Postdoctoral Science Foundation (2016M592827), the Laboratory Foundation of Chinese Universities (SY2015018), and the Training Program of Yangtze Scholars and Innovative Research Team in Universities of China ([2013]373) are acknowledged.

Author details

Long Li[1], Yafei Guo[1,2]* and Tianlong Deng[1]

*Address all correspondence to: guoyafei@tust.edu.cn

1 Tianjin Key Laboratory of Marine Resources and Chemistry, College of Chemical Engineering and Materials Science, Tianjin University of Science and Technology, Tianjin, PR China

2 College of Chemistry and Materials Science, Northwest University, Xi'an, PR China

References

[1] Treptow RS. Lithium batteries: A practical application of chemical principles. Journal of Chemical Education. 2003;**80**:1015–1020. DOI: 10.1021/ed080p1015

[2] Lu L, Han X, Li J, Hua J, Ouyang M. A review on the key issues for lithium-ion battery management in electric vehicles. Journal of Power Sources. 2013;**226**:272–288. DOI: 10.1016/j.jpowsour.2012.10.060

[3] Song JY, Wang YY, Wan CC. Review of gel-type polymer electrolytes for lithium-ion batteries. Journal of Power Sources. 1999;**77**:183–197. DOI: 10.1016/S0378-7753(98)00193-1

[4] Delgado MA, Sánchez MC, Valencia C, Franco JM, Gallegos C. Relationship among microstructure, rheology and processing of a lithium lubricating grease. Chemical Engineering Research & Design. 2005;**83**:1085–1092. DOI: 10.1205/cherd.04311

[5] Delgado MA, Valencia C, Sánchez MC, Franco JM, Gallegos C. Thermorheological behaviour of a lithium lubricating grease. Tribology Letters. 2006;**23**:47–54. DOI: 10.1007/s11249-006-9109-5

[6] Monmaturapoj N, Lawita P, Thepsuwan W. Characterisation and properties of lithium disilicate glass ceramics in the SiO_2-Li_2O-K_2O-Al_2O_3 system for dental applications. Advances in Materials Science & Engineering. 2013;**2013**:680–685. DOI: 10.1155/2013/763838

[7] Huang S, Cao P, Wang C, Huang Z, Gao W. Fabrication of a high-strength lithium disilicate glass-ceramic in a complex glass system. Journal of Asian Ceramic Societies. 2013;**1**:46–52. DOI: 10.1016/j.jascer.2013.02.007

[8] Palacios-Bereche R, Gonzales R, Nebra SA. Exergy calculation of lithium bromide–water solution and its application in the exergetic evaluation of absorption refrigeration systems LiBr-H_2O. International Journal of Energy Research. 2012;**36**:166–181. DOI: 10.1002/er.1790

[9] Hollands KGT. The regeneration of lithium chloride brine in a solar still for use in solar air conditioning. Solar Energy. 1963;**7**:39–43. DOI: 10.1016/0038-092X(63)90003-3

[10] Tobishima SI, Okada T. Effects of quinoneimine dyes on lithium cycling efficiency for $LiClO_4$-propylene carbonate. Journal of Applied Electrochemistry. 1985;**15**:901–906. DOI: 10.1007/BF00614366

[11] Kimura K, Tanaka M, Kitazawa S, Shono T. Highly lithium-selective crown ether dyes for extraction photometry. Chemistry Letters. 1985;**8**:1239–1240. DOI: 10.1246/cl.1985.1239

[12] Dunner DL, Fieve RR. Clinical factors in lithium carbonate prophylaxis failure. Archives of General Psychiatry. 1974;**30**:229–233. DOI: 10.1001/archpsyc.1974.01760080077013

[13] Millard MJ, Kurtis KE. Effects of lithium nitrate admixture on early-age cement hydration. Cement & Concrete Research. 2008;**38**:500–510. DOI: 10.1016/j.cemconres.2007.11.009

[14] Tyrsa VE, Burtseva LP. Generation of thermonuclear energy by fusing hydrogen and lithium atoms. Technical Physics. 2003;**48**:807–812. DOI: 10.1134/1.1593184

[15] Blink JA, Hogam WJ, Hovingh J, Meier ER, Pitts JH. High-yield lithium-injection fusion-energy (HYLIFE) reactor. Journal of Agricultural & Food Chemistry. 2014;**62**:4276–4284. DOI: 10.2172/6124368

[16] Mansfield DK, Strachan JD, Bell MG, Scott SD, Budny R, Marmar ES, et al. Enhanced performance of deuterium–tritium-fueled supershots using extensive lithium conditioning in the Tokamak Fusion Test Reactor. Physics of Plasmas (1994-Present) 1995;**2**:4252–4256. DOI: 10.1063/1.871050

[17] Karki K, Epstein E, Cho JH, Jia Z, Li T, Picraux ST, et al. Lithium-assisted electrochemical welding in silicon nanowire battery electrodes. Nano Letters 2012;**12**:1392–1397. DOI: 10.1021/nl204063u

[18] Swann AC, Koslow SH, Katz MM, Maas JW, Javaid J, Secunda SK, et al. Lithium carbonate treatment of mania. Cerebrospinal fluid and urinary monoamine metabolites and treatment outcome. Archives of General Psychiatry. 1987;**44**:345–354. DOI: 10.1001/archpsyc.1987.01800160057008

[19] Marocho SM, Ozcan M, Amaral R, Bottino MA, Valandro LF. Effect of resin cement type on the microtensile bond strength to lithium disilicate ceramic and dentin using different test assemblies. Journal of Adhesive Dentistry. 2013;**15**:361–368. DOI: 10.3290/j.jad.a28624

[20] Vikström H, Davidsson S, Höök M. Lithium availability and future production outlooks. Applied Energy. 2013;**110**:252–266. DOI: 10.1016/j.apenergy.2013.04.005

[21] Sonoc A, Jeswiet J. A review of lithium supply and demand and a preliminary investigation of a room temperature method to recycle lithium ion batteries to recover lithium and other materials. Procedia Cirp. 2014;**15**:289–293. DOI: 10.1016/j.procir.2014.06.006

[22] Hao H, Liu Z, Zhao F, Geng Y, Sarkis J. Material flow analysis of lithium in China. Resources Policy. 2017;**51**:100–106. DOI: 10.1016/j.resourpol.2016.12.005

[23] Swain B. Recovery and recycling of lithium: A review. Separation and Purification Technology. 2017;**172**:388–403. DOI: 10.1016/j.seppur.2016.08.031

[24] Grosjean C, Miranda PH, Perrin M, Poggi P. Assessment of world lithium resources and consequences of their geographic distribution on the expected development of the electric vehicle industry. Renewable & Sustainable Energy Reviews. 2012;**16**:1735–1744. DOI: 10.1016/j.rser.2011.11.023

[25] Chan LH, Leeman WP, Plank T. Lithium isotopic composition of marine sediments. Geochemistry Geophysics Geosystems. 2006;**3613**:107–109. DOI: 10.1029/2005GC001202

[26] Chitrakar R, Makita Y, Ooi K, Sonoda A. Synthesis of iron-doped manganese oxides with an ion-sieve property: lithium adsorption from Bolivian brine. Industrial & Engineering Chemistry Research. 2014;**53**:3682–3688. DOI: 10.1021/ie4043642

[27] Kesler SE, Gruber PW, Medina PA, Keoleian GA, Everson MP, Wallington TJ. Global lithium resources: relative importance of pegmatite, brine and other deposits. Ore Geology Reviews. 2012;**48**:55–69. DOI: 10.1016/j.oregeorev.2012.05.006

[28] Marinsky, JA, Marcus Y, editors. Ion Exchange and Solvent Extraction: A Series of Advances. Vol. 12. Boca Raton: CRC Press; 1995. p. 93

[29] Yanagase K, Yoshinaga T, Kawano K, Matsuoka T. The recovery of lithium from geothermal water in the Hatchobaru area of Kyushu, Japan. Bulletin of the Chemical Society of Japan. 1983;**56**:2490–2498. DOI: 10.1246/bcsj.56.2490

[30] Yoshinaga T, Kawano K, Imoto H. Basic study on lithium recovery from lithium containing solution. Bulletin of the Chemical Society of Japan. 1986;**59**:1207–1213. DOI: 10.1246/bcsj.59.1207

[31] Epstein JA, Feist EM, Zmora J, Marcus Y. Extraction of lithium from the dead sea. Hydrometallurgy. 1981;**6**:269–275. DOI: 10.1016/0304-386X(81)90044-X

[32] Deng TL, Zhou H, Chen X. Salt-water System Phase Diagrams and Applications. Beijing: Chemical Industry Press; 2013. 11, pp. 203–205

[33] Lepshkov JH, Vopapeva NW. Inorganic Salts Phim. 1958;**3**:278

[34] Lepshkov JH, Vopapeva NW. Inorganic Salts Phim. 1961;**6**:1961

[35] Lepshkov JH, Vopapeva NW. Inorganic Salts Phim. 1962;**7**:1699

[36] Hu KY, Chai WQ. A study of phase equilibria in the quaternary aqueous salt system H^+, Li^+, Mg^{2+}//Cl^-– H_2O at 0°C. Acta Chimica Sinica. 1965 **31**:189–197

[37] Zhang FX, Guo ZJ, Chen PH Ma JH, Chen YS. Studies on the solubility and the isothermal evaporation process in the quinary system Li, Na, K, Mg/Cl-H_2O at 25°C. Chemical Research in Chinese Universities. 1987;**8**:301–305

[38] Zhang FX, Guo ZJ, Chen PH, Ma JH, Chen YS. A study on the ternary system LiCl-$MgCl_2$-H_2O at 25°C. Journal of Northwest University (Natural Science Edition). 1988;**18**:75–78

[39] Ouyang HT, Zeng DW, Zhou HY, Han H, Yao Y. Solubility of the ternary system LiCl–NH_4Cl–H_2O. Journal of Chemical & Engineering Data. 2011;**56**:1096–1104. DOI: 10.1021/je101056t

[40] Yu XD, Jiang DB, Tan Q, Zeng Y. Solid–liquid equilibrium in the aqueous system containing the chlorides of lithium, rubidium and magnesium at 323 K. Fluid Phase Equilibria. 2014;**367**:63–68. DOI: 10.1016/j.fluid.2014.01.037

[41] Meng LZ, Gruszkiewicz MS, Deng TL, Guo YF, Li D. Isothermal evaporation process simulation using the Pitzer model for the quinary system LiCl + NaCl + KCl + $SrCl_2$ + H_2O at 298.15 K. Industrial & Engineering Chemistry Research. 2015;**54**:8311–8318. DOI: 10.1021/acs.iecr.5b01897

[42] Wang SQ, Guo YF, Li DC. Solid-liquid phase equilibria in the ternary systems (LiCl + $MgCl_2$ + H_2O) and (Li_2SO_4 + $MgSO_4$ + H_2O) at 288.15 K. Journal of Chemical & Engineering Data. 2015;**60**:821–827. DOI: 10.1021/je500946w

[43] Li B, Wang QZ, Li J, Fang C, Song P. A study on the ternary systems Li+, K+/SO_4^{2-}-H_2O and Li+, Mg^{2+}/SO_4^{2-}-H_2O at 25°C. Acta Physico-Chimica Sinica.1994;**10**:536–542

[44] Fang CH, Li B, Li J, Wang QZ, Song PS. Studies on the phase diagram and solution properties for the quaternary system Li^+, K^+, Mg^{2+}/SO_4^{2-}-H_2O at 25°C. Acta Chimica Sinica. 1994;**52**:954–959

[45] Sun B, Li B, Fang CH, Du XH, Song PS. Studies on the phase diagram and solution properties for the quinary system, Li^+, K^+, Mg^{2+}/Cl^-, SO_4^{2-}-H_2O at 25°C. Journal of Salt Lake Research. 1995;**3**:50–56.

[46] Li B, Sun B, Fang CH, Du XH, Song PS. Studies on the phase diagram for the quinary system Li^+, Na^+, K^+, Mg^{2+}/SO_4^{2-}-H_2O at 25°C. Acta Chimica Sinica. 1997;**55**:545–552

[47] Wang SQ, Guo YF, Li DC, Tang P, Deng TL. Experimental determination and modelling of the solubility phase diagram of the ternary system (Li_2SO_4 + K_2SO_4 + H_2O) at 288.15 K. Thermochimica Acta. 2015;**601**:75–81. DOI: 10.1016/j.tca.2015.01.003

[48] Liu YH, Guo YF, Wang XK, Deng TL. Phase equilibria in the ternary system (LiCl + Li_2SO_4+ H_2O) at T=308.15 K and p=0.1 MPa: Experimental data and predictions using the Pitzer model. Fluid Phase Equilibria. 2015;**391**:85–89. DOI: 10.1016/j.fluid.2015.02.009

[49] Zeng Y, Yin HA, Tang ML, Wang LS. Studies on the phase diagram and liquid physicochemical properties for the quinary system, Li^+, Na^+, K^+/CO_3^{2-}–H_2O at 298 K. Chemical Engineering(China). 1999;**27**:45–47, 57

[50] Deng TL, Yin HA, Tang ML. A study of solubilities and physicochemistry properties of equilibrium solutions in the reciprocal quaternary system Li^+, K^+/Cl^-, CO_3^{2-}–H_2O at 298 K. Chemical Research in Chinese Universities. 2000;**21**:1572–1574

[51] Deng TL, Yin HA, Tang ML. Experimental and predictive phase equilibrium of the Li^+, $Na^+//Cl^-$, CO_3^{2-}–H_2O system at 298.15 K. Journal of Chemical & Engineering Data. 2002;**47**:26–29. DOI: 10.1021/je000368b

[52] Zeng Y, Yin HA, Tang ML, Feng XX. A study of the phase equilibrium for quinary system Li^+, Na^+, $K^+//CO_3^{2-}$, Cl^-–H_2O at 298.15 K. Chemical Journal of Chinese Universities. 2003;**24**:968–972

[53] Sang SH, Tang ML, Yin HA, Zeng Y. A study on the phase equilibrium of the quaternary system K_2CO_3–Na_2CO_3–Li_2CO_3–H_2O at 288 K. Chinese Journal of Applied Chemistry. 2004;**21**:509–511

[54] Song PS, Du XH. A study of solubilities and physicochemistry properties of equilibrium solutions in the quaternary system $Li_2B_4O_7$–Li_2SO_4–LiCl–H_2O at 25°C. Chinese Science Bulletin. 1986;**3**:209–213

[55] Song PS, Fu HG. Solubilities and properties of solution in the reciprocal system Li^+, Mg^{2+}/ $B_4O_7^{2-}$, SO_4^{2-}–H_2O at 25°C. Chinese Journal of Inorganic Chemistry. 1991;**7**:344–348

[56] Zeng Y, Tang ML, Yin HA, Wang LS. An Experimental study on the equilibrium phase diagram and solution properties of the ternary systems Li^+/CO_3^{2-}, $B_4O_7^{2-}$–H_2O at 298 K. Journal of Mineralogy & Petrology. 1999;**19**:89–92

[57] Zeng Y, Xiao X, Yin HA. A study on the phase equilibrium and solution properties of the quaternary system Li^+, $K^+//CO_3^{2-}$, $B_4O_7^{2-}$–H_2O at 298 K. Journal of Chemical Engineering of Chinese Universities. 2002;**16**:591–596

[58] Yin HA, Hao LF, Zeng Y, Tang ML. Studies on the phase equilibrium and physicochemical properties of solutions for the quinary system Li^+, $Na^+//CO_3^{2-}$, $B_4O_7^{2-}$–H_2O at 298 K. Journal of Chemical Engineering of Chinese Universities. 2003;**17**:1–5

[59] Yin HA, Sang SH, Tang ML Zeng Y. Equilibrium of quaternary system Li^+, $K^+//CO_3^{2-}$, $B_4O_7^{2-}$–H_2O at 288 K. Journal of Chemical Industry and Engineering (China). 2004;**5**:464–467. DOI: 10.3321/j.issn:0438-1157.2004.03.001

[60] Deng TL. Phase equilibrium for the aqueous containing lithium, sodium, potassium, chloride, and borate ions at 298.15 K. Journal of Chemical & Engineering Data. 2004;47:1295–1299. DOI: 10.1021/je049975f

[61] Sang SH, Yin HA, Tang ML. (Liquid + solid) phase equilibria in the quinary system Li^+ + Na^+ + K^+ + CO_3^{2-} + $B_4O_7^{2-}$ + H_2O at 288 K. Physical Inorganic Chemistry. 2005;**36**:1557–1562

[62] Gao DL, Wang Q, Guo YF, Yu XP, Wang SQ, Deng TL. Solid-liquid phase equilibria in the aqueous ternary system Li_2SO_4 + $LiBO_2$ + H_2O at $T = 288.15$ and 298.15 K. Fluid Phase Equilibria. 2014;**371**:121–124. DOI: 10.1016/j.fluid.2014.03.019

[63] Gao DL, Guo YF, Yu XP, Wang SQ, Deng TL. Solubilities, densities, and refractive indices in the salt-water ternary system (LiCl + $LiBO_2$+ H_2O) at $T = 288.15$ K and 298.15 K and $p = 0.1$ MPa. Journal of Chemical & Engineering Data. 2015;**60**:2594–2599. DOI: 10.1021/acs.jced.5b00121

[64] Zhang N, Guo YF, Liu YH, Wang SQ, Deng TL. Thermodynamic phase equilibria of the aqueous ternary system (LiCl + $LiBO_2$ + H_2O) at 308 K: Experimental data and predictions using the Pitzer model. Journal of Chemical Engineering of Japan. 2016;**49**:324–331. DOI: 10.1252/jcej.15we045

[65] Cao LN, Li L, Zhang N, Guo YF, Deng TL. Phase equilibria of quaternary system LiCl–$LiBO_2$–Li_2SO_4–H_2O at 298.15 K. Chinese Journal of Chemical Engineering. 2016;**67**: 1117–1122. DOI: 10.11949/j.issn.0438-1157.20151204

[66] Guo, YF, Li L, Cao LN, Yu XP, Wang SQ, Deng TL. Solubilities, densities and refractive indices in the aqueous quaternary system of lithium sulphate, lithium metaborate, and lithium carbonate at 288.15, 298.15, 308.15 K and 0.1 MPa. Journal of Chemical & Engineering Data. DOI: 10.1021/acs.jced.6b00777

[67] Li L, Guo YF, Zhang SS, Sheng MM, Deng TL. Phase equilibria in the aqueous ternary systems (LiCl + $LiBO_2$ + H_2O) and (Li_2SO_4+ $LiBO_2$ + H_2O) at 323.15 K and 0.1 MPa. Fluid Phase Equilibria. DOI: 10.1016/j.fluid.2016.12.013

[68] Hu KY, Chai WQ, Liu DG. A study of phase equilibria in the quaternary aqueous salt system H^+, Li^+, Mg^{2+}//Cl^- – H_2O at 0°, 20°, 40°. Liu Dagang's Science Works Collection. Beijing: Science Press; 1977. pp. 162–183

[69] Zhang BJ, Sun B, Deng TL. Progresses on meta-stable phase equilibria of salt-water system. Journal of Salt and Chemical Industry. 2007;**36**:39–34. DOI: 10.3969/j.issn.1673-6850.2007.01.014

[70] Zhang BJ. Metastable phase equilibrium of quaternary system Li^+, Na^+, Mg^{2+}//Cl^-–H_2O and ternary system Na^+(K^+), Mg^{2+}//Cl^-–H_2O at 35°. [Dissertation]. Chengdu University of Technology; 2007

[71] Deng TL, Li D. Solid-liquid metastable equilibria in the quaternary system (NaCl + LiCl + $CaCl_2$ + H_2O) at 288.15 K. Journal of Chemical & Engineering Data. 2008;**53**:2488–2492. DOI: 10.1021/je8000798

[72] Zeng Y, Yu XD, Zhang JQ, Li LG. Metastable phase equilibria in the aqueous ternary system containing potassium, lithium and chloride ions at 298.15 K. Advanced Materials Research. 2011;**233–235**:1619–1622. DOI: 10.4028/www.scientific.net/AMR.233-235.1619

[73] Chang WH, Zeng Y, Yan SY, Zhang ZH. Metastable phase equilibrium in the ternary system LiCl + KCl + H_2O at 323 K. Journal of Salt & Chemical Industry. 2012;**41**:26–28

[74] Yu XD, Zeng Y, Yang JY. Sold-liquid isothermal evaporation metastable phase equilibria in the aqueous quaternary system LiCl + KCl + RbCl + H_2O at 298.15 K. Journal of Chemical & Engineering Data. 2012;**57**:172–132. DOI: 10.1021/je200880m

[75] Li ZQ, YU XD, Yin QH, Zeng Y. Thermodynamics metastable phase equilibria of aqueous quaternary system LiCl + KCl + RbCl + H_2O at 323.15 K. Fluid Phase Equilibria. 2013;**358**:131–136. DOI: 10.1016/j.fluid.2013.08.011

[76] Guo ZZ, Liu ZQ, Chen JQ. The metastable phase equilibrium in the system Li^+, Mg^{2+}// Cl^-, SO_4^{2-}–H_2O at 25°. Acta Chimica Sinica. 1991;**49**:937–943

[77] Wang SQ, Deng TL. Metastable phase equilibria of the reciprocal quaternary system containing lithium, sodium, chloride, and sulphate ions at 273.15 K. Journal of Chemical & Engineering Data. 2010;**55**:4213–4215. DOI: 10.1021/je1003669

[78] Li ZY, Deng TL, Liao MX. Solid-liquid metastable equilibria in the quaternary system Li_2SO_4 + $MgSO_4$ + Na_2SO_4 + H_2O at T = 263.15 K. Fluid Phase Equilibria. 2010;**2931**:42–46. DOI: 10.1016/j.fluid.2010.02.016

[79] Deng TL, Yin HJ, Guo YF. Metastable phase equilibrium in the aqueous ternary system (Li_2SO_4 + $MgSO_4$+ H_2O) at 323.15 K. Journal of Chemical & Engineering Data. 2011;**56**:3585–3588. DOI: 10.1021/je800482y

[80] Meng LZ, Yu XP, Li D, Deng TL. Solid-liquid metastable equilibria of the reciprocal quaternary system (LiCl + $MgCl_2$ + Li_2SO_4 + $MgSO_4$ + H_2O) at 323.15 K. Journal of Chemical & Engineering Data. 2011;**56**:4627–4632. DOI: 10.1021/je200563j

[81] Gao J, Deng TL. Metastable phase equilibrium in the aqueous quaternary system (LiCl + $MgCl_2$ + Li_2SO_4 + $MgSO_4$+ H_2O) at 308.15 K. Journal of Chemical & Engineering Data. 2011;**56**:1452–1458. DOI: 10.1021/je1008526

[82] Liu YH, Deng TL Song PS. Metastable phase equilibrium of the reciprocal quaternary system LiCl + KCl + Li_2SO_4 + K_2SO_4 + H_2O at 308.15 K. Journal of Chemical & Engineering Data. 2011;**56**:1139–1147. DOI: 10.1021/je1010888

[83] Wang SQ, Guo YF, Deng TL. Solubility prediction for the reciprocal quaternary system (Li^+, Mg^{2+}//Cl^-, SO_4^{2-}–H_2O) at 273.15 K using Pitzer ion-interaction model. Advanced Materials Research. 2012;**549**:437–440. DOI: 10.4028/www.scientific.net/AMR.549.437

[84] Guo YF, Liu YH, Wang Q, Lin CX, Wang SQ, Deng TL. Phase equilibria and phase diagrams for the aqueous ternary system (Na_2SO_4 + Li_2SO_4 + H_2O) at (288 and 308) K. Journal of Chemical & Engineering Data. 2013;**58**:2763–2767. DOI: 10.1021/je4004146

[85] Guo YF, Gao DL, Han HJ, Shen DL, Wang SQ, Deng TL. Metastable phase equilibria in the aqueous ternary system (Na_2SO_4 + Li_2SO_4 + H_2O) at 308.15 and 348.15 K. Fluid Phase Equilibria. 2013;**358**:56–59. DOI: 10.1016/j.fluid.2013.07.048

[86] Liu YH, Guo YF, Yu XP. Solid-liquid metastable phase equilibria in the five-component system (Li + Na + K + Cl + SO_4 + H_2O) at 308.15 K. Journal of Chemical & Engineering Data. 2014;**59**:1685–1691

[87] Bu BH, Li L, Zhang N, Guo YF, Wang SQ, Sun LY, Deng TL..Solid–liquid metastable phase equilibria for the ternary system ($Li_2SO_4 + K_2SO_4 + H_2O$) at 288.15 and 323.15 K, $p = 0.1$ MPa.Fluid Phase Equilibria. 2015;**402**:78–82. DOI: 10.1016/j.fluid.2015.05.044

[88] Li L, Bu BH, Zhang N, Guo YF, Deng TL. Metastable phase equilibria for the ternary aqueous system of lithium sulphate and potassium sulphate at $T = 308.15$ K: Experimental data and prediction using Pitzer model. Russian Journal of Inorganic Chemistry. 2016;**61**:1169–1174. DOI: 10.1134/S0036023616090138

[89] Sang SH, Yin HA, Zeng Y, Liu FY. Study on metastable equilibria in quaternary system Li^+, Na^+//SO_4^{2-}, CO_3^{2-}–H_2O at 288 K. Acta Chimica Sinica. 2006;**64**:2247–2253

[90] Yan SY, Yin HA, Liang Q, Tang ML, Wang LM. An experimental study on metastable phase equilibria in the quaternary system Li^+, K^+//Cl^-, CO_3^{2-}–H_2O at 298 K. Chemical Research and Application. 2008;**20**:166–168

[91] Sang SH, Yu HY, Cai DZ. Study on metastable equilibria in quaternary system Li^+, Na^+//$SO4^{2-}$, $B_4O_7^{2-}$–H_2O at 288 K. Chinese Journal of Inorganic Chemistry. 2005;**21**: 1316–1320

[92] Sang SH, Yin HA, Tang ML. (Liquid + solid) phase equilibria in the quinary system $Li^+ + Na^+ + K^+ + CO_3^{2-} + B_4O_7^{2-} + H_2O$ at 288 K. Journal of Chemical & Engineering Data. 2005;**50**:1555–1558. DOI: 10.1002/chin.200548015

[93] Zeng Y, Shao M. Liquid-solid metastable equilibria in the quinary system Li^+, K^+//Cl^-, SO_4^{2-}, $B_4O_7^{2-}$–H_2O at 288.15 K. Journal of Chemical & Engineering Data. 2006;**51**:219–222. DOI: 10.1021/je050337m

[94] Sang SH, Yin HA, Lei NF. (Solid + liquid) metastable equilibria in quaternary system $Li_2SO_4 + K_2SO_4 + Li_2B_4O_7 + K_2B_4O_7 + H_2O$ at 288 K. Journal of Chemical Thermodynamics. 2006;**38**:173–178. DOI: 10.1016/j.jct.2005.04.002

[95] Yan SY, Chen T, Liang Q, Yin HA, Zeng Y. A study on metastable equilibrium in the quaternary reciprocal system of Li^+/Cl^-, CO_3^{2-}, $B_4O_7^{2-}$–H_2O at 298 K. Chinese Journal of Inorganic Chemistry. 2006;**26**:95–97

[96] Zeng Y, Lin XF, Meng S. Study on metastable phase equilibrium of quaternary system Li^+, K^+//CO_3^{2-}, $B_4O_7^{2-}$–H_2O at 273 K. Chemical Engineering (2008) 36:47–50

[97] Qu SD. Study on the metastable equilibrium of the quinary system Li^+, K^+//CO_3^{2-}, SO_4^{2-}, $B_4O_7^{2-}$–H_2O at 273 K [Dissertation]. Chengdu University of Technology; 2008

[98] Sang SH, Zhang X, Zhao XP, Xiao LJ. Metastable equilibria in the quinary system of Li^+, K^+//CO_3^{2-}, SO_4^{2-}, $B_4O_7^{2-}$–H_2O at 288 K. Acta Chimica Sinica.2010;**68**:476–480

[99] Gao SY, Song PS, Xia SP, Zheng MP. Salt Lake Chemistry: A New Type Salt Lake of Boron and Lithium. Beijing: Science Press; 2007

[100] Song PS, Yao Y, Li J. Thermodynamics and phase equilibria of the salt lake brine system at 25°. Progress in Chemistry. 2000;**12**:255–267

[101] Pitzer KS. Thermodynamics of electrolytes I: Theoretical basis and general equations. Journal Physical Chemistry. 1973;**77**:268–277. DOI: 10.1021/j100621a026

[102] Pitzer KS. Activity Coefficients in Electrolyte Solutions. 2nd ed. Boca Raton: The Chemical Rubber Company Press; 1991. pp. 75–153

[103] Deng TL, Li J, Li L, Zhang SS, Guo YF. Application of calorimetry to the study of thermodynamic properties of borate systems. Journal of Tianjing University of Science & Technology. 2016; **31**:6–12. DOI: 10.13364/j.issn.1672-6510.20150098

[104] Yao Y, Sun B, Song PS, Zhang Z, Wang RL, Chen JQ. Thermodynamics of aqueous electrolyte solution—Isopiestic determination of osmotic and activity coefficients in $LiCl–MgCl_2–H_2O$ at 25°. Acta Chimica Sinica. 1992;**50**:839–848

[105] Zhang Z, Yao Y, Song PS, Chen JQ. Isopiestic determination of the osmotic and activity coefficients of aqueous mixtures of Li_2SO_4 and $MgSO_4$. Acta Physico-Chimica Sinica. 1993;**9**:366–373

[106] Yang JM, Yao Y, Xia QY, Li BH, Xu SY, Yang JY. Isopiestic determination of the osmotic coefficients and Pitzer model representation for the $Li_2B_4O_7 + LiCl + H_2O$ system at $T = 298.15$ K. Journal of Solution Chemistry. 2008;**37**:377–389. DOI: 10.1007/s10953-007-9244-0

[107] Li J, Song PS, Yao Y, Wang RL. Thermodynamic properties of $KCl–LiCl–H_2O$ at 25°. Acta Physico-Chimica Sinica. 1992;**8**:94–99

[108] Wang DB, Song PS, Yang JZ. Thermodynamics of mixture of boric acid with lithium borate and chloride. Chinese Journal of Chemistry. 1994;**12**:97–104. DOI: 10.1002/cjoc.19940120201

[109] Wu YC, Smith MB, Young TF. Heats of mixing of electrolytes having common ions. Journal of Physical Chemistry. 1965;**69**:1868–1872. DOI: 10.1021/j100890a014

[110] Wood RH, Smith RW. Heats of mixing of aqueous electrolytes. I. Concentration dependence of 1-1 electrolytes. Journal of Physical Chemistry. 2002;**69**:2974–2979. DOI: 10.1021/j100893a025

[111] Leung WH, Millero FJ. The enthalpy of formation of magnesium sulphate ion pairs. Journal of Solution Chemistry. 1975;**4**:145–159. DOI: 10.1007/BF00649155

[112] Apelblat A. Enthalpy of solution of lithium sulphate and lithium sulphate monohydrate in water at 298.15 K. Journal of Chemical Thermodynamics. 1985;**17**:769–773. DOI: 10.1016/0021-9614(85)90108-9

[113] M Bán, Madarász J, Bombicz P, Pokol G, Gál S. Thermal and structural study on the lattice compound 1, 4-diammoniumbutane bis (theophyllinate). Thermochimica Acta. 2004;**420**:105–109. DOI: 10.1016/j.tca.2003.12.033

[114] Li J, Li B, Gao SY. Thermochemistry of hydrated lithium borates. Journal of Chemical Thermodynamics. 1998;**30**:681–688. DOI: 10.1006/jcht.1997.0324

[115] Li P, Liu ZH. Hydrothermal synthesis, characterization, and thermodynamic properties of a new lithium borate, $Li_3B_5O_8(OH)_2$. Journal of Chemical & Engineering Data. 2010;**55**:2682–2686. DOI: 10.1021/je900947d

[116] Li P, Liu ZH. Synthesis, structure and thermodynamic property of a new lithium borate: $Li_4[B_8O_{13}(OH)_2].3H_2O$. Chinese Journal of Chemistry. 2012;**33**:847–853

[117] Zhang ZH, Tao ZC, Yin GY. Heat capacities and thermodynamic functions of the aqueous $Li_2B_4O_7$ solution in the temperature range from 80 K to 355 K. Journal of Chemical & Engineering Data. 2007;**52**:866–870. DOI: 10.1021/je060462d

[118] Yin GY, Yao Y, Jiao BJ, Chen SP, Gao SL. Enthalpies of dilution of aqueous $Li_2B_4O_7$ solutions at 298.15 K and application of ion-interaction model. Thermochimica Acta. 2005;**435**:125–128. DOI: 10.1016/j.tca.2005.04.007

[119] Yin GY, Yao Y, Liu ZH. Enthalpies of dilution of aqueous mixed solutions of LiCl and $Li_2B_4O_7$ at 298.15 K. Journal of Thermal Analysis and Calorimetry. 2009;**95**:377–380. DOI: 10.1007/s10973-008-9239-0

[120] Li JC, Zhai ZX, Zeng ZM, Wang LY. Thermo-chemistry of solution of salt lakes(IV): Heats of dilution, heat capacities and apparent molal enthalpies of $Li_2B_4O_7—Li_2SO_4—LiCl—H_2O$ system at 298.15K. Journal of Salt Lake Research. 1993;**1**:34–38

[121] Li L, Zhang SS, Liu YH, Guo YF, Deng TL. Heat capacities and the ion-interaction of lithium tetraborate aqueous solution system. Chemical Journal of Chinese Universities. 2016;**37**:349–353. DOI: 10.7503/cjcu20150594

2

Membrane Thermodynamics for Osmotic Phenomena

Albert S. Kim and Heyon-Ju Kim

Abstract

In this chapter, we briefly review the thermodynamic ensembles and associated energy functions using the seven thermodynamic variables. The energy E, the entropy S, and the system volume V are used to derive the temperature T and pressure P. The chemical potential μ is derived as the change of the system energy with respect to the number of matters N in the isobaric-isothermal environment. A dilute solution is defined as a homogeneous mixture of solvent and inert solutes, where the total number and volume of solutes are much smaller than those of the solvent. Gibbs free energy of the dilute solution is used to rigorously derive the osmotic pressure by equilibrating chemical potentials of solutes and solvent. Nonequilibrium of the filtration systems is reviewed by introducing the irreversible thermodynamic model with Onsager's reciprocal theorem. Direct applications of the irreversible thermodynamic model are currently limited due to the absence of the exact nonequilibrium statistical mechanics. We hope this chapter, containing a review of statistical mechanics, related to membrane separations and osmosis phenomena, helps researchers and especially graduate students, who seek an in-depth understanding of membrane separation from the theoretical statistical physics as applied to chemical and environmental engineering.

Keywords: membrane thermodynamics, statistical mechanics, thermodynamic ensemble, Gibbs energy function, chemical potential, weak solution, osmotic pressure, Fick's law, solution-diffusion model, thermodynamic irreversible model

1. Introduction

A membrane is a selective barrier between two phases, i.e., a thin layer of material that separates solute and solvent materials when a driving force is applied across it. On membrane surfaces, flows of different thermodynamic phases are introduced and maintained quasi-steady with respect to time. These separation processes require driving forces mainly for mass transfer as gradients of physical quantities associated with the thermodynamic, flowing phases.

Separation implies collecting masses of the same particles/molecules in specific spatial locations, which is strongly correlated to diffusion phenomena. Nature tends to move from a higher energy to a lower energy state, or equivalently highly ordered to randomly disordered phases. The diffusion of solutes in a free (solvent) medium is a spontaneous tendency, which must be well understood to analyze the separation phenomena. On the other hand, if the solutes are spatially confined by permeable interfaces, through which only solvent molecules can pass, the solvents try to move from their high- to low-concentration sides. This phenomena is called *osmosis*, which is equivalent to the solvent moving from low- to high-concentration regions of solutes. Note that in the solute diffusion and the solvent osmosis processes, mass transfer phenomena are from high- to low-concentration regions of the transferring masses. In this light, diffusion and osmosis can be treated equivalently as energy-minimizing and entropy-increasing phenomena of solutes and solvents, respectively.

Most pressure-driven membrane separations aim to produce clean water (solvent) from concentrated solutions. These include ultrafiltration (UF) and microfiltration (MF) for particulate removal and reverse osmosis (RO) and nanofiltration (NF) for ion removal. UF uses a finely porous membrane, which is usually antisymmetric, having a mean pore diameter between 1 and 100 nm. UF aims to separate water and microsolutes from macromolecules and colloids [1–3]. MF uses porous membranes to separate suspended particles with diameters between 0.1 and 10 μm [1, 3, 4]. MF's filtration capacity is therefore between UF and conventional filtration methods. Based on the particle size, dominant diffusion mechanisms of these particulate matters include Brownian diffusion [5–9] and shear-induced diffusion [10–12]. Ballistic motion of non-Brownian particles (usually bigger than 10 μm) in MF can be treated as dynamics of inelastic granules. RO is a desalting process for water production using nonporous membranes that are permeable to water but essentially impermeable to salt. A pressurized feed stream containing dissolved salts contacts the feed-side of the RO membrane, and salt-depleted water is withdrawn as a low-pressure permeate stream [13–16]. NF membranes have lower rejection ratio, i.e., 20–80% of sodium chloride, than that of RO, typically greater than 98–99%. NF resides therefore between UF and RO in terms of salt rejection capability. High hydraulic pressure is an essential component for RO and NF to overcome the osmotic pressure of seawater and brackish water, respectively [17–21].

The concentration (equivalently, osmotic pressure) gradient can be used, however, as a driving force for forward osmosis (FO) and pressure-retarded osmosis (PRO) processes. FO extracts a solvent from the low- to high-concentration sides of the solutes in order to equilibrate the concentrations [22–25]. PRO utilizes the extra gains of hydraulic pressure due to the amount of the transferred solvent for power generation [26–29]. Both pressure-driven and osmosis-driven processes aim to achieve a high flux, i.e., a large amount of water produced per unit time per unit membrane surface area. In order to achieve a steady high flux, increases and decreases in the osmotic pressure gradient need to be prevented in the pressure- and osmosis-driven membrane processes, respectively. Therefore, maintaining a stable osmotic pressure is a primary issue in both types of processes.

The driving forces for membrane separations described above include gradients of the hydraulic pressure, solute concentration, solution temperature, and external electromagnetic field. In statistical mechanics, there are seven primary variables used to explain macroscopic thermodynamic phenomena. An ensemble is made using a set of three selected variables, and a specific energy function of the ensemble is described in terms of the three independent variables. As the membrane separations are coupled phenomena of momentum, mass, and heat transfer, a holistic understanding of statistical mechanics can significantly enhance design, analysis, and optimization of the membrane processes. In this chapter, we explain ensembles and energy functions in statistical mechanics, represent the osmotic pressure using Gibbs energy function of a weak (dilute) solution, and apply statistical laws to explain the separation phenomena using a solution-diffusion model [30].

2. Thermodynamics to statistical mechanics

Statistical mechanics is the microscopic version of thermodynamics [31]. Macroscopic quantities, dealt within thermodynamics, can be fundamentally obtained at the microscopic level in statistical mechanics. There are seven variables in thermodynamics, which are energy E, entropy S, temperature T, pressure P, volume V, number of molecules N, and chemical potential μ. An ensemble is defined as a set, in which three independent variables are used to define a specific form of an energy and the other four variables are represented as functions of the three master variables. For example, the elementary microcanonical ensemble has P, V, S, and μ, represented as functions of three master variables of N, V, and E.

2.1. Primary macroscopic quantities

2.1.1. Temperature

Consider two boxes in contact containing a certain number of particles in equilibrium, forming a closed system. Then, entropy S of the total system has its maximum value for a given system energy, E, i.e.,

$$S = S_{\max} \tag{1}$$

Since the energy is an additive scalar, the total energy of the entire system is the sum of the energies:

$$E = E_1 + E_2 \tag{2}$$

The total entropy can be similarly expressed, knowing that the entropy is a function of the energy:

$$S_{\max} = S_1(E_1) + S_2(E_2) \tag{3}$$

Since the entropy is already maximized in the equilibrium state, it is independent of the energy variation, i.e.,

$$\frac{dS}{dE_1} = \frac{dS_1}{dE_1} + \frac{dS_2}{dE_2}\frac{dE_2}{dE_1} = \frac{dS_1}{dE_1} - \frac{dS_2}{dE_2} = 0 \tag{4}$$

hence, we obtain

$$\frac{dS_1}{dE_1} = \frac{dS_2}{dE_2} \tag{5}$$

The derivative of the entropy S with respect to its energy E is used to define temperature as follows:

$$\frac{dS}{dE} \equiv \frac{1}{T} \quad \left(\rightarrow \frac{1}{k_B T}\right) \tag{6}$$

In the original definition, the magnitude of the temperature is too high so Boltzmann's constant k_B is introduced as shown in the parenthesis of Eq. (6). Temperature T is now represented in terms of the Kelvin unit. Substitution of Eq. (6) into Eq. (5) for each box provides

$$T_1 = T_2 \tag{7}$$

as a condition for the equilibrium. It is worth noting that the internal energy E and entropy S are the basic thermodynamic quantities, and the temperature is a derived variable proportional to the variation of E with respect to S (specifically, in the microcanonical ensemble).

2.1.2. Pressure

In fluid mechanics, pressure is often defined as the ratio of applied force per unit surface area of an object [32]:

$$P = \frac{\langle F_n \rangle}{\text{Area}} \tag{8}$$

where $\langle F_n \rangle$ is the mean normal component of the force vector $\vec{F}$ applied to the object's surface area. A conservative force can be represented as a negative gradient of the total energy $E = K + U$, as a sum of kinetic energy K and potential energy U. Suppose the applied force causes an infinitesimal change in the volume of the body from V to $V + \delta V$ as shown in **Figure 1**. Then, the compressed volume is equal to the surface area multiplied by the thickness variation, i.e., $\delta V = A\,\delta s$, which is in general, $A = \vec{n} \cdot \nabla V$. Using the chain rule, one can represent the normal component of the applied force as a product of the energy density and the total surface area, which is

$$\vec{F} = -\nabla E = -\left(\frac{\partial E}{\partial V}\right)_S \nabla V \tag{9}$$

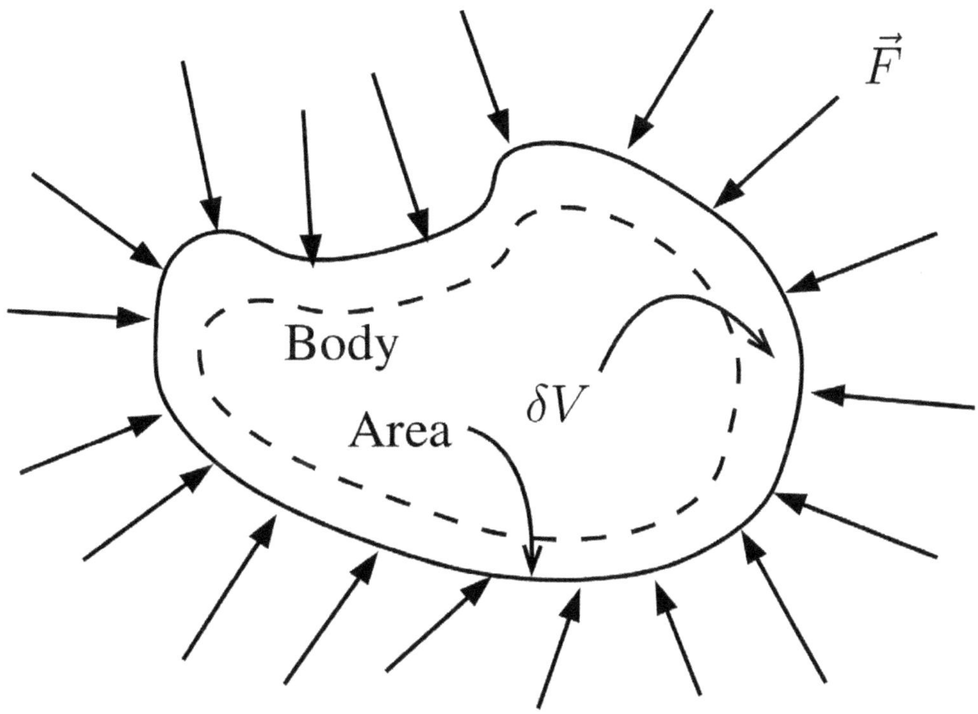

Figure 1. External forces applied to the surface of a body.

where, without losing generality, $(\partial E/\partial V)_S$ can be interpreted as the isentropic (i.e., of constant entropy) energy density inside the body volume V. One can operate the dot product by the normal vector $\vec{n}$ on the left side of Eq. (9) to have

$$\vec{F} \cdot \vec{n} = F_n = -\left(\frac{\partial E}{\partial V}\right)_S A \tag{10}$$

and dividing both sides of Eq. (10) by the area A gives the conceptual definition of the pressure:

$$P = -\left(\frac{\partial E}{\partial V}\right)_S \tag{11}$$

Here, pressure definition can be extended from the normal force per unit area to the energy density in magnitude. Because energy E is a scalar quantity, the direction of the force vector does not need to be considered in the pressure calculation.

2.2. Ensembles and energy functions

2.2.1. Internal energy in microcanonical ensemble

In the previous section, we used three thermodynamic variables of energy E, entropy S, and volume V to generally define temperature T at a constant volume:

$$T = \left(\frac{\partial E}{\partial S}\right)_V \tag{12}$$

and pressure P at a constant entropy:

$$P = -\left(\frac{\partial E}{\partial V}\right)_S \tag{13}$$

Because the derivative operand of both Eqs. (12) and (13) is the internal energy E, the total derivative of E can be written in terms of T and P:

$$\mathrm{d}E = \left(\frac{\partial E}{\partial S}\right)_V \mathrm{d}S + \left(\frac{\partial E}{\partial V}\right)_S \mathrm{d}V = T\,\mathrm{d}S - P\,\mathrm{d}V \tag{14}$$

which indicates that E is an exact function of S and V, i.e., $E = E(S,V)$.

If the system consists of different molecular species, i.e., $k = 1, 2, \ldots, n_K$, where n_K is the total number of species, then the total molecule number N is the sum of the number of molecules of all the species, i.e.,

$$N = \sum_k N_k = N_1 + N_2 + \cdots + N_{n_K} \tag{15}$$

where, for example, N_2 is the total molecule number of species 2. Then, the infinitesimal change of E includes the effect of the particle exchange, using the chemical potential μ_k, as

$$\mathrm{d}E = T\,\mathrm{d}S - P\,\mathrm{d}V + \sum_k \mu_k\,\mathrm{d}N_k \tag{16}$$

In a closed system, the molecule numbers of multiple species can change simultaneously, keeping the total molecule number invariant. If the two systems in contact are at an equilibrium and molecules in the boxes can be exchanged, then the change of energy as per the number of exchanged molecules is equivalent to the chemical potential of the species. From Eq. (16), we can represent an extended version E as an exact function of S, V, and N_k:

$$E = E(S,V,\{N_k\}) \tag{17}$$

If a thermodynamic system is completely controlled by the three variables of N, V, and S, the system is said to be a microcanonical ensemble.

2.2.2. *Helmholtz free energy in canonical ensemble*

Since the temperature is a more convenient variable to measure than the entropy S, one can use the mathematical identity of $T\,\mathrm{d}S = \mathrm{d}(TS) - S\,\mathrm{d}T$ to rewrite Eq. (16) as

$$dE = d(TS) - S\,dT - P\,dV + \mu_k\,dN_k \tag{18}$$

where the notation of the summation over the molecular species k, $\sum_k$ is omitted for simplicity. The total derivative, $d(TS)$, is subtracted from both sides of Eq. (18) to have

$$dA = -S\,dT - P\,dV + \mu_k\,dN_k \tag{19}$$

where A is the Helmholtz free energy defined as

$$A = E - ST \tag{20}$$

If a thermodynamic system is completely described using T, V, and $\{N_k\}$ (for $k = 1, 2\ldots$), this ensemble is called canonical, and the Helmholtz free energy, $A(T,V,\{N_k\})$, is the representative energy function.

2.2.3. *Enthalpy in isentropic-isobaric ensemble*

Similar to how we derived the Helmholtz free energy, we start from the infinitesimal difference of the internal energy E of Eq. (16) using the mathematical identity of $P\,dV = d(PV) - V\,dP$ to have

$$dE = T\,dS - d(PV) + V\,dP + \mu_k\,dN_k \tag{21}$$

We add $d(PV)$ in the both sides of the above equation and obtain

$$dH = T\,dS + V\,dP + \mu_k\,dN_k \tag{22}$$

where

$$H(S,P,\{N_k\}) = E + PV \tag{23}$$

is defined as the enthalpy as a function of S, P, and $\{N_k\}$. Eq. (22) indicates that the enthalpy is independent of T unlike other energy functions (see the next sections for detailed discussion).

2.2.4. *Thermodynamic potential in grand canonical ensemble*

To have an ensemble that is independent of the number of particles, one can start from the infinitesimal change of Helmholtz free energy and use the identity of $\mu_k\,dN_k = d(\mu_k N_k) - N_k\,d\mu_k$ to have

$$dA = -S\,dT - P\,dV + d(\mu_k N_k) - N_k\,d\mu_k \tag{24}$$

Subtracting $d(\mu_k N_k)$ from each side of Eq. (24) gives

$$\mathrm{d}\Phi = -S\,\mathrm{d}T - P\,\mathrm{d}V + N_k\,\mathrm{d}\mu_k \tag{25}$$

where

$$\Phi(T,V,\{\mu_k\}) = A - \mu_k N_k \tag{26}$$

is defined as the thermodynamic potential, varying with respect to T, V, and μ_k. An ensemble described using μ, V, and T is called a grand canonical ensemble. The thermodynamic potential is further derived such that $\Phi = -PV$ if the thermodynamic system is homogeneous.

2.2.5. Gibbs energy in isothermal-isobaric ensemble

Finally, we replace $P\,\mathrm{d}V$ in the infinitesimal change of A in Eq. (19) by $\mathrm{d}(PV) - V\,\mathrm{d}P$ to have

$$\mathrm{d}G = -S\,\mathrm{d}T + V\,\mathrm{d}P + \mu_k\,\mathrm{d}N_k \tag{27}$$

where

$$G(T,P,\{N_k\}) = A + PV = E - TS + PV \tag{28}$$

is defined as the Gibbs free energy varying with respect to T, P, and $\{N_k\}$. Now we assume that G is a homogeneous (i.e., linear) function of N_k such that $G \propto N_k$. In this case, the chemical potential of species k is represented in terms of T and P only as

$$\mu_k = \left(\frac{\partial G}{\partial N_k}\right)_{T,P} = \mu_k(T,P) \tag{29}$$

For the fixed number of particles, the infinitesimal change of the total chemical potential is

$$\mathrm{d}\mu = \frac{1}{N}\,\mathrm{d}G = -\overline{S}\,\mathrm{d}T + \overline{V}\,\mathrm{d}P \tag{30}$$

where $\overline{S} = S/N$ and $\overline{V} = V/N$ are the entropy and the volume per molecule, respectively, of the entire system. In practice, it is often convenient to use the entropy and energy per mole of molecules in engineering applications, but for basic study here we will keep using quantities divided by the number of molecules. For species k, we have the representation of the infinitesimal change in the chemical potential of species k:

$$\mathrm{d}\mu_k = -\overline{S}_k\,\mathrm{d}T + \overline{V}_k\,\mathrm{d}P \tag{31}$$

Keeping the homogeneity assumption, the Gibbs energy function is written as a sum of products of the chemical potentials and the particle numbers:

$$G = \sum_k \left(\frac{\partial G}{\partial N_k}\right)_{T,P} N_k = \sum_k \mu_k(T,P) N_k \tag{32}$$

The thermodynamic potential is generally derived as $\Phi = A - \sum_k \mu_k N_k$ using the Legendre transformation from the previous section. If and only if the Gibbs energy function $G(= A + PV)$ is homogeneous such as Eq. (32), Φ can be further simplified to

$$\Phi = A - G = -PV \tag{33}$$

If the molecular interactions are strong, then Eq. (32) requires an extra coupling term proportional to $N_i N_j$, and Eq. (26) should be revisited as a general definition for Φ (see Section 1.3 for details). Dependences of the energy functions on thermodynamic variables in specific ensembles are summarized in **Table 1**.

Ensemble	Energy functions and relationships
Microcanonical (NVS)	Internal energy $E(N,V,S)$
	$dE = TdS - PdV + \mu dN$
Canonical (NVT)	Helmholtz energy $A(N,V,T) = E - TS$
	$dA = -SdT - PdV + \mu dN$
Grand canonical (μVT)	Thermodynamic potential $\Phi(\mu,V,T) = A - \sum_k \mu_k N_k$
	$d\Phi = -SdT - PdV - Nd\mu$
Isothermal-isobaric (NPT)	Gibbs energy $G(N,P,T) = A + PV = \mu N$
	$dG = -SdT + VdP + \mu dN$
Isentropic-isobaric (NPS)	Enthalpy $H(N,P,S) = E + PV$
	$dH = TdS + VdP + \mu dN$

Table 1. Specific ensembles and associated energy functions.

2.3. Gibbs energy and anisothermal equilibrium

2.3.1. *Thermodynamics variables: extensive and intensive*

Consider a thermodynamic system in equilibrium, shown in **Figure 2**. The system is made by adding two identical systems, which are now in contact with each other. In this case, the seven thermodynamic variables change as follows:

- Additive (extensive): $N \to 2N$, $V \to 2V$, $S \to 2S$, and $E \to 2E$
- Nonadditive (intensive): $T \to T$, $P \to P$, and $\mu_k \to \mu_k$

As expected, the number of particles, volume, entropy, and energy are doubled by adding the two identical systems, and they are called *additive*. On the other hand, temperature, pressure, and chemical potential remain invariant, and they are called *nonadditive*.

The independence of the temperature to the system size can be understood using its basic definition of Eq. (12) as the change ratio of E to S as they are additive quantities. The pressure is defined in Eq. (13) as the negative ratio of changes of E to V. The chemical potential,

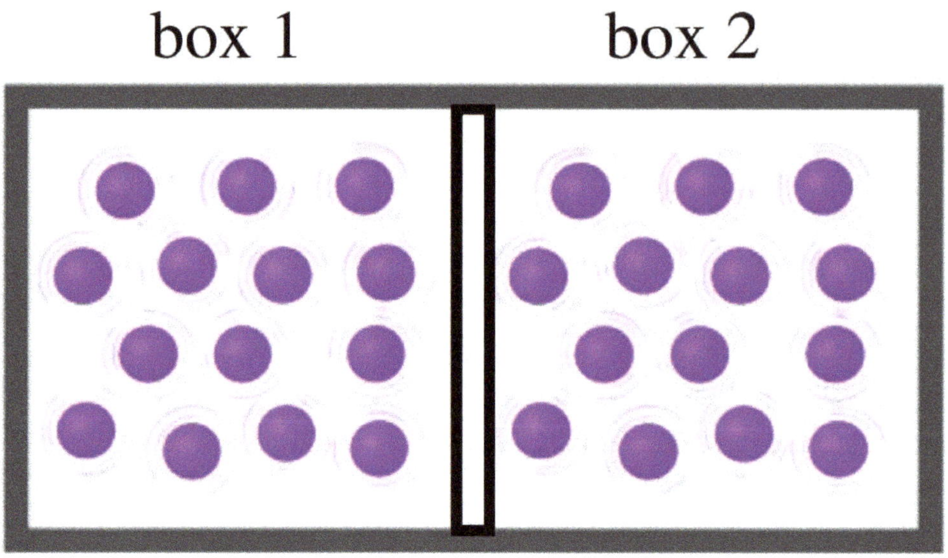

Figure 2. A closed system consisting of two identical boxes in a thermal equilibrium. The outer boundaries (gray) insulate mass and energy transfer from the environment, and each box has the same T, P, and N. If the central wall is removed, then the two identical systems are combined.

interpreted as the ratio of the internal energy change with respect to creation/disappearance of a molecule, must be independent of the number of molecules. Among the seven master variables in thermodynamics, additive quantities are E, S, N, and V, called *extensive*, and nonadditive ones are T, P, and μ_k, called *intensive*. Note that the intensive quantities are defined as ratios of extensive quantities.

In the previous sections, we reviewed the five standard ensembles with their energy functions derived from three independent variables as

- Internal energy $E(S,V,\{N_k\})$
- Helmholtz energy $A(T,V,\{N_k\})$
- Thermodynamic potential $\Phi(T,V,\{\mu_k\})$
- Enthalpy $H(S,P,\{N_k\})$
- Gibbs energy $G(T,P,\{N_k\})$

Among these energy functions, E, A, Φ, and H depend on at least one extensive variable, S or V. Gibbs energy function is the only one that depends on two intensive variables, T and P. Although G basically varies with N_k, if the system is homogeneous, the chemical potential $\mu_k(T,P)$ is independent to the number of particles N_k. In many engineering applications dealing with mass transfer phenomena, temperature and pressure are often maintained as (pseudo-) constants. Molecules and particles translate spatially from one location to other, or are converted to another species (i.e., created or annihilated through physical and chemical reactions). In this light, the Gibbs energy $G(T,P,\{N_k\})$ is the most convenient representation of the system undergoing mass and/or heat transfer in the isobaric and isothermal environment.

Enthalpy $H(S,P,\{N_k\})$ is often used to characterize mass transfer phenomena under an isobaric-isentropic environment between two different temperatures, allowing volume expansion or compression. H is mainly used to link two temperature-dependent quantities such as equilibrium constants of chemical reactions in the NPT ensemble because it does not vary with T.

2.3.2. Anisothermal equilibrium

Consider two heterogeneous systems in equilibrium. This is similar to the case shown in **Figure 2**, but boxes 1 and 2 are not thermodynamically identical. In each box, the internal energy is fully represented using N_i, V_i, and S_i of box i for $i = 1$ and 2. Assume their volumes do not change so that $dV_i = 0$. We express the infinitesimal change of the entropy from Eq. (16) as

$$dS_i = \frac{dE_i}{T_i} - \frac{\mu_i}{T_i} dN_i \tag{34}$$

If the total number of particles $N(= N_1 + N_2)$ is constant, we simply derive

$$\frac{\partial N_2}{\partial N_1} = \frac{\partial}{\partial N_1}(N - N_1) = -1 \tag{35}$$

In equilibrium, the total entropy $S = S_1 + S_2$ must be already maximized, having a constant value $S_{\max}$:

$$\frac{\partial S}{\partial N_1} = \frac{\partial S_{\max}}{\partial N_1} = 0 = \frac{\partial S_1}{\partial N_1} + \frac{\partial N_2}{\partial N_1}\frac{\partial S_2}{\partial N_2} = \frac{\partial S_1}{\partial N_1} - \frac{\partial S_2}{\partial N_2} \tag{36}$$

As the internal energy of each box, E_i, is kept invariant in Eq. (34), we derive

$$\frac{\partial S_1}{\partial N_1} = -\frac{\mu_1}{T_1} \quad \text{and} \quad \frac{\partial S_2}{\partial N_2} = -\frac{\mu_2}{T_2} \tag{37}$$

Substitution of Eq. (37) into (36) gives

$$\frac{\mu_1(T_1,P_1)}{T_1} = \frac{\mu_2(T_2,P_2)}{T_2} \tag{38}$$

which is simplified, if $T_1 = T_2$, to

$$\mu_1(P_1) = \mu_2(P_2) \tag{39}$$

for an isothermal environment. Note that Eqs. (38) and (39) consist of only intensive thermodynamic quantities. The chemical potential can be readily derived using Eq. (29) if the Gibbs energy is known.

3. Dilute solution

3.1. Chemical potentials

Now we consider a dilute (or weak) solution, in which the number of dissolved molecules in the solvent is much less than that of the solvent molecules. Without losing generality for environmental engineering purposes, we set water as the solvent. Gibbs free energy of the weak solution of a single solute species is [31]

$$G = N\mu_0(P,T) + nk_BT\ln\left(\frac{n}{Ne}\right) + n\psi(P,T) \tag{40}$$

where N and n are the numbers of solvent and solute molecules, respectively, μ_0 is the chemical potential of the pure solvent, and $\psi(P,T)$ is an arbitrary function for the chemical potential of the pure solute. Euler's number $e = 2.71828218\ldots$ in the denominator of the logarithmic function on the right-hand side of Eq. (40) stems from Starling's formula, used for entropy calculations: $\ln n! \simeq n\ln n - n = n\ln(n/e)$

If the weak solution contains multiple species of solutes, then the Gibbs energy function is generalized as

$$G = N\mu_0(P,T) + k_BT\sum_i n_i\ln\left(\frac{n_i}{Ne}\right) + \sum_i n_i\psi_i(P,T) \tag{41}$$

One can easily calculate the chemical potentials for the solvent μ_w and solute μ_s as partial derivatives of G in Eq. (41) with respect to N and n, respectively. The former and latter are

$$\mu_w = \frac{\partial G}{\partial N} = \mu_0(P,T) - k_BTx \tag{42}$$

and

$$\mu_s = \frac{\partial G}{\partial n} = \psi(P,T) + k_BT\ln x \tag{43}$$

respectively, where $x\,(= n/N)$ is the number (or mole) fraction of solute molecules to solvent molecules. In a dilute solution, $x \ll 1$.

3.2. Osmotic pressure

Let's consider an isothermal system consisting of two boxes (1 and 2) of the same size in contact. Box 1 (and 2) has the solute mole fraction x_1 (and x_2) and pressure P_1 (and P_2). Since the total system is in isothermal equilibrium, the two boxes have the same temperature: $T_1 = T_2 = T$.

3.2.1. *Using solvent chemical potential*

In this thermodynamic environment, the chemical potentials of water in the two boxes should be equal to each other from Eq. (39):

$$\mu_{w,1} = \mu_{w,2} \tag{44}$$

$$\mu_0(P_1,T) - x_1 k_B T = \mu_0(P_2,T) - x_2 k_B T \tag{45}$$

We assume that the pressure difference is small enough to use the weak solution approximation without drastic thermodynamic changes but large enough to maintain the balance between the two boxes. Then, we expand $\mu_0(P_2,T)$ around P_1 using Taylor's series

$$\mu_0(P_2,T) \simeq \mu_0(P_1,T) + \left(\frac{\partial \mu_0}{\partial P}\right)_T \Delta P \tag{46}$$

at a fixed temperature T. We substitute Eq. (46) into Eq. (45) to obtain

$$\left(\frac{\partial \mu_0}{\partial P}\right)_T \Delta P = (x_2 - x_1) k_B T \tag{47}$$

where $\Delta P = P_2 - P_1$ and $\Delta x = x_2 - x_1$ are differences of pressure and solute mole fraction, respectively, between box 1 and 2. Using Eq. (31), the fundamental representation of the infinitesimal chemical potential, we replaced $\partial\mu_0/\partial P$ with the volume per solvent, V/N. Then, the pressure difference ΔP is calculated as

$$\Delta P = k_B T \frac{N\Delta x}{V} = \mathcal{R}T \frac{\Delta n}{N_A V} \tag{48}$$

and finally denoted as

$$\Delta \pi = \mathcal{R}T\Delta C \tag{49}$$

using

$$\Delta n = N\Delta x = Nx_2 - Nx_1 = n_2 - n_1 \tag{50}$$

$$\Delta C = \Delta n / N_A = C_2 - C_1 \tag{51}$$

where n_i and $C_i\ (= n_i/N_A V)$ are the (absolute) number and the mole concentration of solutes in box i for $i = 1$ and 2, N_A is Avogadro's number, and $\mathcal{R}$ is the universal gas constant. Eq. (49) is called the van't Hoff equation,[1] which resembles the ideal gas law [33]. If the solution contains multiple species of solutes, Eq. (49) can be easily extended to

---

[1]Jacobus H. van't Hoff received the first Nobel Prize in Chemistry in 1901 for the discovery of osmotic pressure in solutions. https://www.nobelprize.org/nobel_prizes/chemistry/laureates/1901/

$$\Delta\pi = \sum_i (C_{i,2} - C_{i,1})\mathcal{R}T = \mathcal{R}T\Delta C \tag{52}$$

where $\Delta C = \sum_i (C_{i,2} - C_{i,1})$ is, in general, the difference of total mole concentration of solutes. *If the total mass concentration of multiple species is known, then it should be carefully converted to total mole concentration using molecular weights of the contained species.* The underlying assumptions of the van't Hoff equation (49) are summarized as follows:

1. The solute concentration is much smaller than the solvent concentration.
2. Temperature gradient between the two boxes is zero.
3. The Gibbs free energy of a dilute solution is described using the weak solution approach.

3.2.2. Using solute chemical potential

If the solvent chemical potentials of boxes 1 and 2 are equal, then the solute chemical potentials should be also the same:

$$\mu_{s,1} = \mu_{s,2} \tag{53}$$

which leads to

$$\psi(P_1,T) + k_B T \ln x_1 = \psi(P_2,T) + k_B T \ln x_2 \tag{54}$$

Using the same approximation for the pressure difference, we derive

$$-\left(\frac{\partial\psi}{\partial P}\right)_T \Delta P = k_B T \Delta(\ln x) \tag{55}$$

$$= k_B T \ln\left(\frac{x_2}{x_1}\right) \tag{56}$$

where $\Delta \ln x = \ln x_2 - \ln x_1$ is the logarithmic difference between concentrations in two boxes. Eq. (56) can further be approximated as follows:

$$-k_B T \ln\left(\frac{x_2}{x_1}\right) = -k_B T \ln\left(1 + \frac{\Delta x}{x_1}\right) \simeq -k_B T \frac{\Delta x}{x_1} \tag{57}$$

We treat the negative derivative of ψ with respect to P as the volume per each solute molecule, i.e.,

$$-\left(\frac{\partial\psi}{\partial P}\right)_T = \frac{V}{n_1} \sim \frac{V}{n_2} \sim \frac{V}{\bar{n}} \tag{58}$$

where $\bar{n} = (n_1 + n_2)/2$, implicitly assuming $N \gg n_i \gg \Delta n$ for $i = 1, 2$. The pressure difference is then calculated as

$$\Delta P = \frac{n_1}{V}\frac{\Delta x}{x_1}k_B T = \frac{n_1/N}{x_1}\frac{N\Delta x}{V}k_B T = 1 \cdot \frac{\Delta n}{V}k_B T = \frac{\Delta n/N_A}{V}\mathcal{R}T \tag{59}$$

which reduces to the identical result of Eq. (49):

$$\Delta\pi = \mathcal{R}T\Delta C \tag{60}$$

The same result can be obtained in a slightly more mathematical way by directly using Eq. (55):

$$\Delta P = \frac{k_B T}{-(\partial\psi[\partial p)_T}\left(\frac{\mathrm{dln}x}{\mathrm{d}x}\right)\Delta x \tag{61}$$

$$= \frac{k_B T}{-(\partial\psi[\partial p)_T}\frac{\Delta x}{x} \tag{62}$$

where

$$\Delta(\ln x) \backsimeq \left(\frac{\mathrm{dln}x}{\mathrm{d}x}\right)\Delta x = \frac{\Delta x}{x} \tag{63}$$

is used. If Δx is finite, a similar approximation can be suggested:

$$\Delta(\ln x) = \left(\frac{\Delta \ln x}{\Delta x}\Delta x\right) = \frac{\Delta x}{\langle x\rangle_{\ln}} \tag{64}$$

where

$$\langle x\rangle_{\ln} = \frac{\Delta x}{\Delta \ln x} \tag{65}$$

is the logarithmic average of the solute mole fraction across the membrane interior. Employing Eq. (58) and $\Delta x/x = \Delta C/C$, we confirm that the osmotic pressure of the dilute concentration is

$$\Delta\pi = \frac{\mathcal{R}T}{-N_A\,(\partial\psi[\partial p)_T}\frac{\Delta x}{x} = \mathcal{R}T\left(\frac{n}{N_A V}\right)\frac{\Delta C}{C} = \mathcal{R}T\Delta C \tag{66}$$

In this section, we mathematically proved that the osmotic pressure (of Eqs. (49), (60), and (66)) is valid for dilute solution consisting of weakly interacting molecules. Without losing generality, the absolute value of the osmotic pressure can be expressed as (similar to the ideal gas law)

$$\pi = C\mathcal{R}T \tag{67}$$

Finally, it is worth noting that in Eq. (58), *the negative sign of the partial derivative indicates that the gradients of solvent and solute concentrations have opposite signs.* If the middle wall between the two boxes in **Figure 3** is partially removed, then solvent and solutes will diffuse in opposite

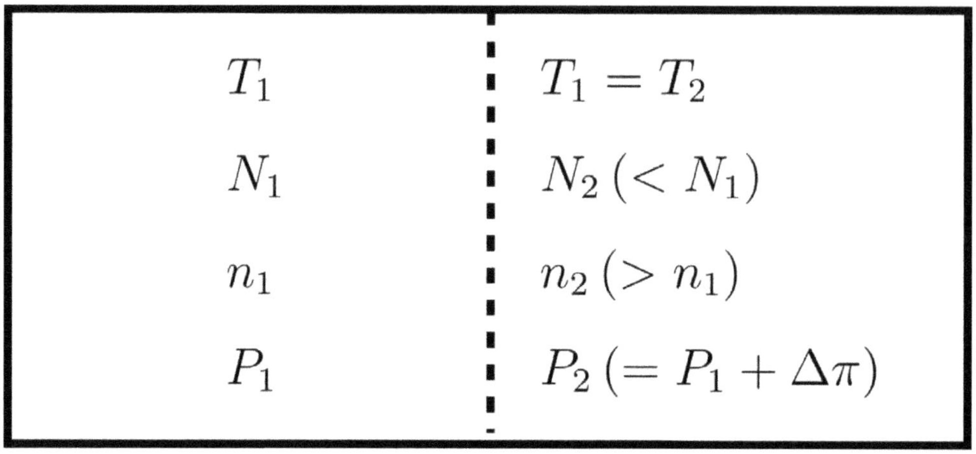

Figure 3. Osmotic pressure schematic: two boxes separated by the semipermeable wall in the isothermal environment.

directions. This should be treated in principle as a binary diffusion of two species (i.e., solvent and solute) by exchanging their positions.

4. Solution-diffusion model revisited

4.1. Solvent (water) transport

For pressure-driven membrane processes such as RO and NF, the applied pressure should overcome the osmotic pressure difference across the membrane. In feed and permeate solutions, salts are dissolved as solutes in the solvent water. The hydraulic pressure generates solvent flow through the membrane, which may contribute to solute transport through the membrane surface. Water molecules, however, dissolve as solutes in the membrane material (as solvent). Due to the high density of the membrane, water molecules can migrate via diffusion from a higher concentration region to a lower concentration region. This normal diffusion is reversed by applying hydraulic pressure to the feed solution with a high concentration such as seawater. Water permeation through a RO membrane can be pictured as diffusion driven by the external hydraulic pressure, which allows us to neglect convective transport of solutes through the membrane. The phenomenological phase of water in the membrane leads to solute transport as Fickian, which is also closely related to the osmotic pressure gradient between two subsystems. The above-mentioned mechanisms are included in *solution-diffusion* model, proposed by Lonsdale et al. [30]. An extensive overview of RO models can be found elsewhere [34–39].

Consider a semipermeable membrane of thickness δ_m, with high and low concentrations on two sides. The solvent flux through the membrane is assumed to be Fickian [40, 41]:

$$J_w = -D_w \frac{dC_w}{dx} \tag{68}$$

where C_w is the concentration of water *dissolved in the membrane*. Assuming that the dissolved water in the membrane material can be treated as a *Henrian* solution, the chemical potential of the pure water (in the membrane solvent) is

$$\mu_w = \text{constant} + \mathcal{R}T\ln C_w \tag{69}$$

In this model, the underlying assumptions are:

1. The water and solute molecules dissolve into a membrane material.
2. The solution is considered as Henrian for water.
3. The feed and permeate streams are immiscible with the membrane.

Substitution of Eq. (69) into Eq. (68) gives

$$J_w = -D_w \frac{C_w}{\mathcal{R}T}\frac{d\mu_w}{dx} \approx \frac{D_w C_w}{\mathcal{R}T\delta_m}\Delta\mu_w \tag{70}$$

where $\Delta\mu_w$ is the transmembrane difference of μ_w at a constant temperature T, which can be written as

$$\Delta\mu_w = \int \frac{\partial\mu_w}{\partial C_s} dC_s + \overline{V}_w \Delta P \tag{71}$$

where C_s is the solute concentration. In the previous section, we proved that the chemical potential difference between two subsystems should vanish in the isothermal equilibrium (i.e., $\Delta T = 0$): $\Delta\mu_w = 0$. In this case, the transmembrane pressure difference is equal to the osmotic pressure difference, i.e., $\Delta P = \Delta\pi$, which gives

$$\int \frac{\partial\mu_w}{\partial C_s} dC_s = -\overline{V}_w \Delta\pi \tag{72}$$

Then, Eq. (71) is simplified to

$$\Delta\mu_w = \overline{V}_w(\Delta P - \Delta\pi) \tag{73}$$

Substitution of Eq. (73) into Eq. (70) provides

$$J_w = A(\Delta P - \Delta\pi) \tag{74}$$

which is the governing equation of solvent transport through the membrane as a medium in which water and solutes can dissolve. Here, A is the water permeability through the membrane:

$$A = \frac{D_w C_w \overline{V}_w}{\mathcal{R}T\delta_m} \tag{75}$$

which is a characteristic value of the membrane. It is challenging to predict or measure the diffusion coefficient D_w and dissolved concentration C_w of water molecules in the membrane material. Therefore, the water permeability A is often experimentally estimated by a linear regression plot of J_w versus ΔP using fresh water as a solute-free feed solution.

4.2. Solute transport

The solute transport through the membrane is also assumed to follow Fick's law:

$$J_s = -D_s \frac{dC_s}{dx} \backsimeq D_s \frac{\Delta C_s}{\delta_m} \tag{76}$$

where C_s and D_s are the concentration and diffusivity of solutes *dissolved* in the membrane, respectively, and ΔC_s is the solute concentration difference across *the membrane interior*. Similarly to C_w, C_s is hard to measure. Therefore, ΔC_s is assumed to be proportional to that between membrane surfaces ΔC_m. The partition coefficient K is then defined as

$$K = \frac{\Delta C_s}{\Delta C_m} \quad (< 1) \tag{77}$$

where ΔC_m is often approximated as the difference between feed concentration C_f and permeate concentration C_p in the RO processes. Substitution of Eq. (77) into Eq. (76) provides

$$J_s = B \Delta C_m \tag{78}$$

where

$$B = \frac{D_s K}{\delta_m} \tag{79}$$

is the solute permeability through the membrane. Note that B conventionally has the same dimension as J_w. Similarly to A, B can be macroscopically measured by independent experiments, providing a J_s versus ΔC_m graph. The slope of the graph, estimated using linear regression, is equal to B.

5. Thermodynamic irreversibility of filtration

A thermodynamic system has three types: open, closed, and isolated. In the open system, mass and heat can pass in and out of the system in contact with the environment. Only heat can be transferred between the closed system and the environment, and neither mass nor heat can be exchanged in the isolated system. Rigorously saying, all the filtration processes are open systems, having entering feed streams to be treated. Temperature gradients across the membrane are often negligible in the pressure- or osmosis-driven filtration processes, but significant in thermal membrane processes such as membrane distillation processes [42–45]. To address the open filtration processes, nonequilibrium statistical mechanics (NESM) should be used [46–48]. To the best of our knowledge, the NESM still burgeons in pure theoretical physics. Therefore, general solutions for irreversible engineering processes are barely found. The minimum condition for us to use equilibrium filtration theory is that the filtration system is already in a steady state, in which no physical quantities vary explicitly with respect to time, i.e., mathematically,

$$\frac{\partial[\ ]}{\partial t} = 0 \tag{97}$$

where [ ] can hold any variables associated to the filtration system. The steady state is, in principle, far away from static equilibrium. Note that the osmotic pressure is derived from a pure equilibrium state, especially for the isobaric-isothermal ensemble. This implies that the solution-diffusion model becomes less accurate if fluid flows in the membrane channels are fast enough or almost turbulent.

To investigate the intrinsically nonequilibrium filtration processes, the irreversible thermodynamic models were developed using the Onsager[2] reciprocal theorem [49]. Kedem and Katchalsky represented the local dissipation rate of free energy per unit volume as *dissipation function* for isothermal, nonelectrolyte systems in a steady state [50]:

$$\phi = \sum_{k=1}^{n} \vec{J}_k \cdot \nabla(-\mu_k) > 0 \tag{80}$$

for species k, having a constant flux $\vec{J}_k$. In the irreversible (i.e., nonequilibrium) process, entropy must increase and therefore the dissipation rate is positive-definite, i.e., $\phi > 0$. The dissipation function for RO is

$$\phi = J_v \Delta P + J_D \Delta\pi \tag{81}$$

where J_v and J_D are the total volumetric flux and the solute velocity relative to the solvent velocity, respectively. One can write

$$J_v = L_{11}\Delta P + L_{12}\Delta\pi \tag{82}$$

$$J_D = L_{21}\Delta P + L_{22}\Delta\pi \tag{83}$$

where L_{ij} are coupling coefficients of the phenomenological fluxes, J_v and J_D. In order to satisfy Eq. (81), the following two conditions must be met

$$L_{11}, L_{22} > 0 \tag{84}$$

and

[2]Lars Onsager received the Nobel Prize in Chemistry in 1968 for the discovery of the reciprocal relations in the fundamental thermodynamics of irreversible processes. http://www.nobelprize.org/nobel_prizes/chemistry/laureates/1968/

$$L_{11}\,L_{22} \geq L_{12}L_{21} = L_{12}^2 \tag{85}$$

where $L_{12} = L_{21}$.

After some theoretical steps, Kedem and Katchalsky [50] derived

$$J_v = L_{11}\,(\Delta P - \sigma \Delta \pi) \tag{86}$$

$$J_s = \overline{C}_s(1-\sigma)J_v + \omega \Delta \pi \tag{87}$$

where $\overline{C}_s$ is the logarithmic average of concentrations on the two membrane sides, $\sigma = -L_{12}/L_{11}$ assuming $L_{11} > 0$ and $L_{12} < 0$, and

$$\omega = \overline{C}_s \frac{L_{11}L_{22} - L_{12}^2}{L_{11}} = \overline{C}_s(L_{22} - \sigma|L_{12}|) \tag{88}$$

Here, σ is defined as the "filtration coefficient," representing the solute rejection property. Kedem and Katchalsky [50] interpret the physical meaning of σ as follows: when $\sigma = 1.0$, the membrane is completely impermeable to solute and rejection is 100%, and when $\sigma = 0.0$, the membrane is completely permeable to solute and rejection is zero. It is worth noting that the irreversible thermodynamic theory includes the solution-diffusion model as a special case. If $\sigma = 1.0$, then Eqs. (86) and (87) reduce to

$$J_v = L_{11}\,(\Delta P - \Delta \pi) \rightarrow A\,(\Delta P - \Delta \pi) \tag{89}$$

$$J_s = \overline{C}_s(L_{22} - |L_{12}|)\Delta \pi \rightarrow B\,\Delta C \tag{90}$$

where $L_{11} = A$ and $B = \omega \Delta \pi / \Delta C$, assuming the osmotic pressure is linearly proportional to the solute concentration.

In our opinion, $\sigma = 1$ can be interpreted in a different way. Because the unity σ in Eqs. (86) and (87) indicates that the effect of $\Delta \pi$ is maximized, the thermodynamic state of the membrane surface is quite close to the static equilibrium state. The solvent flux can be considered as the *barometric diffusion of water* as J_v increases with ΔP, overcoming $\Delta \pi$ across the membrane. The solute flux in this case is purely Fickian, which is dominated by only $\omega \Delta \pi (\propto \Delta C)$ in Eq. (87). The limiting value of $\sigma \rightarrow 1$, however, does not guarantee the perfect rejection of solutes because it does not satisfy $J_s = 0$ in Eq. (90). Knowing $L_{12} < 0$, one can rewrite Eq. (85) to give

$$L_{22} - |L_{12}| \geq \frac{L_{12}^2}{L_{11}} + L_{12} = \sigma^2 L_{11} + L_{12} = L_{11}\sigma(\sigma - 1) \tag{91}$$

which indicates that the solute flux J_s may vanish if $\sigma = 0$ or 1. Here, we have to discard $\sigma = 0$ because J_s in Eq. (87) reaches its maximum at $\sigma = 0$. Then, the condition $\sigma = 1$ applied to Eq. (91) must be only a necessary condition for $J_s = 0$. The inequality relationship in Eq. (91)

indicates that the perfect rejection can be achieved if $L_{22} - |L_{12}| = 0$ in addition to $\sigma = 0$. In membrane separations, the perfect rejection is related not only to the thermodynamic state of the membrane surface, but also to the specific membrane materials having salt rejecting capabilities.

Furthermore, variations of J_v and J_s with respect to σ can be investigated by calculating

$$\frac{1}{L_{11}}\frac{\partial J_v}{\partial \sigma} = -\Delta\pi \tag{92}$$

$$\frac{1}{\overline{C}_s L_{11}}\frac{\partial J_s}{\partial \sigma} = -[\Delta P + \Delta\pi] + \Delta\pi\sigma \tag{93}$$

and substitution of Eq. (92) into Eq. (93) gives

$$\frac{1}{\overline{C}_s L_{11}}\frac{\partial J_s}{\partial \sigma} - \sigma\Delta\pi = \frac{1}{L_{11}}\frac{\partial J_v}{\partial \sigma} - \Delta P \tag{94}$$

which is valid for an arbitrary σ between 0 and 1. Eq. (92) indicates that J_v monotonously decreases with respect to σ. If the filtration system is in a transient, nonequilibrium state far from the pure static equilibrium, the volumetric flux J_v must be higher than that in the quasiequilibrium state. The left-hand side of Eq. (93) is $-\Delta P$ at $\sigma = 1$ and $-[\Delta P + \Delta\pi]$ at $\sigma = 0$: as σ decreases, the magnitude of $\partial J_s/\partial\sigma$ increases.

Overall, σ can be physically interpreted not only as the filtration coefficient, but also as the equilibrium coefficient. When $\sigma \to 1$, the effect of the osmotic pressure difference reaches its maximum of the quasiequilibrium state, but the zero solute flux is not automatically guaranteed. The perfect rejection is achieved if the additional condition $L_{22} = |L_{12}|$ is satisfied, which is, however, independent of σ. The difference of σ indicates how much the filtration system is phenomenologically close to the static equilibrium. On the other side, if $\sigma \to 0$, then the filtration system can be in a steady state, but it is far from the static equilibrium. J_v and J_s approach their theoretical maximum values, and the solute transport is significantly influenced by convection. Although σ is a fundamentally and practically important parameter, to the best of our knowledge, there are no standard theories to directly predict σ. This is because the irreversible thermodynamic model relaxes the equilibrium restriction, but the NESM has not been fully developed yet.

Acknowledgements

This work was financially supported by the R&D project of "Infrastructure Establishment of Thermal Energy Conversion and Desalination using Seawater Thermal Energy(2/3)" (PES9060) from Korea Research Institute of Ships and Ocean Engineering (KRISO).

Author details

Albert S. Kim[1]* and Heyon-Ju Kim[2]

*Address all correspondence to: albertsk@hawaii.edu

1 Civil and Environmental Engineering, University of Hawaii at Manoa, Honolulu, United States of America

2 Seawater Utilization Plant Research Center, Korea Research Institute of Ships and Ocean Engineering, Gangwon-do, Republic of Korea

References

[1] Cheryan M. Microfiltration and Ultrafiltration Handbook. New York, NY: Technomic Presse; 1998.

[2] Shi X, Tal G, Hankins NP, Gitis V. Fouling and cleaning of ultrafiltration membranes: A review. Journal of Water Process Engineering. 2014 Apr;**1**:121–138. Available from: http://dx.doi.org/10.1016/j.jwpe.2014.04.003.

[3] Zeman LJ, Zydney AL. Microfiltration and Ultrafiltration: Principles and Applications. M. Dekker; 1996.

[4] Ho CC, Zydney AL. A Combined pore blockage and cake filtration model for protein fouling during microfiltration. Journal of Colloid and Interface Science. 2000 Dec;**232**(2):389–399. Available from: http://dx.doi.org/10.1006/jcis.2000.7231.

[5] Einstein A. {Ü}ber die von der molekularkinetischen Theorie der W{ä}rme geforderte Bewegung von in ruhenden Fl{ü}ssigkeiten suspendierten Teilchen. Annals d Physik. 1905;**322**(8):549–560. Available from:/journal/112477420/abstract.

[6] Einstein A. On the theory of the Brownian movement. Annals d Physik. 1906;**19**(4):371–381

[7] Einstein A. Zur Theorie der Brownschen Bewegung. Annals d Physik. 1906;**324**(2):371–381. Available from:/journal/112501299/abstract.

[8] Chandrasekhar S. Stochastic problems in physics and astronomy. Reviews of Modern Physics. 1943;**15**(1):1–89. Available from: http://prola.aps.org/abstract/RMP/v15/i1/p1_1.

[9] Chandrasekhar S, Elbert DD. On orthogonal functions which satisfy four boundary conditions.III.TABLES for use in fourier-bessel type expansions. Astrophysical Journal Supplement. 1958;**3**:453–458.

[10] Leighton D, Acrivos A. The shear-induced migration of particles in concentrated suspensions. Journal of Fluid Mechanics. 1987 Sep;**181**(1):415. Available from: http://dx.doi.org/10.1017/S0022112087002155.

[11] Acrivos A, Batchelor GK, Hinch EJ, Koch DL, Mauri R. Longitudinal shear-induced diffusion of spheres in a dilute suspension. Journal of Fluid Mechanics. 1992 Jul;**240**(1):651. Available from: http://dx.doi.org/10.1017/S0022112092000247.

[12] Kim AS, Liu Y. Irreversible chemical potential and shear-induced diffusion in crossflow filtration. Industrial & Engineering Chemistry Research. 2008;**47**(15):5611–5614

[13] Sablani S, Goosen M, Al-Belushi R, Wilf M. Concentration polarization in ultrafiltration and reverse osmosis: A critical review. Desalination. 2001 Dec;**141**(3):269–289. Available from: http://dx.doi.org/10.1016/S0011-9164(01)85005-0.

[14] Lee KP, Arnot TC, Mattia D. A review of reverse osmosis membrane materials for desalination—Development to date and future potential. Journal of Membrane Science. 2011 Mar;**370**(1–2):1–22. Available from: http://dx.doi.org/10.1016/j.memsci.2010.12.036.

[15] Kang Gd, Cao Ym. Development of antifouling reverse osmosis membranes for water treatment: A review. Water Research. 2012 Mar;**46**(3):584–600. Available from: http://dx.doi.org/10.1016/j.watres.2011.11.041.

[16] Pérez-González A, Urtiaga AM, Ibáñez R, Ortiz I. State of the art and review on the treatment technologies of water reverse osmosis concentrates. Water Research. 2012 Feb;**46**(2):267–283. Available from: http://dx.doi.org/10.1016/j.watres.2011.10.046.

[17] Rautenbach R, Linn T. High-pressure reverse osmosis and nanofiltration, a "zero discharge" process combination for the treatment of waste water with severe fouling/scaling potential. Desalination. 1996 Jun;**105**(1–2):63–70. Available from: http://dx.doi.org/10.1016/0011-9164(96)00059-8.

[18] Avlonitis SA, Kouroumbas K, Vlachakis N. Energy consumption and membrane replacement cost for seawater RO desalination plants. Desalination. 2003 Aug;**157**(1–3):151–158. Available from: http://dx.doi.org/10.1016/S0011-9164(03)00395-3.

[19] Mazlan NM, Peshev D, Livingston AG. Energy consumption for desalination — A comparison of forward osmosis with reverse osmosis, and the potential for perfect membranes. Desalination. 2016 Jan;**377**:138–151. Available from: http://dx.doi.org/10.1016/j.desal.2015.08.011.

[20] Werber JR, Deshmukh A, Elimelech M. Can batch or semi-batch processes save energy in reverse-osmosis desalination? Desalination. 2017 Jan;**402**:109–122. Available from: http://dx.doi.org/10.1016/j.desal.2016.09.028.

[21] Lin S, Elimelech M. Kinetics and energetics trade-off in reverse osmosis desalination with different configurations. Desalination. 2017 Jan;**401**:42–52. Available from: http://dx.doi.org/10.1016/j.desal.2016.09.008.

[22] Cath T, AA Childress, Elimelech M. Forward osmosis: Principles, applications, and recent developments. Journal of Membrane Science. 2006 Sep;**281**(1–2):70–87. Available from: http://dx.doi.org/10.1016/j.memsci.2006.05.048.

[23] Lutchmiah K, Verliefde ARD, Roest K, Rietveld LC, Cornelissen ER. Forward osmosis for application in wastewater treatment: A review. Water Research. 2014 Jul;**58**:179–197. Available from: http://dx.doi.org/10.1016/j.watres.2014.03.045.

[24] Chekli L, Phuntsho S, Kim JE, Kim J, Choi JY, Choi JS, et al. A comprehensive review of hybrid forward osmosis systems: Performance, applications and future prospects. Journal of Membrane Science. 2016 Jan;**497**:430–449. Available from: http://dx.doi.org/10.1016/j.memsci.2015.09.041.

[25] Cai Y, Hu XM. A critical review on draw solutes development for forward osmosis. Desalination. 2016 Aug;**391**:16–29. Available from: http://dx.doi.org/10.1016/j.desal.2016.03.021.

[26] Achilli A, Cath TY, Childress AE. Power generation with pressure retarded osmosis: An experimental and theoretical investigation. Journal of Membrane Science. 2009 Nov;**343** (1–2):42–52. Available from: http://dx.doi.org/10.1016/j.memsci.2009.07.006.

[27] Achilli A, Childress AE. Pressure retarded osmosis: From the vision of Sidney Loeb to the first prototype installation — Review. Desalination. 2010 Oct;**261**(3):205–211. Available from: http://dx.doi.org/10.1016/j.desal.2010.06.017.

[28] Yip NY, Elimelech M. Performance limiting effects in power generation from salinity gradients by pressure retarded osmosis. Environmental Science & Technology. 2011 Dec;**45**(23):10273–10282. Available from: http://dx.doi.org/10.1021/es203197e.

[29] Helfer F, Lemckert C, Anissimov YG. Osmotic power with pressure retarded osmosis: Theory, performance and trends —A review. Journal of Membrane Science. 2014 Mar;**453**: 337–358. Available from: http://dx.doi.org/10.1016/j.memsci.2013.10.053.

[30] Lonsdale HK, Merten U, Riley RL. Transport properties of cellulose acetate osmotic membranes. Journal of Applied Polymer Science. 1965 Apr;**9**(4):1341–1362. Available from: http://dx.doi.org/10.1002/app.1965.070090413.

[31] Landau LD, Lifshitz EM. Statistical physics. Oxford, Pergamon Press, 1991. Translated from the Russian by J.B. Sykes and M.J. Kearsley.

[32] Landau LD, Lifshitz EM. Fluid mechanics. Oxford, Pergamon Press (translated from the Russian by J.B. Sykes and W.H. Reid). 1987.

[33] Mulder M. Basic Principles of Membrane Technology. Dordrecht, The Netherlands: Kluwer Academic Publishers; 1996.

[34] Spiegler KS, Kedem O. Thermodynamics of hyperfiltration (reverse osmosis): criteria for efficient membranes. Desalination. 1966 Dec;**1**(4):311–326. Available from: http://dx.doi.org/10.1016/S0011-9164(00)80018-1.

[35] Lee CH. Theory of reverse osmosis and some other membrane permeation operations. Journal of Applied Polymer Science. 1975 Jan;**19**(1):83–95. Available from: http://dx.doi.org/10.1002/app.1975.070190107.

[36] Soltanieh M, Gill WN. Review of reverse osmosis membranes and transport models. Chemical Engineering Communications. 1981 Nov;**12**(4–6):279–363. Available from: http://dx.doi.org/10.1080/00986448108910843.

[37] Potts DE, Ahlert RC, Wang SS. A critical review of fouling of reverse osmosis membranes. Desalination. 1981 Mar;**36**(3):235–264. Available from: http://dx.doi.org/10.1016/S0011-9164(00)88642-7.

[38] Wijmans JG, Baker RW. The solution-diffusion model: a review. Journal of Membrane Science. 1995 Nov;**107**(1–2):1–21. Available from: http://dx.doi.org/10.1016/0376-7388(95)00102-I.

[39] Paul D. Reformulation of the solution-diffusion theory of reverse osmosis. Journal of Membrane Science. 2004 Oct;**241**(2):371–386. Available from: http://dx.doi.org/10.1016/j.memsci.2004.05.026.

[40] Fick A. Neue Ausstellung an dem Begriffe des endosmotischen Aequivalentes. Annalen der Physik und Chemie. 1854;**168**(6):333–335. Available from: http://dx.doi.org/10.1002/andp.18541680612.

[41] Fick A. On liquid diffusion. Journal of Membrane Science. 1995 Mar;**100**(1):33–38. Available from: http://dx.doi.org/10.1016/0376-7388(94)00230-V.

[42] Khayet M, Matsuura T. Membrane Distillation: Principles and Applications. New York: Elsevier; 2011.

[43] Kim AS. A two-interface transport model with pore-size distribution for predicting the performance of direct contact membrane distillation (DCMD). Journal of Membrane Science. 2013;**428**:410–424.

[44] Kim AS. Cylindrical cell model for direct contact membrane distillation (dcmd) of densely packed hollow fibers. Journal of Membrane Science. 2014;**455**:168–186. Available from: http://dx.doi.org/10.1016/j.memsci.2013.12.067.

[45] Kim AS, Ki SJ, Kim HJ. Research perspective of membrane distillation: Multi-scale and multi-physics phenomena. Desalination and Water Treatment. 2017;**58**:351–359. Available from: http://dx.doi.org/10.5004/dwt.2017.11423.

[46] Staverman AJ. Non-equilibrium thermodynamics of membrane processes. Transactions of the Faraday Society. 1952;**48**:176. Available from: http://dx.doi.org/10.1039/TF9524800176.

[47] de Groot SR, Mazur P, King AL. Non-equilibrium thermodynamics. American Journal of Physics. 1963 Jul;**31**(7):558–559. Available from: http://dx.doi.org/10.1119/1.1969680.

[48] De Groot SR, Mazur P. Non-equilibrium Thermodynamics. New York, Dover Publications, Inc.; 2013.

[49] Onsager L. Reciprocal relations in irreversible processes. I. Physical Review. 1931 Feb;**37** (4):405–426. Available from: http://dx.doi.org/10.1103/PhysRev.37.405.

[50] Kedem O, Katchalsky A. Thermodynamic analysis of the permeability of biological membranes to non-electrolytes. Biochimica et Biophysica Acta. 1958 Jan;**27**:229–246. Available from: http://dx.doi.org/10.1016/0006-3002(58)90330-5.

3

Application of Multilayer Thin Film Technology in Desalination Membrane

Syed Javaid Zaidi and Farid Fadhillah

Abstract

Membrane-based desalination is the fastest growing technology in the area of desalination. Reverse osmosis (RO) and nanofiltration (NF) have been established in the last couple of decades; meanwhile, forward osmosis (FO) has begun to find its own place in the field of desalination. Typical commercial polyamide (PA) thin film composite (TFC) membrane has been mostly used in those membrane processes, but it has no drawback. Recently, a versatile, robust technique in preparing ultra-thin films, so-called layer-by-layer assembly (LbL), was adopted in fabrication of desalination membrane. This chapter highlights the most important literatures in the application of LbL assembly for preparing RO, NF and FO membranes, the obstacles and future works, which are essential for those who wish to work in the field.

Keywords: layer-by-layer, dip-LbL, spray-LbL, spin-LbL, reverse osmosis, nanofiltration, forward osmosis, surface modification

1. Introduction

United Nations Environment Programme reported that two out of every three people will live in water-stressed areas by the year 2025. In 2008, 450 million people in 29 countries suffered from water shortages [1]. In addition to that, many more people have been suffering from consuming contaminated water that may cost human lives [2]. These problems just come from domestic uses, which account for 5% of the total water consumption. Meanwhile, 75% of total global water consumption comes from agricultural uses and balance for industrial uses. So far, ground water represents about 90% of the world's readily available freshwater resources, and 1.5 billion people depend on it for their drinking water. Freshwater resources are so unevenly distributed that makes most people lack access to freshwater supply [1].

Desalination has been considered as the only method that can provide freshwater for drinking from traditional sources such as ground water, sea water and other saline aquifers [3]. These traditional sources of water account for 97.5% of all water on the Earth. Thus, capturing even a tiny portion of that water will have a significant impact on reducing water scarcity [3]. As generally known, the main desalination technology can be divided into two main categories, that is thermal-based and membrane-based desalination technology. Membrane technology, particularly RO, has been a dominant technology in the field of desalination membrane. Due to significant improvement of RO process in the last four decades, it has been projected to be the leading desalination technology in general and is mostly used in almost all areas except in the countries having readily available fossil fuels [4].

Since Cadotte developed thin film composite (TFC) membrane polyamide (PA) RO membrane using interfacial polymerization in 1980 [5], it has been mostly used in desalination membrane. However, commercial TFC RO still faces a major problem, particularly bio-fouling. This is a result of hydrophobic and rough nature of the PA membrane itself and also partly due to the fabrication technique. The method is quite powerful to prepare thin films but lacks fine control over surface properties.

In the middle of the 1990s, a rediscovery of the so-called layer-by-layer method has opened a new paradigm in the preparation of an ultra-thin layer. LbL assembly offers nano-level control over several surface properties such as thickness, surface composition, surface roughness and so on. Not to mention, the flexibility in terms of material opened the room for improvement in terms of chemistry of the membrane, a field that has long been considered mature [6].

Therefore, the purpose of this chapter is to review recent activities in the field of LbL assembly, particularly those used in preparing the membrane for desalination. At the beginning, we will briefly highlight some important aspects about LbL assembly in general. We will then focus on method of preparation, some important results, drawbacks and future challenges related to the application of LbL in desalination membrane. In addition to that, we will also highlight some recent works related to the application of this method in the modification of commercially available membranes used in the field of desalination. We limit our discussion to the application of this membrane in separating or removing ions of salt only because there have been many reported works as well in the application of LbL NF, for instance, for removing organic contaminants from water, etc.

2. Layer-by-layer

2.1. Dip-LbL (d-LbL)

The root of LbL assembly might be traced back to 1966 when Iler introduced a novel technique in which colloidal oppositely charged particles can be assembled into layer-by-layer films [7]. However, it was only after Decher reintroduced a similar technique for polyelectrolytes multilayer (PEM) assembly that the technique became very popular in colloidal and interfacial science [8]. After various testing and proofing, particularly for different multilayer precursors [9–13], the systematic way was then reported in 1997 and became the most-cited article in the field of chemistry for 10 years (1998–2008) [14]. The classic approach to assemble PEM thin films is by alternately dipping the substrate into two oppositely charged polyelectrolytes as can be seen in **Figure 1**.

Although LbL assembly was initially invented by making use of electrostatic interaction between the two oppositely charge polyelectrolytes or colloidal particles, nowadays, it can also be formed via donor/acceptor [15, 16], hydrogen bonding [17, 18], covalent bonds [19–21] and stereo-complex formation [22, 23]. The precise structure of each layer depends on a set of control parameters such as polyelectrolyte concentration, adsorption times, ionic strength [24], pH [25] or temperature [26].

2.2. Spray-LbL (Sr-LbL)

Sr-LbL was introduced by Schlenoff [27] by employing poly(diallyl dimethyl ammonium chloride) (PDADMAC) and poly(styrene sulfonate) (PSS) on silicon wafer. Using this technique, similar film structures and properties to d-LbL film can be achieved in shorter deposition time. It was reported that a fully automatic system of this technique can fabricate the film of the same quality as a d-LbL film 25 times faster [28–30]. The main drawbacks of Sr-LbL are still relatively slow polyelectrolyte assembly coupled with inefficient use of polymer solution, which is about 99% of the polymer solution that is rinsed off during film preparation. However, if the size of the substrate is the concern, then, this technique is more suitable than d-LbL [27]. Sr-LbL can also be used to conformally coat individual fibres within a textured surface of hydrophobic textile [29].

2.3. Spin LbL (SA-LbL)

SA-LbL was introduced by Hong *et al.* in 2001 [31, 32] and Chiarelli *et al.* in the same year [33]. Hong *et al.* successfully fabricated very smooth thin layers with controllable thickness that comprised of the combination of nanoparticles and polyelectrolyte. The major difference

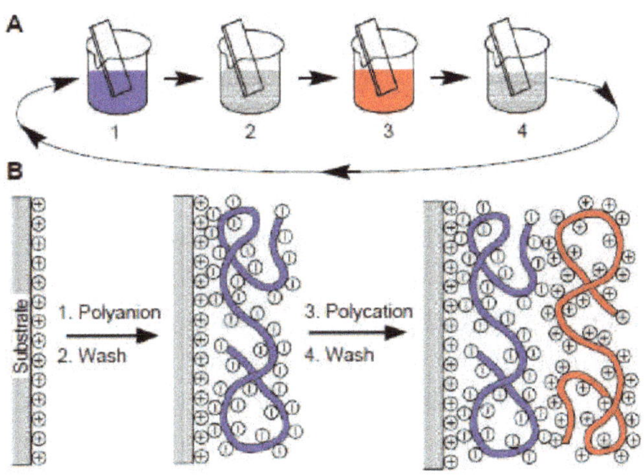

Figure 1. A schematic diagram of dip-layer-by-layer assembly (adapted from Ref. [14]).

between d-LbL and SA-LbL is the way polyelectrolytes are deposited on the substrate. As it is generally known, in the case of d-LbL, the substrate is immersed in polyelectrolyte solutions, while in the case of SA-LbL, only a small amount of polyelectrolyte is injected onto the spinning substrate.

It was reported that the SA-LbL film is several times thicker than d-LbL film for the same number of bilayers, which is due to different adsorption mechanisms. In d-LbL, the polymer chain will be adsorbed on the substrate under influence of electrostatic force and then followed by chain rearrangement. Meanwhile, in SA-LbL, due to high-speed spinning, the adsorption, the rearrangement of polyelectrolyte chains and water removal occur simultaneously. Quick water removal increases polyelectrolyte concentration in a very short time and also removes screening effect by water molecules. This in turn promotes faster adsorption and stronger electrostatic force; hence, more polymer chains are adsorbed within short time and result in a thicker film than d-LbL film [31].

Air shear force that occurs due to the relative movement between spinning substrate and air enhances the planarization of multilayer film, significantly reduces the surface roughness of the film and enhances mechanical integrity [34]. The above features are indirect evidences that SA-LbL film has a highly ordered internal structure [31], and experimentally, it has been revealed by neutron reflectivity study. Because a highly ordered internal structure can be obtained in much shorter time than conventional d-LbL or Sr-LbL, SA-LbL is considered to be more "technologically friendly" [34].

3. Layer-by-layer application

3.1. LbL for separation membrane

3.1.1. Nanofiltration

Due to its charged characteristic, polyelectrolyte LbL membrane has been found to be promising for NF application. There have been many works done and are ongoing in this area. Some representative works are listed in **Table 1**. We selected the best performance in terms of flux and selectivity of Cl^-/SO_4^{2-} from each work so they can be compared to each other. All those reported works used d-LbL approach unless it is stated differently.

The earliest notable work in d-LbL NF membrane perhaps was done by Krasemann and Tieke in 1999 [35]. They suggested the mechanism of salt rejection by LbL membrane similar to that of bipolar membrane as can be seen in **Figure 2**. However, some fundamental studies showed that PEM films do not exhibit an internal structure as originally projected and commonly depicted. There is significant intertwine between polyelectrolytes as a consequence of charge compensation. In d-LbL assembly, this charge compensation is most likely attained since the contact time between adjacent layers is sufficient to achieve it [36]. In accordance with this statement, our work using SA-LbL assembly also showed no significant increase in rejection

LbL film substrate	Main result	Testing condition	Ref
(PAH/PSS)-PAN/PET with 20–200 nm pore size[1]	$(PAH/PSS)_{60}$ showed selectivity of Na^+/Mg^{2+} = 15.1; Cl^-/SO_4^{2-} = 9.9	0.1 M for each of NaCl, $MgCl_2$ and $NaSO_4$, membrane active area: 4.53 cm^2, dead end cell	[35]
(PAH/PSS)-porous alumina with 20 nm pore size	$(PAH/PSS)_5$ showed selectivity of Cl^-/SO_4^{2-} = 7	0.1 M for each of KCl and K_2SO_4, membrane active area: 2 cm^2, dead end cell	[38]
(PAH/PSS)-porous alumina with 20 nm pore size	$(PAH/PSS)_{4.5}$ showed individual rejection of $MgSO_4$ = 96%, NaCl = 29% and selectivity Cl^-/SO_4^{2-} = 30	1000 ppm for each of NaCl, $MgSO_4$; P = 4.8 bars; membrane active area: 1.5 cm^2; cross flow cell.	[39]
(PSS/PDADMAC)-porous alumina	$(PSS/PDADMAC)_{4.5}$ showed WF = 2.4 m^3/m^2.day; rejection of SO_4^{2-} = 92.3; selectivity Cl^-/SO_4^{2-} = 15	1000 of Cl^-, 1000 ppm of SO_4^{2-} ; P = 4.8 bars; membrane active area: 1.5 cm^2; cross flow cell.	[42]
$(PAH/PSS)_5$ $(PAH/PAA)_{2.5}$ Support: Porous alumina with 20 nm pore size After heat-induced crosslinking at 115°C	Selectivity of Cl^-/SO_4^{2-} = 150 Selectivity of Cl^-/SO_4^{2-} = 360	0.1 F for each of KCl and K_2SO_4 Membrane active area: 2 cm^2.	[47]
(PDADMAC/PSS)-PES UF 50 kDa MWCO[2]	$(PSS/PDADMAC)_{4.5}$ showed WF = 1.6 m^3/m^2.day; rejection of SO_4^{2-} = 96%; selectivity Cl^-/SO_4^{2-} = 32	1000 of Cl^-, 1000 ppm of SO_4^{2-}; P = 4.8 bars; membrane active area: 1.5 cm^2; cross flow cell.	[48]
(PAH/PSS)/(PAH/PSSMA)-PAN UF[3]	$(PAH/PSS)_1/(PAH/PSSMA)_1$ showed WF = 28.6 L/m^2.h, rejection of Na_2SO_4 = 91.4%	1000 ppm Na_2SO_4, P = 2 bars membrane active area: 23.75 cm^2	[40]

[1] PAN/PET = polyacrylonitrile/polyether terephthalate.
[2] PES = polyethersulfone.
[3] PSSMA = poly (4-styrenesulfonic acid-co-maleic acid).

Table 1. Selected work for NF prepared from polyelectrolyte multilayer membrane using LbL assembly.

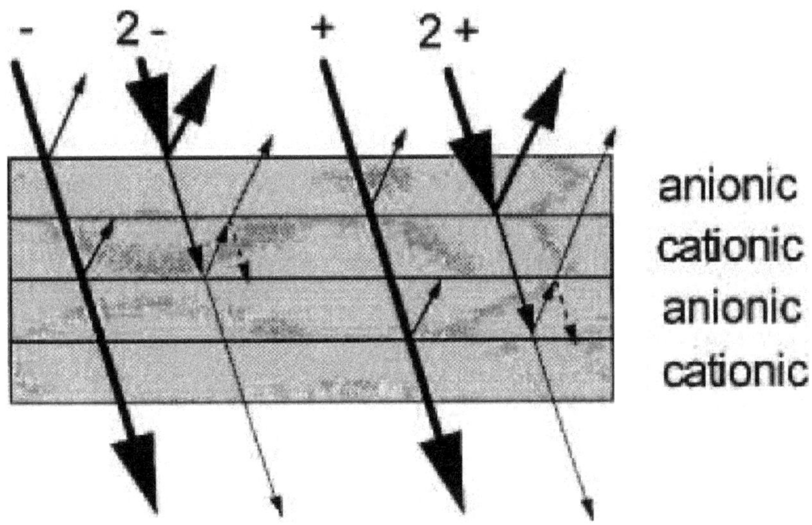

Figure 2. A rejection model of multi-bipolar membrane (adapted from Ref. [35]).

by doubling the number of layers, even though it is well known that the contact between layers is very short in this type of LbL assembly. This is evidence that even in short time, charge compensation is attained to some extent. It is also evidence that a bipolar mechanism may not be suitable, and this type of membrane rather follows a typical mechanism for NF or RO membrane [37]. It was also investigated that by changing the outermost layer, the anion flux changes significantly suggest that only uncompensated charges lie near the film/solution interface that have significant effect in ion rejection [38, 39].

It is widely known that ion rejection is one of the most importance performance indicators of a membrane. Aside employing highly charged polyelectrolytes which can improve rejection from the outermost layer, using dynamic LbL assembly seems quite helpful in this aspect. In this type of LbL assembly, polyelectrolyte solution is forced to pass the support layer and leaves behind the polyelectrolyte on the surface of the support layer. Dynamic LbL produces a more compact and tighter LbL membrane, which in turn increases the rejection [40]. There is also so-called semi-dynamic LbL assembly, which involves the dynamic introduction/replacement of solution into the fibre lumen by syringes followed by static contact for desired time. This technique is quite useful for hollow fibre membrane as the support [41].

In addition to ion rejection, flux is also considered as the most important performance indicator. Many membranes have high flux due to swelling which in turn will reduce the rejection. From investigation of several frequently used PEM pairs, it was reported that solution fluxes decrease in the following order: PDADMAC/PSS > Poly(allylamine hydrochloride) PAH/PSS > PDADMAC/Poly(acrylic acid) PAA >> PAH/PAA [42]. PDADMAC/PSS, for instance, permits higher flux than PAH/PSS due to higher degree of swelling [43]. In addition to swelling, highly porous support can also contribute to higher flux because it helps in reducing membrane resistance. Several methods have been used to produce highly porous support such as using electrospun nanofiber [44] or incorporating nanoparticles such as montmorillonite (MMT) or silica gel (SG). For example, polyvinyl alcohol (PVA) mixed with 1 wt% montmorillonite (MMT) can achieve porosity of 80% [45]. Likewise, PAN support mixed with 1 wt% SG can also achieve porosity of 80% [46].

3.1.2. Forward osmosis

The application of LbL assembly for NF is quite promising as can be seen in the previous section. There have been many works done and are ongoing in that area; meanwhile, the use of LbL assembly for fabricating FO is still at its infancy. The main results from FO mode of testing of several LbL FO membranes can be seen in **Table 2**. It is important to note that the information given in **Table 2** is only part of the works that showed the best performance in terms of water flux (WF) and salt to water flux ratio (SWFR) with active layer facing draw solution (ALDS) testing mode. It is known that ALDS mode shows higher initial water flux compared to ALFS mode, but it is also prone to internal fouling because foulants enter the porous support easily. This type of fouling in fact is much more difficult to be cleaned. In order to overcome this problem, a double-skinned design was proposed [49]. Double-skinned layer approach uses active layer deposited on both surfaces with the support layer, which will be in between as can be seen in **Figure 3**. Hence, it will prevent the foulants from entering the membrane from the support layer side.

As previously discussed, swelling of the membrane can result in higher flux due to expansion of the pores. For NF application, in which the membrane is exposed to relatively low salt concentration solution, swelling may not cause serious problems, but in the case of FO or RO, swelling degree can be very high and significantly decreases the salt rejection. Several attempts have been done to enhance LbL film stability, for example, by using crosslinking. Glutaraldehyde (GA) is the mostly used chemical crosslinker as can be seen in **Table 2**. With this crosslinking, LbL membrane can maintain its structure and performance and showed quite comparable performance with the commercial membrane [50]. Besides GA, UV light can also be employed to further increase the extent of crosslinking [51]. The mechanism of GA and UV light crosslinking can be seen in **Figure 4**. Some polyelectrolytes that contain carbonyl and amine functional group can be crosslinked simply by heating at 180°C, and this results in a polyamide-like layer which has better stability and lower swelling degree [52].

So far, the preparation of LbL membrane is mainly from polyelectrolytes. Recently, the researcher has started utilizing the technique for fabrication of LbL membrane using methylphenylene diamine (MPD) and trimesoyl chloride (TMC), two most commonly used

LbL film substrate	LbL assembly	Result/testing condition[4]	Ref.
(PAH/PSS)-PAN	Dip, crosslinked using GA	$(PAH/PSS)_3$ showed WF of 105.4 L/m^2.h and SWFR of 2.8 mM	[50]
	Dip, crosslinked using GA, double-skinned layer	$(PAH/PSS)_{3\text{-}3}$ showed WF of 35.5 L/m^2.h and SWFR of 4.86 mM	[49]
	Dip, crosslinked using GA, embedded silver nanoparticle	$(PAH/PSS)_{2.5}$ showed WF = 42 L/m^2.h and SWFR = 0.84 mM / FS: 10 mM NaCl	[55]
	Dip, crosslinked using combined GA and UV light	$(PAH/PSS)_3$ with UV exposure = 2 h showed WF = 15 L/m^2.h and SWFR = 0.8 mM	[51]
(PAH/PSS)-hollow fibre PES	Dip	$(PAH/PSS)_6$ showed WF = 40.5 L/m^2.h and SWFR = 2.1 mM	[56]
	Semi-dynamic	$(PAH/PSS)_2$ inner deposited layers showed WF = 73.5 L/m^2.h and SWFR = 0.6 mM	[41]
(PAH/PAA-PSS)-PEI	Dip, crosslinked using GA	$(PAH/PAA\text{-}PSS)_3$ showed water flux of 28 L/m^2.h and SWFR of 0.74 mM/dead end cell	[57]
(CHI/PAA)–(PVA+MMT–TA+ LiCl)	Dip	$(CHI/PAA)_3$ with MMT = 1 wt% showed WF = 19 L/m^2.h and SWFR of 0.09 mM/DS: 0.5 M NaCl, dead end cell	[45]
(PAH/PSS)–(PAN+SG)	Dip	$(PAH/PSS)_3$ with SG = 1 wt% showed WF = 77.9 L/m^2.h and SWFR = 0.94 mM	[46]
(MPD/TMC)–(PEI/PAA coated PAN)	Dip	10 layers of molecular LbL showed WF of 33 L/m^2.h and SWFR of 1.7 mM/DS: 0.5 M NaCl	[53]

[4]All reported default testing condition is DS = 0.5 M $MgCl_2$, FS = pure water, cross flow system, unless it is mentioned differently. π_{NaCl} at 0.5 M = 22.74 atm and π $MgCl_2$ at 0.5 M = 32.65 atm.

Table 2. The work progress for FO prepared from polyelectrolyte multilayer membrane using LbL assembly.

Figure 3. A conceptual illustration of doubled-skin LbL membrane (adapted from Ref. [49]).

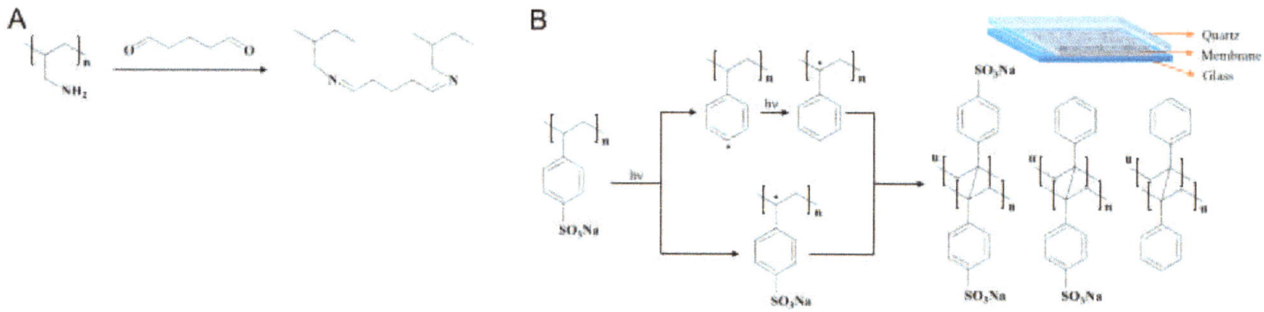

Figure 4. A schematic mechanism of crosslinking (A) using GA linker and (B) using UV light (adapted from Ref. [51]).

monomers for PA. This method is known as molecular layer-by-layer assembly (m-LbL) [53]. The mechanism of m-LbL is shown in **Figure 5**. Compared to several typical LbL FO and commercial FO membranes, m-LbL FO membrane showed superior performance for both ALDS and ALFS modes. Having said that, m-LbL FO membrane still has problems with stability and irreversible defect, especially after exposing it to very high salt concentration solution, for example, at 2 M NaCl. This instability is most likely due to relatively weak interaction between m-LbL film and the support. One of the approaches to enhance this interaction is by employing interlayers that will be discussed in more detail in the RO section.

3.1.3. Reverse osmosis

The use of LbL assembly for fabricating RO membrane is one of the most challenging applications because a typical RO membrane is tested at relatively high salt concentration up to 32,000 ppm with high operating pressure. Many polyelectrolyte multilayer films are not stable when exposed to such conditions.

The earliest work in this field to the author's knowledge was done by Jin *et al.* in 2003 [58]. Polyvinyl Amine (PVAm)/Polyvinyl sulphate (PVS) was used in their work and deposited on PAN/PET support using d-LbL assembly. Membrane performance was examined at a pressure of 40 bars, with NaCl concentration of 584.4 ppm using dead end cells with membrane active area of 36.8 cm^2. The result showed salt rejection of 93.5% and permeability of around 0.11 $L/m^2.h.bar$. Similar work using thermally induced PAH/PAA deposited onto

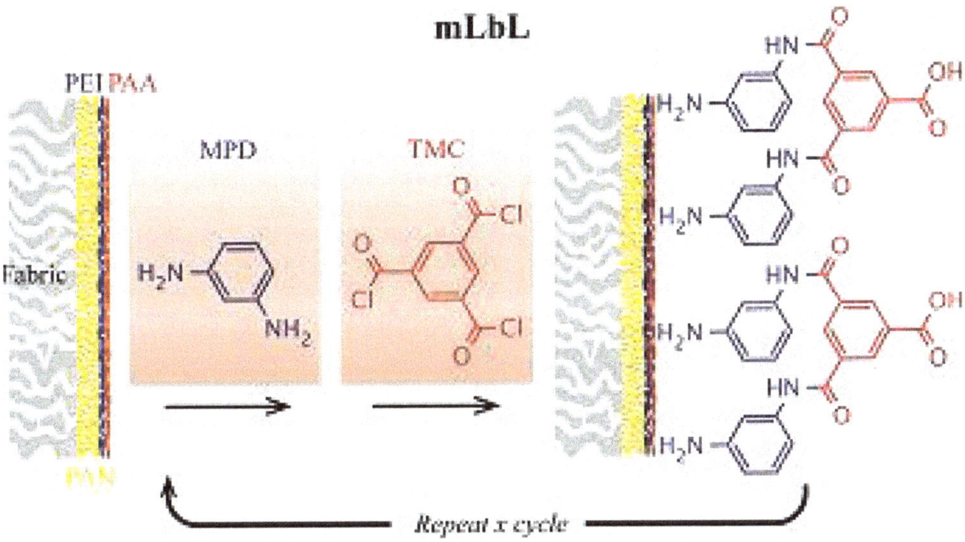

Figure 5. A schematic diagram for m-LbL process (adapted from Ref. [54]).

PSF supports showed salt rejection of around 81% with water permeability of 0.35 L/m^2.h.bar tested at a pressure of 20 bars and NaCl concentration of 2000 ppm using cross flow cells with active area of 13.85 cm^2 [59].

From the above literature review, almost all RO and FO LbL membranes were tested at low salt concentrations. Meanwhile, RO and FO membranes are normally used to treat water with rather high salt concentrations. There has been an issue with LbL film stability when it is exposed to such solution. Some PEMs are quite stable even without crosslinking. For instance, it was reported that 35 layers of PAH/PAA were successfully deposited on PSF support using SA-LbL and tested at pressure of 48.6 bars and NaCl concentration of 15,000 ppm using cross flow cells with active area of 42 cm^2. The test showed very stable performance with salt rejection of around 88% and water permeability of 0.22 L/m^2.h.bar. This work was considered the first attempt to apply SA-LbL for fabricating RO membrane [60]. Besides that, the use of Sr-LbL to fabricate RO membrane was also investigated, for example, assembly of clay (laponite (LAP)) and polyelectrolyte multilayers on the top of PSF support. (PAH/PAA) (PAH/LAP) deposited at pH 5 and tested at pressure of 18 bars, with NaCl concentration of 10,000 ppm using dead end permeation cells showed salt rejection of 89% with water permeability of 2.82 x 10^{-13} m^2/Pa.s. [61].

As previously mentioned, the use of m-LbL for RO membrane was also studied (see **Figure 5**). Using m-LbL, the membrane performance can be finely tuned by simply varying the number of layers. In general, as the number of layers increases, the flux decreases and rejection increases as can be seen in **Figure 6**. For instance, with 10 layers of m-LbL only, salt rejection of around 96% coupled with flux of 82% higher than IP TFC PA membrane was attained. In addition to this remarkable result, the surface roughness of this m-LbL membrane was only 3.4 nm, which was much smoother than surface roughness of IP PA TFC (i.e. 45.1 nm in this study) [54].

Recently, the same group of researchers investigated the role of interaction between m-LbL as active layer and the support layer, that is, PAN by showing the performance difference between hydrolyzed PAN (HPAN) and non-hydrolyzed PAN (see **Figure 7**) [62].

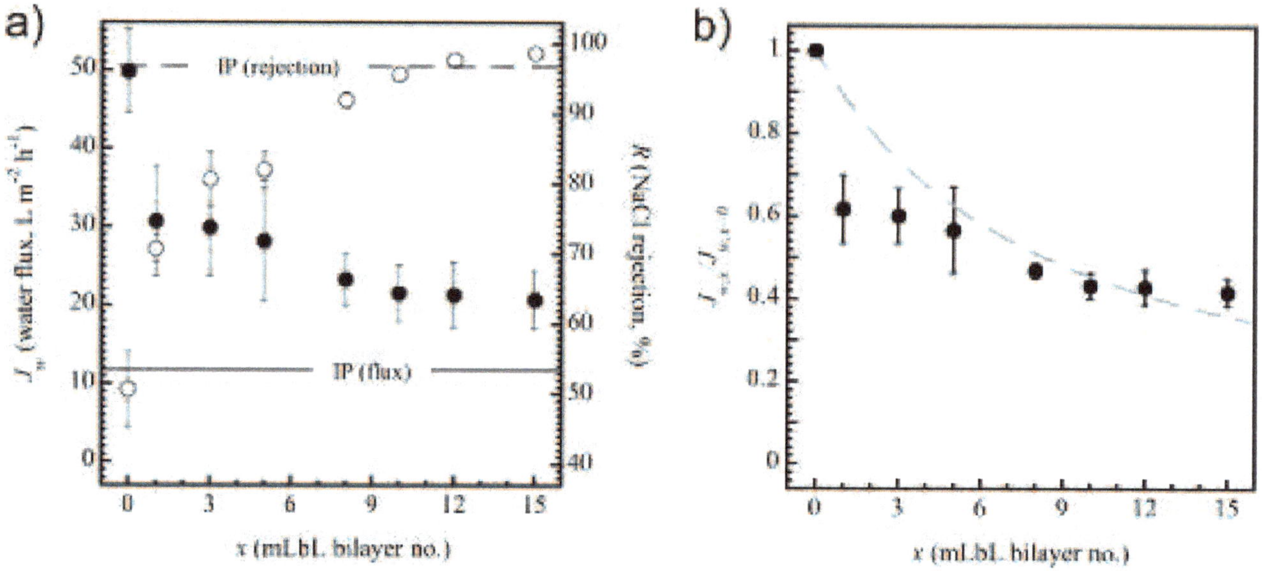

Figure 6. (a) Water flux (Jw, filled symbols) and NaCl rejection (unfilled symbols) as functions of a number of layers. (b) Normalized flux as a function of a number of bilayers (membrane was tested at NaCl concentration of 2000 ppm and pressure of 15.5 bars; adapted from Ref. [54]).

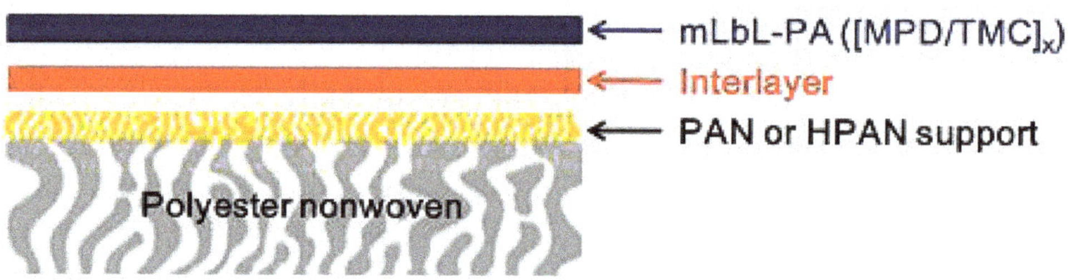

Figure 7. A conceptual design of m-LbL TFC PA membrane with interlayer (adapted from Ref. [62]).

As can be seen in **Figure 8**, deposition of m-LbL PA on the top of HPAN resulted in much higher rejection than that on the top of PAN. This improvement occurred due to more uniform active layers deposited on HPAN as a result of better interaction between carboxylate group of HPAN with amine group of MPD. Even though the interaction was better, in general, the active layer was still found to be insufficiently dense and selective towards NaCl so the rejection was still lower than IP PA TFC. The extent of uniformity then was further increased by additional interlayers. PEI/PAA interlayer on the top of HPAN showed by far the best performance. This was due to more carboxylate groups provided by PAA as the outermost layer prior to the deposition of MPD/TMC. Carboxylate groups of PAA will serve as seeding sites for subsequent deposition of MPD/TMC. The membrane with PEI/PAA interlayer showed remarkable results, that is, rejection of 98.7% coupled with water flux of 20.7 L/m^2.h which is 75% higher than IP PA TFC (the membrane was tested at pressure of 15.5 bars and NaCl concentration of 2000 ppm). This is indeed a very promising result in the field of RO.

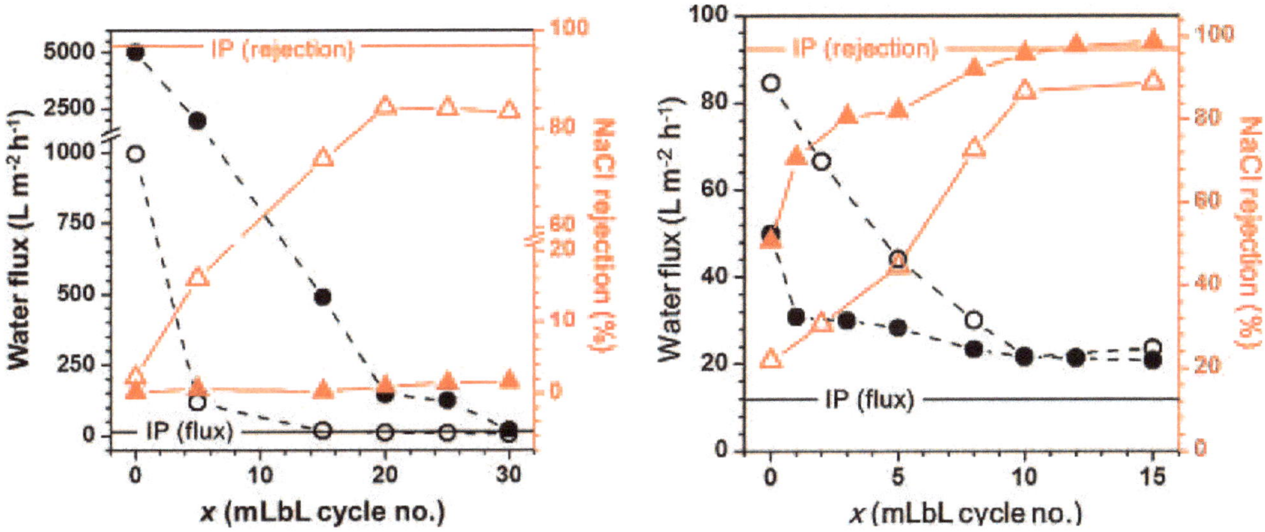

Figure 8. m-LbL PA TFC performance (represented by a circle for water flux and triangle for rejection). PAN support (left figure, filled symbol), HPAN (left figure, unfilled symbol), PEI/PAA-HPAN support (right figure, filled symbol) and PEI-HPAN (right figure, unfilled symbol) (adapted from Ref. [62]).

3.2. Membrane modification

In this section, the use of LbL assembly to modify the surface of existing RO, NF or FO membranes will be discussed. One of the most important surface modification purposes is to protect the membrane from being directly exposed to harsh environment such as chlorine or oxidants content in water, foulants and pH of the water at which the membrane life will be shortened. There have been tremendous works in the modification of commercially available membranes using different approaches such as coating, blending, incorporating nanomaterials, functionalization, grafting, etc. [63]. This section is devoted only for membrane modification using LbL assembly, which is mostly a sort of coating technique that can be simply a physical process or chemically bonded to the membrane surface.

The idea behind modification of the RO membrane using LbL is basically to alter surface properties that can reduce fouling tendency for example by reducing the roughness and hydrophobicity. The study showed that additional LbL films improve, significantly, the smoothness of the surface, improve salt rejection but sacrifice the flux. However, as far as fouling is concerned, this surface modification is very promising as no flux decrease was noticed when the membrane was exposed to foulants containing water [64].

Recently, graphene oxide (GO), a single-sheet functionalized graphene with oxygen-rich functional group, has attracted attention from researchers in the field of water treatment. This material provides fast water transport, hydrophilicity as well as excellent chemical stability. The main objectives of employing GO or GO/polyelectrolyte are to improve chlorine resistance and reduce the bio-fouling without sacrificing the flux (see **Figure 9** for the schematic illustration of GO-modified membrane) [65, 66]. Biofouling test using bovine serum albumin (BSA) showed that GO-modified PA TFC membrane can maintain the flux constant after approximately 6 h of testing; meanwhile, pristine PA TFC membrane kept showing

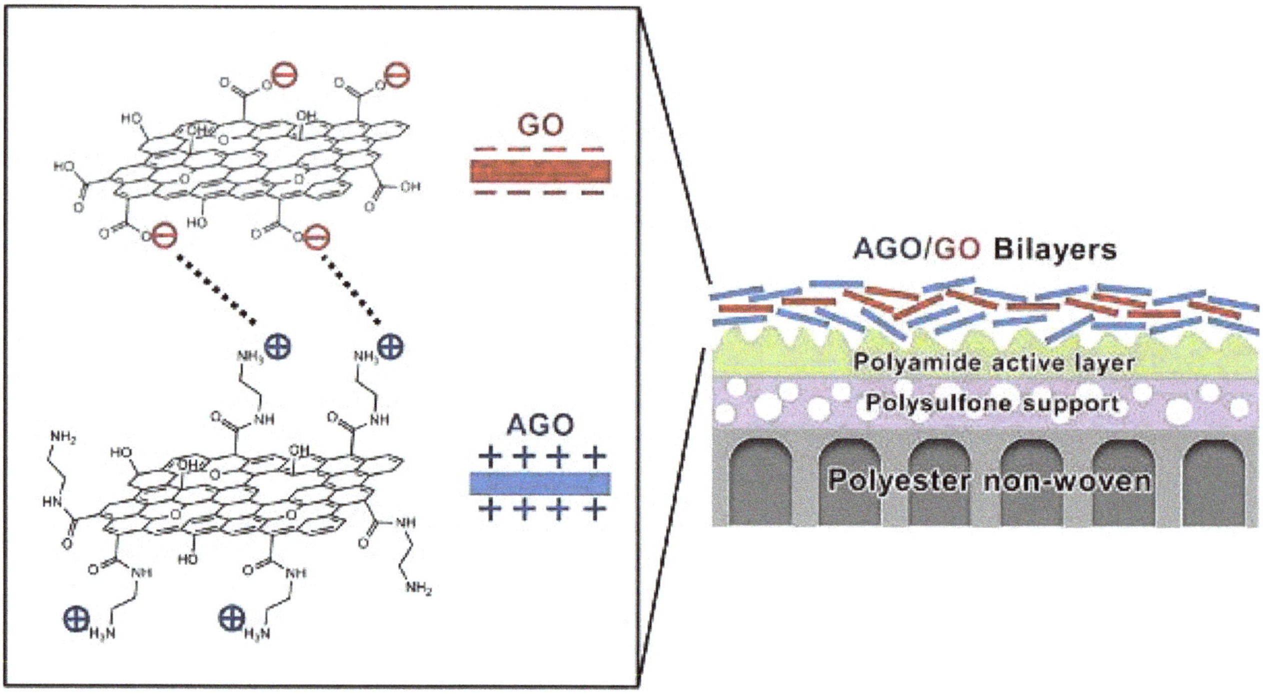

Figure 9. A schematic illustration of a multilayer GO on PA TFC membrane (adapted from Ref. [65]).

linear flux reduction even after 12 h of testing. This result showed that GO-modified PA TFC improves the fouling resistance but does not completely remove the fouling as indicated by flux decrease for the first 6 h of the experiment [65].

4. Conclusion and future work

IP PA TFC membrane has been extensively used in the area of RO, NF and FO. However, the nature of the process as well as the properties of the polyamide creates several problems such as chlorine and fouling resistance that are likely difficult to solve as long as the same fabrication technique is used. This is because the IP itself does not provide fine control over the film properties, not to mention its limited applicability only to few types of polymers such as polyamide and polycarbonate. Meanwhile, polyamide itself is naturally hydrophobic which causes severe fouling problems and has weak resistance against chlorine and oxidants. Relatively new technology, LbL assembly, offers flexibility and great control over the film properties which are the main keys to overcome aforementioned problems associated with PA TFC.

Based on literature study, most of the work in LbL desalination membrane still focuses in employing polyelectrolyte to form an active layer. Meanwhile, it has been investigated that many polyelectrolytes are highly hydrophilic in nature, they swell a lot when exposed to water and even swelling can be aggravated by salt infiltration [67] while those are two main compounds in saline water. Some attempts have been done to maintain stability of polyelectrolyte such as heat-induced, chemical or UV light crosslinking. However, only few studies have been done to investigate long-term stability of those membranes and even to the best of our knowledge most of studies were done under very soft conditions such as low salt concentration, typically 1000–2000 ppm. Thus, testing in harsh conditions for example at NaCl concentration of 32,000 ppm for longer duration must be done particularly for those LbL membranes intended to be used in RO or FO applications. The result of this study will definitely drive the research in finding the best protocol to create the most stable LbL membrane.

Also, from literature study, we found that most of the works were done using traditional dip-LbL. This method is quite difficult to be brought to an industrial scale as the adsorption process limited by diffusion of polyelectrolyte onto the support surface is a time-consuming process. Several researchers have migrated to other LbL assemblies such as dynamic LbL, semi-dynamic LbL, Sr-LbL and SA-LbL that utilize external force to speed up the adsorption process. Surely, more extensive works are urgently required for those LbL assemblies.

One of the advantages of LbL assembly that has not been thoroughly observed was the flexibility and applicability of this technique to create ultra-thin films from various materials. To the best of our knowledge, most of the works are still focused on polyelectrolyte. It is true some researchers have started introducing some inorganic nanomaterials such as silver nanoparticles, graphene oxide and clay, but still there is huge space available to do research in terms of membrane material. There were many polymers that have never been investigated because no appropriate technology was applicable to prepare ultra-thin films using those polymers. After the rediscovery of LbL, the door has opened. Using LbL assembly, one can create the film either from the polymer itself or from the monomers as in the case of m-LbL. One can also combine organic and inorganic materials with nano-level control easily to fabricate highly resistant membranes towards chlorine and foulants and, at the same time, give high flux and high rejection or selectivity for instance. Tailoring the film properties is one of the strength and advantages of LbL assembly that has not been deeply investigated but for preparing separation membrane.

As a new emerging technology, LbL must still go a long journey; it is going to face many challenges in the future. However, with all of its strength, versatility and robustness, we believe that LbL membrane one day will dominate the desalination membrane just as IP PA TFC did in the last couple of decades.

Author details

Syed Javaid Zaidi[1*] and Farid Fadhillah[2]

*Address all correspondence to: szaidi@qu.edu.qa

1 Center for Advanced Materials, Qatar University, Qatar

2 Chemical Engineering Department, Al Imam Mohammad Ibn Saud Islamic University (IMSIU), Riyadh, Kingdom of Saudi Arabia

References

[1] Executive Summary of Vital Water Graphics: An overview of the state of the world's Fresh and Marine Waters, 2nd Edition, 2008. Available from: http://new.unep.org/dewa/vitalwater/article186.html

[2] Dringking Water, Fact sheet, 2016, available from: http://www.who.int/mediacentre/factsheets/fs391/en/

[3] Shannon MA, Bohn PW, Elimelech M, Georgiadis JG, Marinas BJ, Mayes AM. Science and technology for water purification in the coming decades. Nature. 2008;**452**(7185):301-310

[4] Greenlee LF, Lawler DF, Freeman BD, Marrot B, Moulin P. Reverse osmosis desalination: Water sources, technology, and today's challenges. Water Research. 2009;**43**(9):2317-2348

[5] Cadotte JE. Interfacially synthesized reverse osmosis membrane [Internet]. Google Patents, 1981. Available from: https://www.google.com/patents/US4277344

[6] Petersen RJ. Composite reverse osmosis and nanofiltration membranes. Journal of Membrane Science. 1993;**83**:81-150

[7] Iler RK. Multilayers of colloidal particles. Journal of Colloid and Interface Science. 1966; **21**(6):569-594

[8] Decher G, Hong JD, Schmitt J. Buildup of ultrathin multilayer films by a self-assembly process: III. Consecutively alternating adsorption of anionic and cationic polyelectrolytes on charged surfaces. Thin Solid Films. 1992;**210-211**(Part 2):831-835

[9] Lvov Y, Decher G, Mohwald H. Assembly, structural characterization, and thermal-behavior of layer-by-layer deposited ultrathin films of poly(vinyl sulfate) and poly(Allylamine). Langmuir. 1993;**9**(2):481-486

[10] Schmitt J, Grunewald T, Decher G, Pershan PS, Kjaer K, Losche M. Internal structure of layer-by-layer adsorbed polyelectrolyte ilms-a neutran and X-ray reflectivity study. Macromolecules. 1993;**26**(25):7058-7063

[11] Lvov Y, Haas H, Decher G, Mohwald H, Mikhailov A, Mtchedlishvily B, et al. Successive deposition of alternate layers of polyelectrolytes and a charged virus. Langmuir. 1994;**10**(11):4232-4236.

[12] Schmitt J, Decher G, Dressick WJ, Brandow SL, Geer RE, Shashidhar R, et al. Metal nanoparticle/polymer superlattice films: Fabrication and control of layer structure. Advanced Materials. 1997;**9**(1):61-65.

[13] Sukhorukov GB, M○hwald H, Decher G, Lvov YM. Assembly of polyelectrolyte multilayer films by consec1utively alternating adsorption of polynucleotides and polycations. Thin Solid Films. 1996;**284**(285):220-223

[14] Decher G. Fuzzy nanoassemblies: Toward layered polymeric multicomposites. Science. 1997;**277**(5330):1232-1237

[15] Shimazaki Y, Mitsuishi M, Ito S, Yamamoto M. Preparation and characterization of the layer-by-layer deposited ultrathin film based on the charge-transfer interaction in organic solvents. Langmuir. 1998;**14**(10):2768-2773

[16] Shimazaki Y, Mitsuishi M, Ito S, Yamamoto M. Preparation of the layer-by-layer deposited ultrathin film based on the charge-transfer interaction. Langmuir. 1997;**13**(6):1385-1387

[17] DeLongchamp DM, Hammond PT. Highly ion conductive poly(ethylene oxide)-based solid polymer electrolytes from hydrogen bonding layer-by-layer assembly. Langmuir. 2004;**20**(13):5403-5411

[18] Lee H, Mensire R, Cohen RE, Rubner MF. Strategies for hydrogen bonding based layer-by-layer assembly of poly(vinyl alcohol) with weak polyacids. Macromolecules. 2012;**45**(1):347-355

[19] Tian Y, He Q, Tao C, Li J. Fabrication of fluorescent nanotubes based on layer-by-layer assembly via covalent bond. Langmuir. 2005;**22**(1):360-362

[20] Liang Z, Dzienis KL, Xu J, Wang Q. Covalent layer-by-layer assembly of conjugated polymers and CdSe nanoparticles: multilayer structure and photovoltaic properties. Advanced Functional Materials. 2006;**16**(4):542-548

[21] Bechler SL, Lynn DM. Reactive polymer multilayers fabricated by covalent layer-by-layer assembly: 1,4-conjugate addition-based approaches to the design of functional bio-interfaces. Biomacromolecules. 2012;**13**(5):1523-1532

[22] Hamada K, Serizawa T, Kitayama T, Fujimoto N, Hatada K, Akashi M. Stepwise stereocomplex assembly of isotactic poly(methyl methacrylate) and syndiotactic poly(alkyl methacrylate)s on surfaces. Langmuir. 2001;**17**(18):5513-5519

[23] Serizawa T, Hamada K, Kitayama T, Katsukawa K, Hatada K, Akashi M. Stepwise assembly of isotactic poly(methyl methacrylate) and syndiotactic poly(methacrylic acid) on a substrate. Langmuir. 2000;**16**(18):7112-7115

[24] Helm C, Lösche M, Möhwald H, Decher G, Schmitt J. Fine-Tuning of the film thickness of ultrathin multilayer films composed of consecutively alternating layers of anionic and cationic polyelectrolytes. In: Trends in Colloid and Interface Science VI. Berlin/Heidelberg: Springer; 1992. pp. 160-164.

[25] Shiratori SS, Rubner MF. pH-dependent thickness behavior of sequentially adsorbed layers of weak polyelectrolytes. Macromolecules. 2000;**33**(11):4213-4219

[26] Büscher K, Graf K, Ahrens H, Helm CA. Influence of adsorption conditions on the structure of polyelectrolyte multilayers. Langmuir. 2002;**18**(9):3585-3591

[27] Schlenoff JB, Dubas ST, Farhat T. Sprayed polyelectrolyte multilayers. Langmuir. 2000;**16**(26):9968-9969

[28] Nogueira GM, Banerjee D, Cohen RE, Rubner MF. Spray-layer-by-layer assembly can more rapidly produce optical-quality multistack heterostructures. Langmuir. 2011;**27**(12):7860-7.

[29] Krogman KC, Zacharia NS, Schroeder S, Hammond PT. Automated process for improved uniformity and versatility of layer-by-layer deposition. Langmuir. 2007;**23**(6):3137-3141

[30] Izquierdo A, Ono SS, Voegel JC, Schaaf P, Decher G. Dipping versus spraying: Exploring the deposition conditions for speeding up layer-by-layer assembly. Langmuir. 2005;**21**(16): 7558-7567

[31] Cho J, Char K, Hong JD, Lee KB. Fabrication of highly ordered multilayer films using a spin self-assembly method. Advanced Materials. 2001;**13**(14):1076-1078

[32] Lee S-S, Hong J-D, Kim CH, Kim K, Koo JP, Lee K-B. Layer-by-layer deposited multilayer assemblies of ionene-type polyelectrolytes based on the spin-coating method. Macromolecules. 2001;**34**(16):5358-5360

[33] Chiarelli PA, Johal MS, Casson JL, Roberts JB, Robinson JM, Wang HL. Controlled fabrication of polyelectrolyte multilayer thin films using spin-assembly. Advanced Materials. 2001;**13**(15):1167-1171

[34] Kharlampieva E, Kozlovskaya V, Chan J, Ankner JF, Tsukruk VV. Spin-assisted layer-by-layer assembly: Variation of stratification as studied with neutron reflectivity. Langmuir. 2009;**25**(24):14017-14024

[35] Krasemann L, Tieke B. Selective ion transport across self-assembled alternating multilayers of cationic and anionic polyelectrolytes. Langmuir. 1999;**16**(2):287-290

[36] Farhat TR, Schlenoff JB. Ion transport and equilibria in polyelectrolyte multilayers. Langmuir. 2001;**17**(4):1184-1192. Available from: http://dx.doi.org/10.1021/la001298+

[37] Fadhillah F, Zaidi SMJ, Khan Z, Khaled MM, Rahman F, Hammond PT. Development of polyelectrolyte multilayer thin film composite membrane for water desalination application. Desalination. 2013;**318**:19-24

[38] Harris JJ, Stair JL, Bruening ML. Layered polyelectrolyte films as selective, ultrathin barriers for anion transport. Chemistry of Materials. 2000;**12**(7):1941-1946

[39] Stanton BW, Harris JJ, Miller MD, Bruening ML. Ultrathin, multilayered polyelectrolyte films as nanofiltration membranes. Langmuir. 2003;**19**(17):7038-7042

[40] Deng H-Y, Xu Y-Y, Zhu B-K, Wei X-Z, Liu F, Cui Z-Y. Polyelectrolyte membranes prepared by dynamic self-assembly of poly (4-styrenesulfonic acid-co-maleic acid) sodium salt (PSSMA) for nanofiltration (I). Journal of Membrane Science. 2008;**323**(1): 125-133

[41] Liu C, Shi L, Wang R. Enhanced hollow fiber membrane performance via semi-dynamic layer-by-layer polyelectrolyte inner surface deposition for nanofiltration and forward osmosis applications. Reactive and Functional Polymers. 2015;**86**:154-160

[42] Hong SU, Malaisamy R, Bruening ML. Optimization of flux and selectivity in Cl-/SO42- separations with multilayer polyelectrolyte membranes. Journal of Membrane Science. 2006;**283**(1-2):366-372

[43] Hong SU, Ouyang L, Bruening ML. Recovery of phosphate using multilayer polyelectrolyte nanofiltration membranes. Journal of Membrane Science. 2009;**327**(1-2):2-5

[44] Ritcharoen W, Supaphol P, Pavasant P. Development of polyelectrolyte multilayer-coated electrospun cellulose acetate fiber mat as composite membranes. European Polymer Journal. 2008;**44**(12):3963-3968

[45] Pardeshi P, Mungray AA. Synthesis, characterization and application of novel high flux FO membrane by layer-by-layer self-assembled polyelectrolyte. Journal of Membrane Science. 2014;**453**:202-211

[46] Lee J-Y, Qi S, Liu X, Li Y, Huo F, Tang CY. Synthesis and characterization of silica gel–polyacrylonitrile mixed matrix forward osmosis membranes based on layer-by-layer assembly. Separation and Purification Technology. 2014;**124**:207-216. Available from: http://www.scopus.com/inward/record.url?eid=2-s2.0-84894071531&partnerID=tZOtx3y1

[47] Stair JL, Harris JJ, Bruening ML. Enhancement of the ion-transport selectivity of layered polyelectrolyte membranes through crosslinking and hybridization. Chemistry of Materials. 2001;**13**(8):2641-2648

[48] Malaisamy R, Bruening ML. High-flux nanofiltration membranes prepared by adsorption of multilayer polyelectrolyte membranes on polymeric supports. Langmuir. 2005;**21** (23):10587-10592

[49] Qi S, Qiu CQ, Zhao Y, Tang CY. Double-skinned forward osmosis membranes based on layer-by-layer assembly-FO performance and fouling behavior. Journal of Membrane Science. 2012; **405**:20-9

[50] Qiu C, Qi S, Tang CY. Synthesis of high flux forward osmosis membranes by chemically crosslinked layer-by-layer polyelectrolytes. Journal of Membrane Science. **381**(1-2):74-80

[51] Duong PHH, Zuo J, Chung T-S. Highly crosslinked layer-by-layer polyelectrolyte FO membranes: Understanding effects of salt concentration and deposition time on FO performance. Journal of Membrane Science. 2013; **427**:411-21.

[52] Harris JJ, DeRose PM, Bruening ML. Synthesis of passivating, nylon-like coatings through cross-linking of ultrathin polyelectrolyte films. Journal of the American Chemical Society. 1999;**121**(9):1978-1979. Available from: http://dx.doi.org/10.1021/ja9833467

[53] Kwon S-B, Lee JS, Kwon SJ, Yun S-T, Lee S, Lee J-H. Molecular layer-by-layer assembled forward osmosis membranes. Journal of Membrane Science. 2015;**488**:111-120

[54] Gu J-E, Lee S, Stafford CM, Lee JS, Choi W, Kim B-Y, et al. Molecular layer-by-layer assembled thin-film composite membranes for water desalination. Advanced Materials. 2013;**25**(34):4778-4782. Available from: http://dx.doi.org/10.1002/adma.201302030

[55] Liu X, Qi S, Li Y, Yang L, Cao B, Tang CY. Synthesis and characterization of novel antibacterial silver nanocomposite nanofiltration and forward osmosis membranes based on layer-by-layer assembly. Water Research. 2013;**47**(9):3081-3092. Available from: http://www.scopus.com/inward/record.url?eid=2-s2.0-84876699477&partnerID=tZOtx3y1

[56] Liu C, Fang W, Chou S, Shi L, Fane AG, Wang R. Fabrication of layer-by-layer assembled FO hollow fiber membranes and their performances using low concentration draw solutions. Desalination. 2013;**308**:147-153. Available from: http://www.scopus.com/inward/record.url?eid=2-s2.0-84870728544&partnerID=tZOtx3y1

[57] Cui Y, Wang H, Wang H, Chung T-S. Micro-morphology and formation of layer-by-layer membranes and their performance in osmotically driven processes. Chemical Engineering Science. **101**:13-26

[58] Jin W, Toutianoush A, Tieke B. Use of polyelectrolyte layer-by-layer assemblies as nanofiltration and reverse osmosis membranes. Langmuir. 2003;**19**(7):2550-2553

[59] Park J, Park J, Kim SH, Cho J, Bang J. Desalination membranes from pH-controlled and thermally-crosslinked layer-by-layer assembled multilayers. Journals of Materials Chemistry. 2010;**20**(11):2085-2091. Available from: http://dx.doi.org/10.1039/B918921A

[60] Fadhillah F, Zaidi SMJ, Khan Z, Khaled MM, Hammond PT, Javaid Zaidi SM, et al. Reverse osmosis desalination membrane formed from weak polyelectrolytes by spin assisted layer by layer technique. Desalination and Water Treatment. 2011;**34**(1-3):44-49. Available from: http://www.tandfonline.com/doi/abs/10.5004/dwt.2011.2856%5Cnpapers2://publication/doi/10.5004/dwt.2011.2856

[61] Kovacs JR, Liu C, Hammond PT. Spray layer-by-layer assembled clay composite thin films as selective layers in reverse osmosis membranes. ACS Applied Materials and Interfaces. 2015;**7**(24):13375-13383. Available from: http://www.scopus.com/inward/record.url?eid=2-s2.0-84932643621&partnerID=tZOtx3y1

[62] Gu J-E, Lee JS, Park S-H, Kim IT, Chan EP, Kwon Y-N, et al. Tailoring interlayer structure of molecular layer-by-layer assembled polyamide membranes for high separation performance. Applied Surface Science. 2015;**356**:659-667. Available from: http://www.sciencedirect.com/science/article/pii/S0169433215019327

[63] Rana D, Matsuura T. Surface modifications for antifouling membranes. Chemical Reviews. 2010;**110**(4):2448-2471. Available from: http://dx.doi.org/10.1021/cr800208y

[64] Ishigami T, Amano K, Fujii A, Ohmukai Y, Kamio E, Maruyama T, et al. Fouling reduction of reverse osmosis membrane by surface modification via layer-by-layer assembly. Separation Purification Technology. 2012;**99**:1-7. Available from: http://www.sciencedirect.com/science/article/pii/S1383586612004133

[65] Choi W, Choi J, Bang J, Lee J-H. Layer-by-layer assembly of graphene oxide nanosheets on polyamide membranes for durable reverse-osmosis applications. ACS Applied Material Interfaces. 2013;**5**(23):12510-12519. Available from: http://pubs.acs.org/doi/abs/10.1021/am403790s

[66] Hu M, Zheng S, Mi B. Organic fouling of graphene oxide membranes and its implications for membrane fouling control in engineered osmosis. Environmental Science & Technology. 2016;**50**(2):685-693. Available from: http://www.scopus.com/inward/record.url?eid=2-s2.0-84955322151&partnerID=tZOtx3y1

[67] Dubas ST, Schlenoff JB. Swelling and smoothing of polyelectrolyte multilayers by salt. Langmuir. 2001;**17**(25):7725-7727. Available from: http://dx.doi.org/10.1021/la0112099

4

On the Purification of Agro-Industrial Wastewater by Membrane Technologies: The Case of Olive Mill Effluents

Javier Miguel Ochando-Pulido and
Antonio Martinez-Ferez

Abstract

The olive oil production is one of the main industrial activities in the Mediterranean Basin: Italy, Portugal, Greece, and Northern African countries—Syria, Algeria, Turkey, Morocco, Tunisia, Libya, Lebanon, and Egypt. Also, France, Serbia and Montenegro, Macedonia, Cyprus, Turkey, Israel, and Jordan produce a considerable annual yield. Moreover, it is an emergent agro-food industry in China, the USA, Australia, the Middle East, and China, which is expected to develop a considerable production potential. Hence, the treatment of olive mill effluents is a task of global concern. In this context, advanced separation technologies comprising membranes and adsorption resins have been a breakthrough in terms of advanced separation and purification technologies, but many aspects are still in development or under investigation. In this chapter, a focus on the use of membrane and ion adsorption technologies for the purification of these wastewaters will be given. The effect of different factors comprising the type of membrane, i.e., ultrafiltration, nanofiltration, and reverse osmosis; the type of adsorbent (waste material, resins); and the operating conditions will be addressed. Conventional treatments are not able to abate the high concentration of dissolved species present in these effluents. The use of these technologies can be a feasible solution if properly engineered.

Keywords: olive mill effluents, membranes, adsorption, ion exchange, wastewater treatment

1. Introduction

The scarcity of water particularly concerns agricultural irrigation, which demands more than 70% of the total water consumption worldwide. In this scenario, however, there is a big potential to use regenerated wastewater for irrigation purposes. Reuse of used regenerated wastewater could be a very positive solution regarding environmental and economic impacts.

In the last decades, new advanced separation technologies, less intensive in terms of specific energy consumption than conventional separation procedures and "greener" regarding the minor use of chemicals and reagents to achieve the desired separation, have been more and more implemented. Concretely, membrane technology, adsorption, and ion-exchange (IE) processes can take the lead for these purposes.

Adsorption technology implies relatively low operation costs and long service life because adsorption materials can be regenerated and reused or raw materials can be used. Indeed, adsorption is attractive for its simplicity of design, avoidance of toxic solvents, and minimization of the transformation of the target substances [1]. On another hand, ion exchange is considered a green and simple technology, capable of achieving high removal efficiencies and targeted selectivity, with application for species from wastewaters [2].

Otherwise, membrane technology is modular, easy to design, thus easily scalable to the industry, requires low maintenance, and is environmentally friendly, providing high purifying capacity and selectivity [3–10]. There has been a significant boost in the use of membranes for a plethora of applications and particularly in the field of water and wastewater treatments in the last years. This impulse has been a result of the new membrane materials, modules designs, and optimization of the operating conditions, in specific those for the minimization of fouling issues.

In fact, membrane processes are becoming increasingly used in purifying processes for water and groundwater, to replace classic separation processes, as well as for the reclamation of effluents of different origins, especially those by-produced in agro-industries [6–15].

Among agro-industrial effluents, olive mill wastewater (OMW), generated during the production of olive oil in factories called "mills," is one of the most hardly polluted, depending on the procedure used, reaching chemical oxygen demand (COD) which values up to 100,000 mg/L. The volumes of these effluents have increased markedly in the last decades. Currently, average-sized modern olive oil mills generate several tenths of cubic meters of OMW daily, which raises several millions of cubic meters a year. This can be explained by two linked facts: the increase in the demand of olive oil worldwide due to its health-promoting features (nutritional, antioxidant, anti-inflammatory, cosmetic) and the adoption of continuous production procedures as a response to cope with that demand. Now, the affected countries are not only those in the Mediterranean region, where these industries are ancestral and represent an important sector of the industrial economy, i.e., Spain, Italy, Portugal, Greece, and Northern African countries—Syria, Algeria, Turkey, Morocco, Tunisia, Libya, Lebanon, and Egypt, but also France, Serbia and Montenegro, Macedonia, Cyprus, Turkey, Israel, and Jordan, as well as the USA, the Middle East, and China, with important production capacities nowadays.

The treatment of the olive mill effluents includes wastewater from the washing of the olives (OWW), as well as olive mill wastewater (OMW-3, for three phases of mills), wastewater from olive oil washing (OMW-2), as well as from other activities in the facility, including cleaning and sanitation. OWW includes high concentration of suspended solids (mainly, peel, pulp, ground, branches, and leaves debris) dragged during the olive fruit washing process but low concentration of dissolved organic matter—depending on the water flow exchange rate in the washing machines during the fruit cleaning procedure—usually lower than the standardized limits to discharge the effluent on superficial land suitable for the purpose.

OMW is characterized by strong odor nuisance, acid pH, intensive violet-dark color, and high saline toxicity, exhibiting considerable electroconductivity (EC) values [16]. Uncontrolled disposal of these effluents represents an environmental hazard, causing soil contamination, underground leakage, and water body pollution. Due to this presence of high COD load including refractory compounds, as well as fats and lipids, direct discharge of these wastewaters to the municipal sewage treatment plants is forbidden. Legal limits are set to prevent difficulties to keep the biomasses alive on which the municipal sewer treatment plants rely.

Discharge of untreated OMW to the ground fields and superficial waters bodies is currently prohibited in Spain, whereas in Italy as well as in other European countries, only partial discharge on suitable terrains is allowed; otherwise, in Portugal OMW can be stored and used for irrigation of arbustive cultures under control manner. Straight discharge of OMW has been reported to cause strong odor nuisance, soil contamination, plants growth inhibition, underground leaks, water body pollution, and hindrance of self-purification processes, as well as severe impacts to the aquatic fauna and to the ecological status [16–20].

The two-phase system appears to be more ecological, thus has been strongly promoted in Spain, and is now being implemented in Portugal and Greece. Nevertheless, the three-phase system is still surviving in other countries where scarcity of financial support has not permitted the technological switch. Considering that in the two-phase extraction water injection is only practiced in the final vertical centrifugation step, the volume of liquid effluent derived from the decanting process (OMW-2) is reduced by one third on average if compared to the amount required for the three-phase system. Moreover, much of the organic matter remains in the solid waste, which contains more humidity than the pomace from the three-phase system (60–70% in two-phase systems versus 30–45% in three-phase ones, OMW-3) and hence OMW-2 exhibits lower pollutants degree, too (**Table 1**).

In this chapter, a focus on the use of membrane and adsorption technologies for the purification of OMW will be given. The effect of different factors comprising the type of membrane, that is, microfiltration (MF), ultrafiltration (UF), nanofiltration (NF), and reverse osmosis (RO), the type of adsorbent (waste material, resins), and the operating conditions will be addressed. The problem of membrane fouling for these systems will also be covered.

Parameter	OMW-P	OMW-3	OMET-2	OMW-2	OWW
pH	4.5	5.4	7.2	4.9	6.3
Moisture, %	93.0	93.4	99.4	99.3	99.7
Total solids, %	12.0	6.6	0.59	0.6	0.27
Organic matter, %	10.5	5.8	0.39	0.49	0.10
Ashes, %	1.5	0.9	0.21	0.11	0.17
BOD_5, mg/L	90.0	42.0	0.29	0.79	0.50
COD, mg/L	180.0	151.4	7.1	7.8	0.8
Total phenols, mg/L	2,400	921.0	86.0	157.0	4.0
EC, $mS \cdot cm^{-1}$	9.0	7.9	1.9	1.3	0.9

OWW: olive washing wastewater; OMW-P, OMW-3 and OMW-2: olive mill wastewater from batch press process as well as three-phase and two-phase continuous extraction procedures, respectively; OMET-2: mixture of all effluents produced in the olive mill, including OWW, OMW, and from other activities in the facility (e.g., cleaning and sanitation); COD: chemical oxygen demand; BOD_5: biological oxygen demand; EC: conductivity.

Table 1. Average physicochemical composition of the different types of olive mill effluents [28].

Conventional physicochemical treatments are not effective to eliminate the significant salinity of these wastewaters, reflected in their high electric conductivity values. These treatments are not able to abate the high concentration of dissolved species present in these effluents, which present hazardous salinity levels per guideline established by the FAO for irrigation purposes or discharge into public waterways. The use of membrane and adsorption technologies can be a feasible solution if properly engineered.

2. Membrane processes for olive mill effluent purification

Membrane technology, in particular pressure-driven membrane processes comprising MF, UF, NF, and RO, can offer a series of advantages in contrast with other separation processes, making them very promising and environmentally friendly for the purification and remediation of OMW: needless of chemicals and reagents, e.g., solvents—to accomplish the desired separation and concentration; minor costs (capex and opex) and specific energetic requirements than most conventional separation alternatives, upon significant purification potentiality, selectivity, and recovery factors; and also easy scaling to the industry, low space necessities as they are modular, design simplicity, easy operation, and low maintenance needs [5–10].

Fouling mechanisms are very important to fully understand what is taking place between the membrane and the effluent, in view of the adoption and implementation of adequate decisions for the successful design of the membrane plant. This comprises the setup of specifically tailored pretreatment process and optimized operating conditions. Irreversible fouling arises quickly on the membranes due to the high concentration of pollutants when wastewater is purified without any pretreatment [5–12]. Therefore, adequate and optimally designed pretreatment processes on each particular feedstock, in other words, pretreatment tailoring of membrane processes, must be developed in order to maximize productivity and minimize fouling.

2.1. Microfiltration and ultrafiltration membranes

Canepa et al. [21] conducted a study on OMW-3 (average COD = 90 g/L) treatment based on a combined UF membrane followed by adsorption and RO process at pilot scale. The UF membrane was made of polysulfone, the permeate was treated with adsorbing polymers, and the resins eluate with RO polypiperazine-amide membranes. A rapid decline of the UF permeate flux was observed by the authors, and a daily washing was necessary. The fact that there was a flux decline of 80% after 20 h of UF and 70% after 15 h for the RO membrane was a capital handicap for the feasibility of the proposed process. UF was also examined by Borsani and Ferrando [22]. They proposed a pretreatment consisting of oil removal and suspended solid settling, achieving a final 50% organic matter removal with respect to the raw OMW-3 (COD = 70 g/L). Turano et al. [23] proposed centrifugation as pretreatment, simple, and mechanical operation which does not need addition of chemicals and is available in the production line, before UF—polysulfone, molecular weight cutoff (MWCO) 17 kDa—for the treatment of OMW-3. The centrifugation pretreatment removed 80% of the suspended solid concentration, and 90% COD reduction was achieved at the outlet of the integrated process. Paraskeva et al. [24] examined the treatment/fractionation of OMW-3 by a combination of polypropylene filter screening (80 μm), UF (zirconia, mean pore size 100 nm) followed by NF (polymeric, 200 Da) and RO. In particular, UF removed suspended solids and solid fat/lipid components (90%).

Stoller and Chianese [14] studied the purification of olive washing wastewater (OWW) to comply with municipal sewer discharge standards in Italy. In contrast with OMW, this effluent presents moderate organic pollutant load but high concentration of suspended solids. The authors proposed a coagulation-flocculation pretreatment with polyelectrolytes (aluminum sulfate (AS) or aluminum hydroxide (AH)) before UF and NF in series (composite thin-film spiral-wound membranes). Both pretreatment processes yielded similar COD and biological oxygen demand (BOD_5) rejection efficiencies, but higher productivity was observed in the subsequent membrane-in-series process after flocculation with AS. The same procedure was applied by these researchers to OMW-3 [8, 10]. In this case, they also compared the results with two additional pretreatments, heterogeneous photocatalysis with titania nanocatalysts and aerobic digestion. They noted that major pollutant removal by the pretreatment, e.g., COD, is not sufficient to state the suitability of the adopted pretreatment for the downstream membrane operation, but the fact that the pollutant particle size is shifted away from the size of the membrane pores is relevant to enhance steady-state flux.

On another hand, Akdemir and Ozer [25] applied pH adjustment (acidic or alkaline) and cartridge filtration (20 μm) as pretreatment before UF (MWCO 30–100 kDa) and reported the best performance for the 100 kDa UF membrane, pointing for 1 bar as the critical pressure to reduce fouling issues. Also, Coskun et al. [26] examined UF as part of an integral membrane process for OMW-3 reclamation, but did not consider the entity of fouling in the performance of the membranes. Zirehpour et al. [27] examined a lab-made UF membrane for the purification of OMW-3, preceded by a MF membrane (50, 5, and 0.2 μm). The commercial UF membrane provided higher flux than the lab-made one, but the antifouling properties and rejection efficiency of the latter were better.

On another hand, Ochando-Pulido et al. [28, 29] tested a polymeric UF membrane for the reclamation of both OWW and OMW-2, separately or mixed. As pretreatments, three different processes were examined: (i) first, gridding (cut size equal to 300 μm) of the raw effluent was carried out in order to remove the coarse particles; (ii) then, pH-temperature (T) flocculation (pH-T F) was performed by adding HNO_3 (70% w/w) under continuous stirring (320 rpm); and (iii) the supernatant phase of the pH-T F process after the separation of the mud was either directly conducted to the UF unit and thereby referred to as OMWW-F or further pretreated by photocatalysis (PhC) under ultraviolet irradiation (UV) with lab-made ferromagnetic-core TiO_2 nanoparticles (UV/TiO_2 PhC), thus named OMWW-F/PhC. Finally, the differently pretreated streams were driven to the UF and NF membrane units. The lab-made ferromagnetic photocatalyst nanoparticles were produced in three consecutive steps, following the procedure fully reported elsewhere [30]. Briefly, in the first place, magnetite was synthesized by a sol-gel process in a spinning disk reactor, using tetraethylorthosilicate and tetraisopropoxide as precursors. Moreover, threshold flux-based methods, as previously pointed by Stoller [9], were applied for to optimize the design of the plant and control of the operation.

The first studies on the experimental and theoretical behavior of fouling in MF and UF membranes were performed by Belfort et al., Field and Aimar [31, 32]. The first theoretical model that shed light on membrane transport phenomena of colloidal particles was elaborated by the research groups of Field et al. [3, 32], Bacchin et al. [33], and Mänttäri and Nystörm, [34]. The concept of the critical flux, as the flux which can be successfully attained by a given membrane without incurring in fouling formation during operation time, was explained and proven by these scientists for MF membranes. The insight into the different fouling typologies was addressed by Bacchin et al. and Ognier et al. Fouling was explained because of local conditions that may trigger liquid/gel phases over the membrane layer and in the membrane pores caused by concentration polarization profiles [35, 36]. Later, this concept was also extended to UF and NF membranes [4] and complemented with the concept of the threshold flux. In many cases, it was observed that it was not possible to completely inhibit fouling during the operation of some liquid-liquid membrane systems, as is the case of wastewater [4–7, 37]. The threshold flux divides a low fouling region from a high fouling region. Stoller et al. [38] worked on a reliable methodology to convert previously measured critical flux data into threshold flux values in his studies on OMW-3 purification with polymeric NF membranes. Stoller underlined the key necessity to develop quick and reliable methods capable to allow the estimation of threshold flux conditions and for the conversion of previously available critical flux measurement data into threshold flux data. Finally, both concepts were merged by Stoller and Ochando [39] into a new concept, the boundary flux. The knowledge of real-time boundary flux values is a key factor to design stable control systems for membrane processes (**Figure 1**).

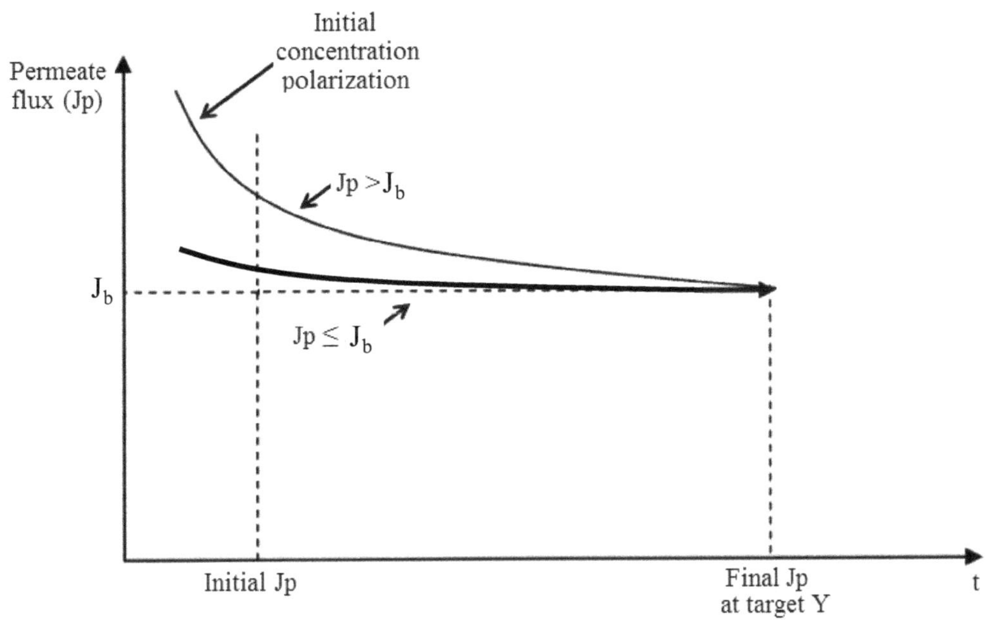

Figure 1. Operation strategies for membranes: above (curve line) or upon boundary flux (J_b) (straight line) conditions; J_p: permeate flux at any time t [29].

2.2. Nanofiltration and reverse osmosis membranes

The specific selectivity toward small solutes and the lower energy consumption of NF membranes have boosted their use as tertiary treatment in integrated wastewater treatment processes. On the other hand, RO membranes permit complying with the most stringent regulations for public health and environment protection. Both types of membranes are already applied in the management of industrial effluents of very different sectors, such as stainless steel [40], energy cogeneration [41], nuclear power [42], textile and tannery [43–47], coking [48, 49], carwash [50], pulp and paper [34, 51], pharmaceutical [52], and agro-food industries, such as dairy [13], tomato [12], and olive oil [5–10, 24], among others. However, again, the problem of fouling is always present in the treatment of any kind of effluent by NF and RO membranes. The critical, threshold, and boundary flux theories are, nonetheless, applicable as well for the design and control of the membrane plant.

Paraskeva et al. [24] used NF after MF/UF for the purification of OMW-3, achieving flux values up to 100–120 L/h and 95% phenol removal. Moreover, better efficiency was achieved by applying RO after NF, ensuring a significant conductivity, salinity, and turbidity decrease and nearly 30 L/h flux and 75–80% recovery of the initial feed volume. However, the permeate flux data regarding fouling was not discussed by the authors. The post-treated effluent was suitable for disposal in aquatic receptors or for irrigation purposes, and the inorganic part (including nitrogen, phosphorus, magnesium, potassium, and metal traces) and the organic fraction (hydrocarbons, nitrogenous compounds, organic acids, polyalcohols) may be potential plant nutrients, mixed with other inorganic or organic fertilizers, for instance, manure or sludge from biological treatments of other sorts of wastes. Zirehpour et al. (2012) evaluated the performance of a NF membrane at the end of a treatment line comprising MF and UF for OMW-3 purification. A specific arrangement of the integrated membrane system was found to be the UF membrane followed by two-step NF membranes in series, the first NF step providing high flux while the second one providing high rejection. Analysis of the fouling behavior of the used membranes was performed, per the recovery ratio and degree of the total flux loss during volume reduction factor experiments.

Di Lecce et al. [53] used polyamide thin-film composite spiral-wound NF at pilot scale for the fractionation of OMW-3 after MF. The purified NF permeate obtained presented COD and phenolic loads in the range of the standards for discharge into surface waters, with a rejection of 98% for COD, dry matter, and phenolic compounds. However, again, the performance of the NF membrane regarding its fouling was not reported.

On another hand, Ochando-Pulido et al. [28, 29] tested polymeric NF and RO membranes for the reclamation of OWW and OMW-2. The NF membrane was downstream of an UF one, and the effluent was initially pretreated by flocculation and/or heterogeneous photocatalysis (pH-T F and/or UV/TiO_2 PhC). This treatment sequence helped reducing the required membrane area for both membranes, and the applied UV/TiO_2 photocatalysis process enhanced the productivity and longevity of both membranes, achieving a final treated effluent stream compatible for irrigation. Moreover, the threshold flux theory was applied successfully to determine the threshold operating conditions for the process. In a similar line, Stoller [10] worked during 3 years continuously with a NF membrane for the treatment of OMW-3 with the goal to check for the reliability of threshold flux optimization methodologies for the fouling control on the used membrane in the long run. The author reported the successful operation of the NF membrane for the used period of pilot-scale work. Stoller highlighted that this was possible if adequately targeted fouling minimization control methods were carried out, based on the estimation of the critical flux for the optimal operation of the plant. These theories were successfully applied to this system, which could be explained by the threshold flux model.

Ochando-Pulido et al. [6, 7] studied the reclamation of OMW-2 (Spain) by RO, with the goal of "closing the loop," as is the trend of the current "circular economy." the quality to recirculate the final effluent to the olive washing machines of the manufacture process to finally close the loop. The effluent was firstly treated by means of an advanced oxidation process (AOP) which consisted in homogeneous catalysis with Fenton's reagents [16]. The authors noticed the importance of an operating variable of the RO process, the recirculation of a fraction of the permeate. The adoption of this operating sequence permitted enhancing the permeate production of the membrane, which was also stabilized, and resulted as well in a steady rejection of the target species, comprising suspended solids, phenols, and iron removal total rejection, as high as 99.4 and 98.2% of the COD and conductivity, respectively.

3. Adsorption processes for olive mill effluent purification

Adsorption technology has several advantages such as simple handling, relatively low operation costs, and long service lifetime because adsorption materials can be regenerated and reused. Also, adsorption is attractive for its simplicity of design, avoids toxic solvents, and minimizes the transformation of the target substances [1]. On the contrary, it can present limited purification efficiency and the loss of capacity depending on the feed solution [54].

Among the adsorption methodologies tested with OMW, there are some studies using so-called biosorbents (those of biological origin) as alternatives to other expensive adsorbents. Achak et al. [55] investigated the performance of a low-cost biosorbent made of banana peel. They examined various parameters including adsorbent dosage, pH, and contact time. After optimization, adsorption of phenols from 60 to 88% was achieved upon the equilibrium reached after 3 h of contact time. These authors also tested low-cost biosorbent, wheat bran, which is inexpensive and an available biomaterial. This biosorbent material yielded adsorption efficiencies toward phenolic compounds up to 67%, and the equilibrium was reached in 4 h of contact time. Moreover, both materials were found to provide better performances upon alkaline pH environments.

Another biosorbent, olive stone biomass, by-product of olive oil industry, was addressed by as adsorbent for iron from secondary-treated OMW [18, 56, 57]. The OMW effluent was pre-treated by an advanced oxidation process based on Fenton's reaction, which uses iron-based catalysts. The equilibrium adsorption capacity was found to increase when the particle size decreased (from < 1 to 4.8 mm). The percentage of iron adsorption increased from 30 to 70% when the initial concentration of biomass increased from 25 to 125 g/L. These authors demonstrated that direct reuse of olive pitches or a simple washing with cold and hot water can be sufficient for the adsorption process, which was fast and spontaneous within the first 10–20 min. The experimental data supports both pseudo-first and pseudo-second order models. Also, the adsorption capacity became incremented with the temperature, pin pointing for an endothermic adsorption process. In addition, the values of thermodynamic parameters of the process were reported: the activation energy ($Ea = 8.04$ kJ mol^{-1}), the standard enthalpy ($\Delta H_0 = 8.86$ kJ mol^{-1}), the standard free energy ($\Delta G_0 = -19.51$ kJ mol^{-1}), and the standard entropy ($\Delta S_0 = 91.4$ J $mol^{-1}K^{-1}$). Furthermore, column regeneration studies were carried out for various adsorption-desorption cycles [56], and the column performance was confirmed to permit multiple service and regeneration cycles, thus enabling the treatment of considerably large volumes of wastewater. A negligible loss in bed height and olive stone mass was observed. After adsorption, the olive stone biosorbent could be used as a biofuel for domestic or industrial uses. The present process was highlighted to be environmentally friendly and capable to reduce the iron load from other different effluents, providing an affordable technology for these small- and medium-scale industries.

Dried *Azolla caroliniana*, a freshwater aquatic fern, and granular activated carbon (AFC-LS activated with phosphoric acid) were studied to evaluate their adsorption and desorption capacities by Ena et al. [58]. Granular activated carbon (GAC) showed better efficiency, although total phenols were more than twice as concentrated in Azolla as those found in the GAC powder. Other studies with activated carbon, in this case in powdered form, for OMW treatment were performed by Sabbah et al. [59], after sand filtration, achieving 95% removal of the phenolic compounds present in OMW. Also, Azzam et al. [60] examined powdered activated carbon in one-stage, two-stage, and three-stage countercurrent adsorption system series in stirred batch vessels. The three-stage countercurrent adsorption process could reduce the phenol concentration in 96%, and the equilibrium was reached within 2–3 h.

Another adsorption material used is adsorption resins. Novel ion-exchange (IE) resins developed over the past few decades have promoted this technology as a suitable separation and purification process for wastewater treatment and fractionation purposes. In this sense, IE is considered a green and simple technology, capable of achieving high removal efficiencies and targeted selectivity [2].

IE technology is very attractive because of its relative simplicity of application as well as due to the low cost and the effectiveness to remove target species from wastewaters, particularly from diluted solutions. IE resins have also found an increasing application in the drinking water treatment sector over the last few decades, especially when there is a high concentration of natural organic matter (NOM) since high percentages on the removal efficiency of NOM by IE process are found (Comstock and Boyer) [61]. The efficient use of IE depends on several operating factors such as the contact time, operating temperature, pH, flow rate, initial pollutant concentration, and resin characteristics [62].

Canepa et al. [21] used a resin-bed column for the final purification of OMW-3 previously treated by UF, and performed good regarding the retention of polyphenols, and could treat around 3000 L until saturation.

Mineral resins, such as zeolites, have been reported to be useful mineral materials for reducing the phenolic compound concentration from OMW compared to other substrates, such as clay soil and bentonite. The regeneration of zeolite was easy after treatment either by simple settling or light centrifugation procedures. Moreover, this material could be easily regenerated by a very interesting eco-friendly technique known as low-temperature ashing (LTA) [63].

On another hand, Amberlite XAD16, a nonpolar resin, was used by Scoma et al. [64] as the adsorbent phase and ethanol as biocompatible desorbing phase. This process could remove until the 60% of the phenolic compounds from OMW. Ferri et al. [65] studied the adsorption and desorption efficiency of five resins with different physical properties (Amberlite XAD4, XAD7, XAD16, IRA96, and ISOLUTE ENV+) toward an aqueous solution of 10 typical phenolic compounds occurring in OMW. The desorption solvents tested were water, methanol, and ethanol under basic and acidified conditions. IRA96 polar resin reached the highest phenol adsorption (76%). However, the maximum desorption ratios achieved (60%) was achieved with ENV + resin and ethanol as the desorbing phase.

Zagklis et al. [66] tested the fractionation of OMW-3 exiting a previous membrane process comprising NF and RO by a three-step resin process. The nonionic XAD4, XAD16, and XAD7HP resins were selected. The proposed three-step resin process enabled at high extent the separation of phenols from carbohydrates, which hinder further concentration of phenols from the RO concentrate. After this, vacuum distillation was implemented for the concentration of phenols, facilitated by the carbohydrate removal and the change of solvent from water to ethanol. The final product had phenol concentration of 378 g/L in gallic acid equivalents, from an initial content in the raw OMW of 2.64 g/L. Kaleh and Geißen [54] studied the behavior of 16 commercial resins, and they demonstrated that the reduction and selective uptake of phenols from OMW are feasible by choosing the appropriate sorbents, conditions, and pretreatment.

Víctor-Ortega et al. [67] studied the removal of phenols from synthetic olive mill aqueous solutions with two different anionic resins: a strong-base resin (Amberlyst A26) and a weak-base one (Amberlite IRA-67). The effect of the phenolic load in the raw effluent and the recirculation time were studied. The equilibrium data were modeled with Langmuir, Freundlich, and Temkin isotherms, and the best correlated one was found to be Langmuir isotherm for both resins. Kinetics of the IE process was examined with the pseudo-first order, pseudo-second order, and intraparticle diffusion models. Both second-order and intraparticle diffusion models could describe the IE mechanism accurately. On the other hand, results revealed that phenol uptake is a spontaneous process for Amberlyst A26 anionic resin ($\Delta G° = -1.55$ kJ mol^{-1}), whereas it was found to be non-spontaneous for Amberlite IRA-67 ($\Delta G° = 3.06$ kJ mol^{-1}). Finally, Amberlyst A26 resin was confirmed to be considerably more efficient (80–98%) than Amberlite IRA-67 for the potential removal of phenols from olive mill industrial effluents.

In addition to this, Víctor-Ortega et al. [68] examined a continuous-flow IE process for the purification of OMW-2. They compared the performance of strong-base and weak-base anionic resins. They found that the pH of the feed affected the IE process. This permitted to achieve up to 98% removal of phenols upon an increment of the pH up to 7 with the strong-base resin. The efficiency was noted to be maintained constant at higher pH value. Otherwise, with the other resin, a similar behavior was observed, and the phenolic concentration could be removed up to 57% at pH around 7. With the aim to predict the performance of the resins bed break breakthrough curves, the experimental data were fitted to various models for varying effluent concentrations (5–100 mg/L) under the optimal simulated operating conditions. The authors found an enhancement of the IE efficacy with major concentration of phenols in the feed. For the different model equations tested, Thomas model yielded utmost accuracy, above that of Yoon-Nelson and Clark models. To sum up, and as a very important fact, column regeneration studies showed that almost 100% phenol recovery efficiencies were ensured. The proposed IE process led to a phenol solution susceptible to be concentrated and used in food, cosmetic, or pharmaceutical sectors and a purified effluent for irrigation purposes.

4. Conclusions

The production of olive oil is becoming global, and thus the treatment of the derived effluents from this industry includes wastewater from olive washing (OWW), olive mill wastewater (OMW, for three phases of mills), and wastewater from olive oil washing (OOW).

Conventional physicochemical treatments are not effective to eliminate the significant salinity of OME, reflected in their high electric conductivity values. In this context, advanced separation technologies comprising membranes and adsorption resins have been a breakthrough in terms of advanced separation and purification technologies, but many aspects are still in development or under investigation.

In this chapter, a focus on the use of membrane and ion adsorption technologies for the purification of OMW is reported. The effect of different factors comprising the type of membrane, i.e., ultrafiltration (UF), nanofiltration (NF), and reverse osmosis (RO); the type of adsorbent (waste material, resins); and the operating conditions will be addressed. The problem of membrane fouling for these systems will also be covered. The use of membrane and adsorption technologies can be a feasible solution if properly engineered, to comply with the guidelines established, e.g., by the Food and Agricultural Organization (FAO) for irrigation purposes or discharge into public waterways.

Author details

Javier Miguel Ochando-Pulido* and Antonio Martinez-Ferez

*Address all correspondence to: jmochandop@gmail.com

Department of Chemical Engineering, University of Granada, Granada, Spain

References

[1] Soto ML, Moure A, Domínguez H, Parajó JC. Recovery, concentration and purification of phenolic compounds by adsorption: A review. Journal of Food Engineering. 2011; **105**:1-27

[2] Caetano M, Valderrama C, Farran A, Cortina JL. Phenol removal from aqueous solution by adsorption and ion exchange mechanisms onto polymeric resins. Journal of Colloid and Interface Science. 2009;**338**:402-409.

[3] Field RW, Wu D, Howell JA, Gupta BB. Critical flux concept for microfiltration fouling. Journal of Membrane Science. 1995;**100**:259-272

[4] Field RW, Pearce GK. Critical, sustainable and threshold fluxes for membrane filtration with water industry applications. Advances in Colloid and Interface Science. 2011;**164**: 38-44

[5] Le-Clech P, Chen V, Fane TAG. Fouling in membrane bioreactors used in wastewater treatment. Journal of Membrane Science. 2006;**284**(1-2):17-53

[6] Ochando-Pulido JMS, Rodriguez-Vives JM, Martinez-Ferez A. The effect of permeate recirculation on the depuration of pretreated olive mill wastewater through reverse osmosis membranes. Desalination. 2012;**286**:145-154

[7] Ochando-Pulido JM, Hodaifa G, Rodriguez-Vives S, Martinez-Ferez A. Impacts of operating conditions on reverse osmosis performance of pretreated olive mill wastewater. Water Research. 2012;**46**(15):4621-4632

[8] Stoller M. On the effect of flocculation as pretreatment process and particle size distribution for membrane fouling reduction. Desalination. 2009;**240**:209-217

[9] Stoller M. Effective fouling inhibition by critical flux based optimization methods on a NF membrane module for olive mill wastewater treatment. Chemical Engineering Journal. 2011;**168**:1140-1148

[10] Stoller M. A three-year long experience of effective fouling inhibition by threshold flux based optimization methods on a NF membrane module for olive mill wastewater treatment. Membranes. 2013;**32**:37-42

[11] Bódalo A, Gómez JL, Gómez E, Máximo F, Hidalgo AM. Application of reverse osmosis to reduce pollutants present in industrial wastewater. Desalination. 2003;**155**:101-108

[12] Iaquinta M, Stoller M. Optimization of a nanofiltration membrane for tomato industry wastewater treatment. Desalination. 2009;**245**:314-320

[13] Luo J, Ding L, Wan Y, Jaffrin MY. Threshold flux for shear-enhanced nanofiltration: Experimental observation in dairy wastewater treatment. Journal of Membrane Science. 2012;**409**:276-284

[14] Stoller M, Chianese A. Optimization of membrane batch processes by means of the critical flux theory. Desalination. 2006;**191**:62-70

[15] Stoller M, Bravi M. Critical flux analyses on differently pretreated olive vegetation waste water streams: some case studies. Desalination. 2010;**250**:578-582

[16] Martínez NL, Hodaifa G, Rodríguez VS, Giménez JA, Ochando J. Degradation of organic matter in olive-oil mill wastewater through homogeneous Fenton-like reaction. Chemical Engineering Journal. 2011;**173**:503-510

[17] Danellakis D, Ntaikou I, Kornaros M, Dailianis S. Olive oil mill wastewater toxicity in the marine environment: Alterations of stress indices in tissues of mussel *Mytilus galloprovincialis*. Aquatic Toxicology. 2011;**101**(2):358-366

[18] Hodaifa G, Ben Driss Alami G, Ochando-Pulido JM, Víctor-Ortega MD, Martinez-Ferez A. Kinetic and thermodynamic parameters of iron adsorption onto olive stones. Ecological Engineering. 2014;**73**:270-275

[19] Karaouzas I, Skoulikidis NT, Giannakou U, Albanis TA. Spatial and temporal effects of olive mill wastewaters to stream macroinvertebrates and aquatic ecosystems status. Water Research. 2011;**45**(19):6334-6346

[20] Ntougias S, Gaitis F, Katsaris P, Skoulika S, Iliopoulos N, Zervakis GI. The effects of olives harvest period and production year on olive mill wastewater properties – Evaluation of Pleurotus strains as bioindicators of the effluent's toxicity. Chemosphere. 2013;**92**:399-405

[21] Canepa P, Marignetti N, Rognoni U, Calgari S. Olive mills wastewater treatment by combined membrane processes. Water Research. 1988;**22**(12):1491-1494

[22] Borsani R, Ferrando B. Ultrafiltration plant for olive vegetation waters by polymeric membrane batteries. Desalination. 1996;**108**:281-286

[23] Turano E, Curcio S, De Paola MG, Calabrò V, Iorio G. An integrated centrifugation–ultrafiltration system in the treatment of olive mill wastewater. Journal of Membrane Science. 2002;**206**:519-531

[24] Paraskeva CA, Papadakis VG, Tsarouchi E, Kanellopoulou DG, Koutsoukos PG. Membrane processing for olive mill wastewater fractionation. Desalination. 2007;**213**:218-229

[25] Akdemir EO, Ozer A. Investigation of two ultrafiltration membranes for treatment of olive oil mill wastewater. Desalination. 2009;**249**:660-666

[26] Coskun T, Debik E, Demir NM. Treatment of olive mill wastewaters by nanofiltration and reverse osmosis membranes. Desalination. 2010;**259**:65-70

[27] Zirehpour A, Jahanshahi M, Rahimpour A. Unique membrane process integration for olive oil mill wastewater purification. Separation and Purification Technology. 2012;**96**:124-131

[28] Ochando-Pulido JM, Hodaifa G, Victor-Ortega MD, Rodriguez-Vives S, Martinez-Ferez A. Effective treatment of olive mill effluents from two-phase and three-phase extraction processes by batch membranes in series operation upon threshold conditions. Journal of Hazardous Material. 2013;**263**(1):168-176

[29] Ochando-Pulido JM, Verardo V, Segura-Carretero A, Martinez-Ferez A. Technical optimization of an integrated UF/NF pilot plant for conjoint batch treatment of two-phase olives and olive oil washing wastewaters. Desalination. 2015;**364**:82-89

[30] Ochando-Pulido JM, Stoller M. Boundary flux optimization of a nanofiltration membrane module used for the treatment of olive mill wastewater from a two-phase extraction process. Separation and Purification Technology. 2014;**130**:124-131

[31] Belfort G, Davis RH, Zydney AL. The behavior of suspensions and macromolecular solutions in crossflow microfiltration. Journal Membrane Science. 1994;**96**:1-58

[32] Field RW, Aimar P. Ideal limiting fluxes in ultrafiltration: Comparison of various theoretical relationships. Journal of Membrane Science. 1993;**80**:107-115

[33] Bacchin P, Aimar P, Sanchez V. Influence of surface interaction on transfer during colloid ultrafiltration. Journal of Membrane Science. 1996;**115**:49-63

[34] Mänttäri M, Nystörm M. Critical flux in NF of high molar mass polysaccharides and effluents from the paper industry. Journal of Membrane Science. 2000;**170**:257-273

[35] Bacchin P, Aimar P, Field RW. Critical and sustainable fluxes: Theory, experiments and applications. Journal of Membrane Science. 2006;**281**:42-69

[36] Ognier S, Wisniewski C, Grasmick A. Membrane bioreactor fouling in sub-critical filtration conditions: A local critical flux concept. Journal of Membrane Science. 2004;**229**:171-177

[37] Espinasse B, Bacchin P, Aimar P. On an experimental method to measure critical flux in ultrafiltration. Desalination. 2002;**146**:91-96

[38] Stoller M, Bravi M, Chianese A. Threshold flux measurements of a nanofiltration membrane module by critical flux data conversion. Desalination. 2012;**315**:142-148

[39] Stoller M, Ochando Pulido JM. The boundary flux handbook: A comprehensive database of critical and threshold flux values for membrane practitioners. Netherlands: Elsevier; 2015

[40] Lee J, Kwon T, Moon I. Performance of polyamide reverse osmosis membranes for steel wastewater reuse. Desalination. 2006;**189**:309-322

[41] Samodurov AN, Lysenko SE, Gromov SL, Panteleev AA, Fedoseeva EB. The use of reverse osmosis technology for water treatment in power engineering. Thermal Engineering. 2006;**53**(6):439-443

[42] Epimakhov VN, Oleinik MS, Moskvin LN. Reverse osmosis filtration based water treatment and special water purification for nuclear power systems. Atomic Energy. 2004;**96**(4): 234-240

[43] Ellouze E, Tahri N, Ben Amar R. Enhancement of textile wastewater treatment process using nanofiltration. Desalination. 2012;**286**:16-23

[44] Gönder ZB, Arayici S, Barlas H. Advanced treatment of pulp and paper mill wastewater by nanofiltration process: Effects of operating conditions on membrane fouling. Separation and Purification Technology. 2011;**76**(3):292-302

[45] Liu M, Lü Z, Chen Z, Yu S, Gao C. Comparison of reverse osmosis and nanofiltration membranes in the treatment of biologically treated textile effluent for water reuse. Desalination. 2011;**281**:372-378

[46] Stoller M, Sacco O, Sannino D, Chianese A. Successful integration of membrane technologies in a conventional purification process of tannery wastewater streams. Membranes. 2013;**3**:126-135

[47] Srisukphun T, Chiemchaisri C, Urase T, Yamamoto K. Foulant interaction and RO productivity in textile wastewater reclamation plan. Desalination. 2010;**250**:845-849

[48] Korzenowski C, Minhalma M, Bernardes AM, Ferreira JZ, de Pinho MN. Nanofiltration for the treatment of coke plant ammoniacal wastewaters. Separation and Purification Technology. 2011;**76**(3):303-307

[49] Yin N, Yang G, Zhong Z, Xing W. Separation of ammonium salts from coking wastewater with nanofiltration combined with diafiltration. Desalination. 2011;**268**:233-237

[50] Boussu K, Kindts C, Vandecasteele C, Van der Bruggen B. Applicability of nanofiltration in the carwash industry. Separation and Purification Technology. 2007;**54**(2):139-146

[51] Pizzichini M, Russo C, Di Meo C. Purification of pulp and paper wastewater with membrane technology for water reuse in a closed loop. Desalination. 2005;**178**:351-359

[52] Wei X, Wang Z, Fan F, Wang J, Wang S. Advanced treatment of a complex pharmaceutical wastewater by nanofiltration: Membrane foulant identification and cleaning. Desalination. 2010;**251**:167-175

[53] Di Lecce G, Cassano A, Bendini A, Conidi C, Giorno L, Gallina T. Characterization of olive mill wastewater fractions treatment by integrated membrane process. Journal of Science of Food and Agriculture. 2014;**94**:2935-2942

[54] Kaleh Z, Geißen SU. Selective isolation of valuable biophenols from olive mill wastewater. Journal of Environmental Chemical Engineering. 2016;**4**:373-384

[55] Achak M, Hafidi A, Mandia L, Ouazzani N. Removal of phenolic compounds from olive mill wastewater by adsorption onto wheat bran. Desalination and Water Treatment. 2014;**52**:2875-2885

[56] Hodaifa G, Ochando-Pulido JM, Alami SBD, Rodriguez-Vives S, Martinez-Ferez A. Iron removal from liquid effluents by olive stones on adsorption column: Breakthrough curves. Ecological Engineering. 2014;**73**:270-275

[57] Martínez Nieto L, Ben Driss Alami S, Hodaifa G, Faur C, Rodríguez S, Giménez JA, Ochando J. Adsorption of iron on crude olive stones. Industrial Crops and Products. 2010;**32**:467-471

[58] Ena A, Pintucci C, Carlozzi P. The recovery of polyphenols from olive mill waste using two adsorbing vegetable matrices. Journal of Biotechnology. 2012;**157**:573-577

[59] Sabbah I, Marsook T, Basheer S. The effect of pretreatment on anaerobic activity of olive mill wastewater using batch and continuous systems. Process Biochemistry. 2004;**39**:1947-1951

[60] Azzam MOJ, Al-Malah K, Al-Gazzawi Z, Al-Omari SA. Dynamic treatment response of olive mills wastewater using series of adsorption steps. Clean – Soil, Air, Water. 2010;**38**:822-830

[61] Comstock SEH, Boyer TH. Combined magnetic ion exchange and cation exchange for removal of DOC and hardness. Chemical Engineering Journal. 2014;**241**:366-375

[62] Víctor-Ortega MD, Ochando-Pulido JM, Hodaifa G, Martinez-Ferez A. Final purification of synthetic olive oil mill wastewater treated by chemical oxidation using ion exchange: Study of operating parameters. Chemical Engineering Process. Process Intensification. 2014;**85**:241-247

[63] Santi CA, Cortes S, D'Acqui LP, Sparvoli E, Pushparaj B. Reduction of organic pollutants in olive mill wastewater by using different mineral substrates as adsorbents. Bioresources Technology. 2008;**99**:1945-1951

[64] Scoma A, Bertin L, Zanaroli G, Fraraccio S, Fava F. A physicochemical-biotechnological approach for an integrated valorization of olive mill wastewater. Bioresources Technology. 2011;**102**:10273-10279

[65] Ferri F, Bertin L, Scoma A, Marchetti L, Fava F. Recovery of low molecular weight phenols through solid-phase extraction. Chemical Engineering Journal. 2011;**166**:994-1001

[66] Zagklis DP, Vavouraki AI, Kornaros ME, Paraskeva C. Purification of olive mill wastewater phenols through membrane filtration and resin adsorption/desorption. Journal of Hazardous Materials. 2015;**285**:69-76

[67] Víctor-Ortega MD, Ochando-Pulido JM, Martínez-Férez A. Phenols removal from industrial effluents through novel polymeric resins: Kinetics and equilibrium studies. Separation and Purification Technology. 2016a;**160**:136-144

[68] Víctor-Ortega MD, Ochando-Pulido JM, Martínez-Férez A. Performance and modeling of continuous ion exchange processes for phenols recovery from olive mill wastewater. Process Safety and Environment Protection. 2016b;**100**:242-251

5

Desalination: A Means of Increasing Irrigation Water Sources for Sustainable Crop Production

OrevaOghene Aliku

Abstract

Globally, water resources for agricultural production have been on the decline. This is associated with increase in water demand over limited resources and poor quality water that adversely affects crop quality and yield and deteriorates soil properties. Even though soil salinity has been affecting agriculture for thousands of years, significant research has been conducted only in the past 100 years. Desalination, which is the process of reducing the salt content in water to an acceptable level, could be an alternative for improving water quality, thereby increasing water sources and reducing the competition among various users of water. Thus, desalination could lead to improved crop quality, improved crop yield, enhanced all-year round crop production, and as such become an important tool for effective agricultural water management.

Keywords: salinity, reverse osmosis, forward osmosis, irrigation water, crop production

1. Introduction

As more than 60% of rainfall, the primary source of water for agriculture is lost to evapotranspiration [1], with the continuous increase in human population and its resultant increase in water demand which is expected to nearly double its size in the next 50 years, the exploitation of the available water resources and the advent of climate change with its global warming effect on available water for crop production, the search for new, sustainable and drought-proof water resources is inevitable [2]. He further stated that since agricultural activities consume more than 60% of the total water demand, using treated wastewater for irrigation can reduce depletion of groundwater significantly. In Refs. [1, 3], it was stated that water-scarce countries especially the Middle East countries located in the arid and semi-arid zones will have to rely more on the use of non-conventional irrigation water resources such as saline aquifers to partly alleviate water scarcity. Although, the present freshwater resources may soon be insufficient to meet the growing demand for food [4], most of these drought-proof water resources contain dissolved solids and chemicals such as salts. The application of these water resources for irrigation purposes often result to the detrimental effect of salinization of soils, environmental degradation and low crop yield.

Salinization is one of the land degradation processes rendering millions of hectares of land unproductive for crop cultivation. It was stated in Ref. [5] that salinization is one of the most serious land degradation problems facing the world. According to El-Swaify [6], salinity is when an 'excessive' amount or concentration of soluble salts occurs in the soil, either naturally or as a result of mismanaged irrigation water. Although, he further reported that salt-affected soils are most abundant in arid regions worldwide, the extent of saline soils is variable [7], whereas Yan et al. [5] stated that soil salinity vary in time and space. Salts are often introduced into soil and water systems via the use of excessive inorganic fertilizers which are leached or washed away as runoff into underground water bodies used for irrigation purposes. According to El-Swaify [6], salts in soil and irrigation water may be either naturally present as products of geochemical weathering of rocks and parent materials or derived directly from sea water flooding, spray or intrusion into groundwater sources and/or caused by irrigation mismanagement, particularly when internal soil drainage is impeded. Due to the presence of salts, most saline lands are virtually uncultivated in the dry season because of strong salinity and lack of water in good quality and quantity [7]. According to Gleick [8], almost half of the human population suffers insufficient access to portable water, and water scarcity in agriculture has been considered to be a global crisis [9].

Hence, desalination, which is any process that removes salt from water [10] to produce desalinized water, is increasingly considered a source of water for agriculture [4]. Even though soil salinity has been affecting agriculture for thousands of years, significant research has been conducted only in the past 100 years [11]. Thus, this review highlights some of the effects of salinity on soil and crop growth and yield, and some possible methods of desalinization of water and soil resources for optimum utilization in a crop production system.

2. Effects of salinity on soil and crop

Different salts, cations and anions vary in their effects on plants and soils, and as such differences in ionic compositions of soil solutions and waters with similar electrical conductivity values may lead to dissimilar effects [6]. Salinity may adversely affect soil structure and other physical properties, and this could finally be transmitted to crop growth and development. For instance, the breakdown of soil structure can exacerbate salt effects on crops through increased surface crusting, germination inhibition and reduced permeability, porosity and aeration [6]. In Ref. [12], it was reported that soil infiltration rate was greatly affected by sodicity and electrolyte concentration of the irrigation water. In Ref. [13], it was reported that increasing salinity and sodicity resulted in a progressively smaller, more stressed microbial community which was less metabolically efficient. Saline soils have been reported to contain sufficient salts at the root zone to impair crop growth [7]. Also, Corwin et al. [14] noted salinity as one of the most significant soil properties influencing cotton yield in a response model. In Ref. [15], it was reported that the emergence of sunflower and maize was affected by salinity and that the higher the salinity, the lower the leaf area and the dry matter production.

2.1. Desalination

Desalination describes a range of processes which are used to reduce the amount of dissolved solids in water [16]. Also, Nofal [17] defined desalination as the removal of excess salt and other minerals from water in order to get fresh water suitable for drinking water, animal consumption and irrigation purposes. It is used to produce clean water from water sources containing dissolved chemicals, and in most cases, water sources are salty, producing fresh water from sea water or brackish water [16]. They further stated that natural waters may be classified approximately according to their total dissolved solid (TDS) values as listed in **Table 1**. Desalination is a water saving alternative to brackish water irrigation even though its diffusion as a viable method of water treatment has been limited by high costs and concern about the lack of plant nutrients in desalinated water [17]. In Ref. [4], it was also confirmed that desalination not only separates the undesirable salts from the water but also removes ions that are essential to plant growth. Although, a recent report concludes that the costs of desalination remain prohibitively expensive for full use by irrigated agriculture [18], for high value cash crops like green-house vegetables and flowers, its use may be economically feasible [4]. According to Smith and Shaw [16], low-cost methods of desalination by distillation are also available.

2.2. Prospects of desalinized water for agriculture

Due to the impact of climate change which has led to uncertainty in the amount and duration of rainfall for crop production, 69% of global water supply is being channelled for irrigation purpose [19]. As a result, present fresh water resources may soon be insufficient to meet the growing demand for food [4]. Although, at present, sea water desalination provides 1% of the world's drinking water, desalinized water is increasingly considered a source of water for agriculture [4]. In some countries, farmers have already adopted the use of desalinized

Type of water	Total dissolved solids (mg/L)
Sweet waters	0–1000
Brackish waters	1000–5000
Moderately saline waters	5000–10,000
Severely saline waters	10,000–30,000
Seawater	More than 30,000

Source: Smith and Shawerji [16].

Table 1. Classes of natural waters.

brackish water for crop production. For instance, Mechell and Lesikar [20] reported that ~22% of water desalinated in Spain are used for agricultural irrigation purposes, whereas an Australian survey found that 53% of the population envisioned desalinated water usage for irrigation of vegetables as highly likely.

2.3. Benefits of desalination

Desalination is a water saving alternative to brackish water irrigation [17]. By implication, it could increase the possible sources of water for irrigation, and as such enhance sustainable all-year round crop production. According to Ref. [4], the low level of salinity of desalinized water is an extra benefit, because the salts [especially Sodium (Na^+) and Chlorine (Cl^-)] damage soils, stunt plant growth and harm the environment. Hence, desalinized water could improve the quality of irrigation water thereby reducing the possibilities of the incidence of soil salinity with its consequent adverse effect on crop growth and yield via its deteriorating effects on soil properties. Furthermore, desalination could increase the size of land area for cultivation, the number of crops (including salt sensitive crops) cultivated, improve crop quality, increase crop productivity and increase the broad band of water use for other purposes [17]. Desalination has been reported to improve farmers' income [17].

2.4. Techniques in desalination

According to Refs. [10, 21], techniques used in a desalination process essentially separates saline water into two parts, hence, two streams of water are produced.

(a) Treated water that has low concentrations of salts and minerals.

(b) Concentrate or brine, which has salt and mineral concentrations higher than that of the pre-treated water.

It is often associated with electrical generation plants, from which both electricity and waste heat are available [16]. Some of these desalination methods could be relatively expensive, whereas others such as desalination by distillation could be low-cost methods. According to Refs. [10, 21], the two major types of technologies used for desalination can be broadly classified into thermal technologies (multi-stage flash distillation, multi-effect distillation and vapour compression distillation) and membrane technologies (electrodialysis/electrodialysis reversal and reverse osmosis), with reverse osmosis, and distillation followed by condensation being two main desalination methods [16]. In Ref. [10], it was stated that both technologies need energy to operate and produce fresh water. However, the most appropriate method can be selected on the basis of the total dissolved solids (TDS) value of the raw water (**Table 2**).

2.5. Thermal technologies

These technologies involve the eating of saline water and collecting the condensed vapour distillate to produce pure water [10]. In Ref. [21], it was reported that thermal distillation technologies are widely used in the Middle East, primarily because the region's petroleum reserves keep

Process	Total dissolved solid value (mg/L)
Ion exchange (not described here)	500–1000
Electrodialysis	500–3000
Reverse osmosis (standard membranes)	500–5000
Reverse osmosis (high-resistance membranes)	Over 5000
Distillation	Over 30,000

Source: Smith and Shawerji [16].

Table 2. Suitability of desalination process based on the total dissolved solids.

energy cost low. However, thermal technologies have rarely been used for brackish water desalination, because of the high cost involved [10]. According to Refs. [6, 21], thermal technologies are grouped into three major large scale processes, i.e., multi-stage flash distillation (MSF), multi-effect distillation (MED) and vapour compression distillation (VCD). They stated that solar distillation, which is another thermal technology, is typically used for very small production rates.

2.5.1. *Multi-stage flash distillation (MSF)*

This process of distillation involves the use of several (multi-stage) chambers [10]. According to Ref. [21], this process sends the pre-treated saline water through multiple chambers as illustrated in **Figure 1** [22]. In the MSF process, each successive stage of the plant operates at progressively lower pressures. In Ref. [21], it was explained that the pre-treated saline water is heated and compressed to a high temperature and high pressure, and the pressure is reduced as the water progressively passes through the chambers, causing the water to rapidly boil. In other words, the pre-treated water is first heated under high pressure as it is passed into the first 'flash chamber', where the pressure is released, causing the water to boil rapidly, resulting in sudden evaporation or 'flashing', which continues in each successive stage, because the pressure at each stage is lower than that of the previous stage [10]. The vapour produced by the flashing is then condensed on a heat exchanger tubing that runs through each stage and collected as fresh water. Generally, only a small percentage of the pre-treated saline water is converted into vapour and condensed [10].

2.5.2. *Multi-effect distillation (MED)*

The MED process has been used since the late 1950s and the early 1960s [10]. According to Ref. [21], the MED employs the same principles as the MSF process except that instead of using multiple chambers of a single vessel, MED uses successive vessels (**Figure 2**), i.e., MED occurs in a series of vessels, using the principles of evaporation and condensation at reduced ambient pressure [21]. Here, water is produced by a series of evaporator vessels at progressively lower pressures. Water boils at lower temperatures as pressure decreases, such that the water vapour of the first vessel serves as the heating medium for the second, and so on [10]. According to Ref. [21], the multiple vessels make the MED process more efficient, while [10] stated that the more the vessels, the higher the performance ratio of the MED.

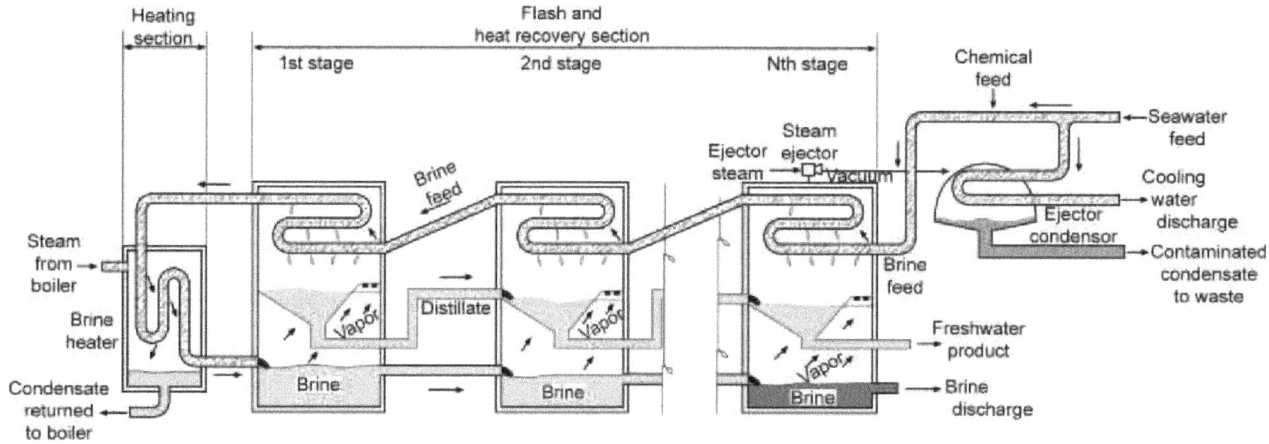

Figure 1. An illustration of the multi-stage flash distillation (MSF) process (Source: Buros, 1990).

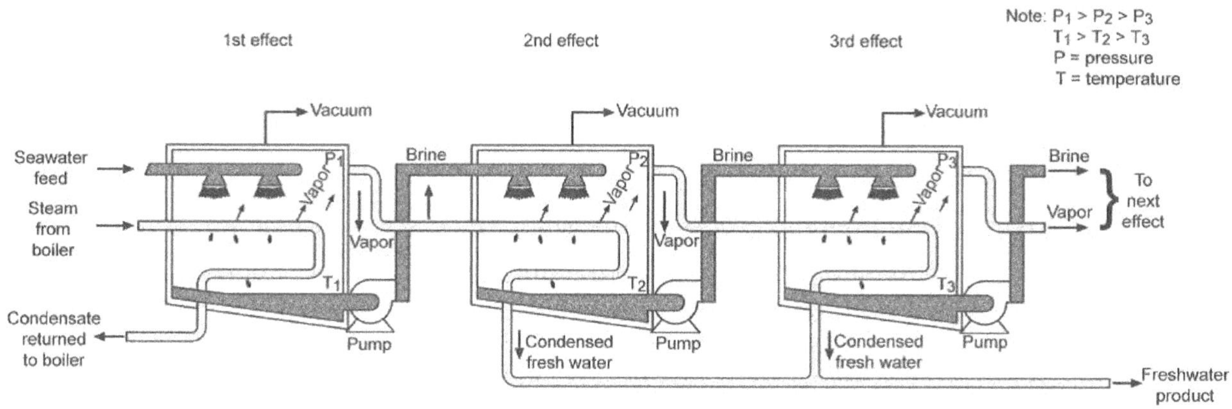

Figure 2. A schematic diagram of a multi-effect distillation (MED) process (Source: [22]).

2.5.3. Vapour compression distillation (VCD)

The VCD can function independently or in combination with other thermal distillation processes such as the MED [10, 21]. According to Ref. [23], the heat for evaporating the pretreated saline water comes from the compression of vapour, rather than the direct exchange of heat from steam produced in a boiler (**Figure 3**). It usually involves the use of a mechanical compressor to generate heat for evaporation [10]. Vapour compression distillation unit are commonly used to produce fresh water for small- to medium-scale purposes such as resorts, hotels and industrial applications [21].

2.5.4. Solar distillation

This involves the use of solar energy for water desalination as shown in **Figure 4**. Also, Buros [21] stated that although the designs of solar distillation units vary greatly, the basic principles are the same. They explained that the sun provides the energy to evaporate the saline water, and the water vapour formed from the evaporation process then condenses on a clear glass covering before it is collected as fresh water in the condensate trough. The clear glass or

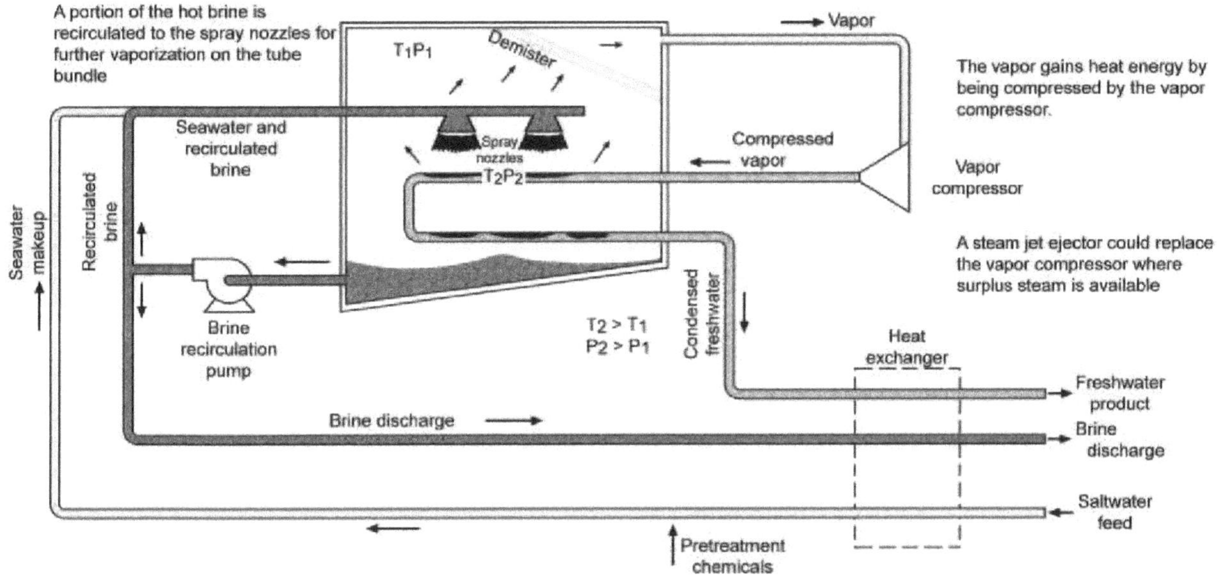

Figure 3. An example of a vapour compression distillation (VCD) process (Source: [22]).

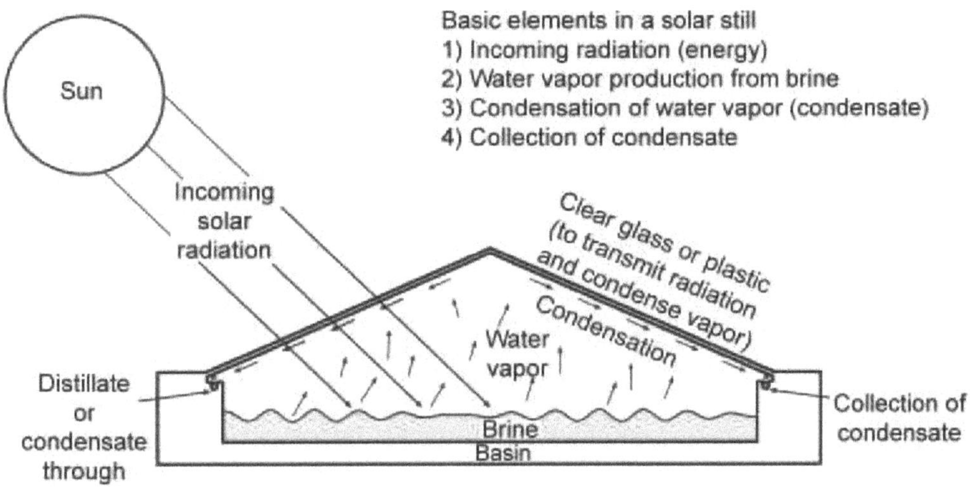

Figure 4. An example of a solar still distillation process (Source: [22]).

plastic covering is used to transmit radiant energy and also to allow water vapour to condense on its interior surface before it is collected as fresh water. Alike VCD, solar desalination is generally used for small-scale operations [21].

2.6. Membrane technologies

According to Ref. [21], there are several membrane treatment processes, including reverse osmosis, nanofiltration, ultrafiltration and microfiltration. These processes involve the use of a barrier, which is a membrane, and a driving force. The membranes contain pores which differ in sizes according to the type of process (**Figure 5**). It was explained in Ref. [21] that membrane technologies often require that the water undergo chemical and physical pre-treatment

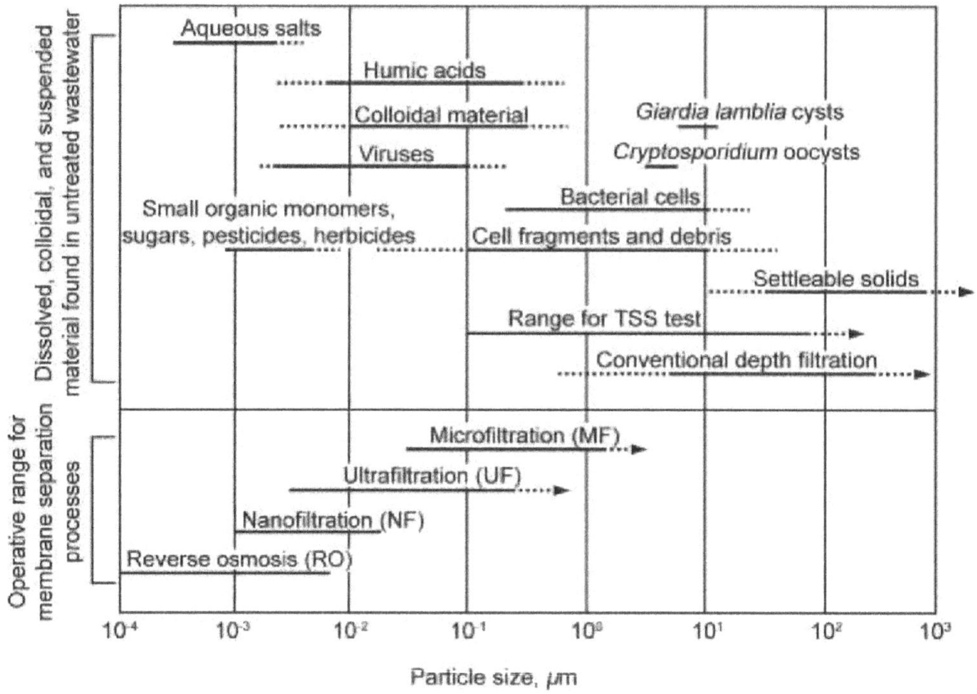

Figure 5. An illustration of the range of nominal membrane pore sizes for reverse osmosis (RO), nanofiltration (NF), ultrafiltration (UF), and microfiltration (MF) (Source: [23]).

to limit blockage by debris and scale formation on the membrane surfaces. The general characteristics of membrane processes are presented in **Table 3**. Membrane technologies can be subdivided into two broad categories: electrodialysis/electrodialysis reversal (ed/edr) and reverse osmosis (RO) [10]. According to Ref. [21], the driving force used in electrodialysis or electrodialysis reversal is an electrical potential, whereas that used in reverse osmosis is a pressure gradient.

2.6.1. Electrodialysis and electrodialysis reversal (ED/EDR)

This is a voltage-driven membrane process in which an electrical potential is used to move salts through a membrane, leaving fresh water behind as product water [10]. In Ref. [21], it was explained that the membrane used for ED/EDR are built in such a way that they only allow passage of either positively or negatively charged ions, but not both. Here, ionic molecules, such as sodium, chloride, calcium and carbonate in saline water, that are known to cause adverse effects on soil and crop productivity are removed from the treated water as the cations are attracted to the negative electrode, whereas the anions are attracted to the positive electrode while passing through selected membranes. According to Ref. [10], the membranes are usually arranged in an alternate pattern, with anion-selective membrane followed by a cation-selective membrane. He further explained that during this process, the salt content of the water channel is diluted, while concentrated solutions are formed at the electrodes. Concentrated and diluted solutions are created in the spaces between the alternating membranes, and these spaces bound by two membranes are called cells [10]. The pre-treated saline

Membrane process	Membrane driving force	Typical separation mechanism	Operating structure (pore size)	Typical operating range (μm)	Permeate description	Typical constituents removed
Microfiltration	Hydrostatic pressure difference or vacuum in open vessels	Sieve	Macropores (>50 nm)	0.08–2.0	Water + dissolved solutes	TSS, turbidity, protozoan oocysts and cysts, some bacteria and viruses
Ultrafiltration	Hydrostatic pressure difference	Sieve	Mesopores (2–50 nm)	0.005–0.2	Water + small molecules	Macromolecules, colloids, most bacteria, some viruses, proteins
Nanofiltration	Hydrostatic pressure difference	Sieve + solution/ diffusion + exclusion	Micropores (<2 nm)	0.001–0.01	Water + very small molecules, ionic solutes	Small molecules, some hardness, viruses
Reverse osmosis	Hydrostatic pressure difference	Solution/ diffusion + exclusion	Dense (<2 nm)	0.0001–0.001	Water + very small molecules, ionic solutes	Very small molecules, colour, hardness, sulfates, nitrate, sodium, other ions
Dialysis	Concentration difference	Diffusion	Mesopores (2–50 nm)	–	Water + small molecules	Macromolecules, colloids, most bacteria, some viruses, proteins
Electrodialysis	Electromotive force	Ion exchange with selective membranes	Micropores (<2 nm)	–	Water + ionic solutes	Ionized salt ions

Source: Metcalf and Eddy [22].

Table 3. General characteristics of membrane processes.

water passes through all the cells simultaneously to provide a continuous flow of desalinated water and a steady stream of concentrate from the stack [10]. Although the ED was originally conceived as a seawater desalination process, it has generally been used for brackish water desalination [10].

According to Refs. [10, 21], the EDR functions in a similar way as the ED. However, El-Swaify [6] explained that the only exception to the EDR operating on the same general principle as the ED unit is that both the product and the concentrate channels are identical in the EDR, whereas Buros [21] also explained that the polarity or charge of the electrodes is switched periodically in the reverse process. Immediately following reversal, the product water is removed until the lines are flushed out and the desired water quality restored [10]. They explained that the reversal in flow of ions helps to remove scaling, slimes and other debris from the membranes before they accumulate in large amount, thus extending the system's operating life.

2.6.2. Reverse osmosis

In relation to thermal processes, reverse osmosis is a relatively new process that was commercialized in the 1970s [10, 24]. Currently, it is the most widely used method for desalination in the United States [10]. This process of desalination uses a pressure gradient as the driving force to move high pressure pre-treated saline water through a membrane that prevents the salt ions from passing, thus, yielding the product water stream and a concentrated brine stream as shown in **Figure 6**, respectively [10, 21]. In other words, reverse osmosis utilizes hydraulic pressure to offset osmotic pressure and induces mass transport of water across a semi-permeable membrane [25]. This is simply applying pressure (in excess of the osmotic pressure) to the saline water [16]. Osmotic pressure (π) is calculated using the Van't Hoff equation:

$$\pi = MRT \tag{1}$$

where M is the molar concentration of dissolved species; R is the ideal gas constant and T is the temperature on the Kelvin scale.

According to Ref. [10], high pressure pumps supply the pressures between the range of 150 psi for slightly brackish water to 800–1000 psi for salt water, to enable the water to pass through the membrane and have the salt rejected. It is worthy to note that the membrane is easily torn and needs to be supported carefully [16]. Due to the fact that the membrane of the reverse osmosis process consists of small pores, the salt water needs to be filtered first to remove particles which might damage the membranes, while chemical additives may be added to prevent biological growth and scaling [16, 21]. This is very important as the membrane surfaces must remain clean [10].

The individual spiral reverse osmosis membrane element through which the high pressure pre-treated saline water flows are constructed in a concentric spiral pattern that allow alternating layers of pre-treated water and brine spacing, reverse osmosis membrane and a porous product water carrier (**Figure 7**) [21]. The porous product water carrier allows the fresh water to flow into the centre of the membrane element to be collected in the product water tube. According to Ref. [10], the reverse osmosis processes are used for desalinating brackish water (TDS > 1500 mg/L) and seawater. Although membrane desalination processes using reverse osmosis or nanofiltration are diffusion-controlled membrane processes [25], also, Krishna [10]

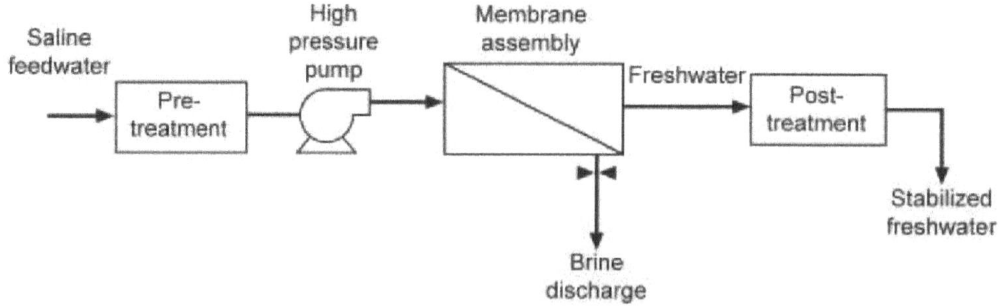

Figure 6. Basic components of a reverse osmosis membrane treatment process (Source: [21]).

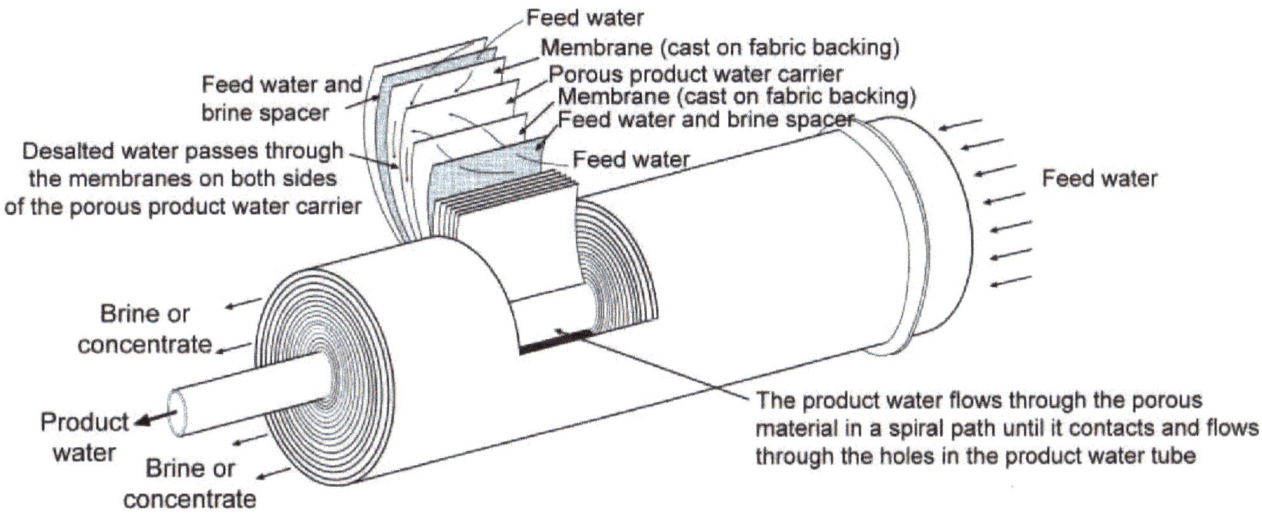

Figure 7. Dissected view of a spiral reverse osmosis membrane element (Source: [22]).

explained that unlike nanofiltration, which is a membrane process that is used for the removal of divalent salt ions such as calcium, magnesium and sulphate, reverse osmosis is used for the removal of sodium and chloride.

According to Ref. [26], following mass balance equations are commonly used to describe reverse osmosis and nanofiltration membrane process performance. Equation (2) indicates mass balance for water flow:

$$Q_f \rho_f = Q_p \rho_p + Q_c \rho_c \tag{2}$$

where Q_f is feedwater flow rate (m^3/d); ρ_f is density of feedwater; Q_p is permeate flow rate (m^3/d); ρ_p is density of permeate; Q_c is concentrate flow rate (m^3/d) and ρ_c is density of concentrate.

Equation (3) describes mass balance for solute flux:

$$Q_f C_f = Q_p C_p + Q_c C_c \tag{3}$$

where C_f is feedwater solute concentration, units of mass per volume (mg/L); C_p is permeate solute concentration, units of mass per volume (mg/L) and C_c is concentrate solute concentration, units of mass per volume (mg/L).

2.6.3. *Forward osmosis*

Forward osmosis is used to describe the use of osmosis as a salt-water separation mechanism through an engineered membrane. It is an emerging membrane treatment process that belongs to the class of osmotically driven membrane processes [25]. It was first presented by Cath et al. [27] and could also be called direct osmosis. Unlike reverse osmosis where pressure is applied to the pre-treated saline water and a low salinity permeate is produced, forward osmosis involves a semi-permeable membrane which separates a high osmotic pressure 'draw' solution from the pre-treated saline water with relatively lower salinity and osmotic pressure. Here, water is drawn across the membrane by natural osmosis, restricting the passage of salts at the membrane surface. In Ref. [25], it explained that when equal volumes of a dilute feed solution and a concentrated draw solution are separated by a semi-permeable membrane, water flows into the concentrated draw solution, which has a higher osmotic pressure. This flow continues until chemical equilibrium is reached. The increase in water column height in the high osmotic pressure chamber at equilibrium equates to the difference in osmotic pressure between the dilute and concentrated solutions. Thus, forward osmosis uses the osmotic pressure differential ($\Delta\pi$) across the membrane, rather than the hydraulic pressure differential as in reverse osmosis, as the driving force for transport of water through the membrane. The transport of water in forward osmosis is described in Eq. (4):

$$J_W = K_W(\sigma\Delta\pi - \Delta P) \tag{4}$$

where J_W = water flux; $\Delta\pi$ *is* differential osmotic pressure across the membrane; K_W is water permeability coefficient of the membrane; σ is reflection coefficient (a measure of the relative permeability of a particular membrane to a particular solute) and ΔP is differential applied pressure across the membrane.

Past research has shown that forward osmosis membranes are good barriers to a broad range of contaminants, including bacteria, protozoa, viruses and other dissolved organic and inorganic constituents in contaminated water [27]. Also, in comparison to other desalination processes such as the multi-stage flash, multi-effect distillation and reverse osmosis, McGinnis and Elimelech [28] estimated that the forward osmosis has relatively lowest relative energy consumption (**Figure 8**). The authors estimated that forward osmosis with a thermally decomposing draw solution [such as in the forward osmosis low temperature distillation (FO-LT) process which incorporates the use of low-quality heat for thermal decomposition of the draw solution and recovery using distillation columns] would use less than one-third the work energy of reverse osmosis for desalination.

2.7. Application of desalinated water in irrigated agriculture

According to Ref. [1], the amount of fresh groundwater or agricultural activities is negligible and exists only in some locations. He further stated that desalination of brackish and saline water seems to be promising, especially in the absence of any other alternative. In spite of this, the cost of desalinated water are still too high for full use of this resource in irrigated agriculture, with the exception of intensive horticulture or high-value cash crops, such as vegetables and flowers grown in greenhouses [29]. In Refs. [1, 29], reverse osmosis was reported to be the preferred desalination technology for agricultural uses because of the cost reductions driven by improvements in membranes in recent years. An example of countries that have adopted the application of desalinated water for irrigated agriculture is Spain. According to Ref. [30], Spain has more than 300 treatment plants with most of the plants processing brackish water, and located in coastal areas or within 60 km of the sea. It was also noted in Ref. [29] that small and medium size brackish water desalination plants, with a capacity of less than 1000 m^3/d (11.6 L/s), are common because they adapt better to individual farmer requirements and to

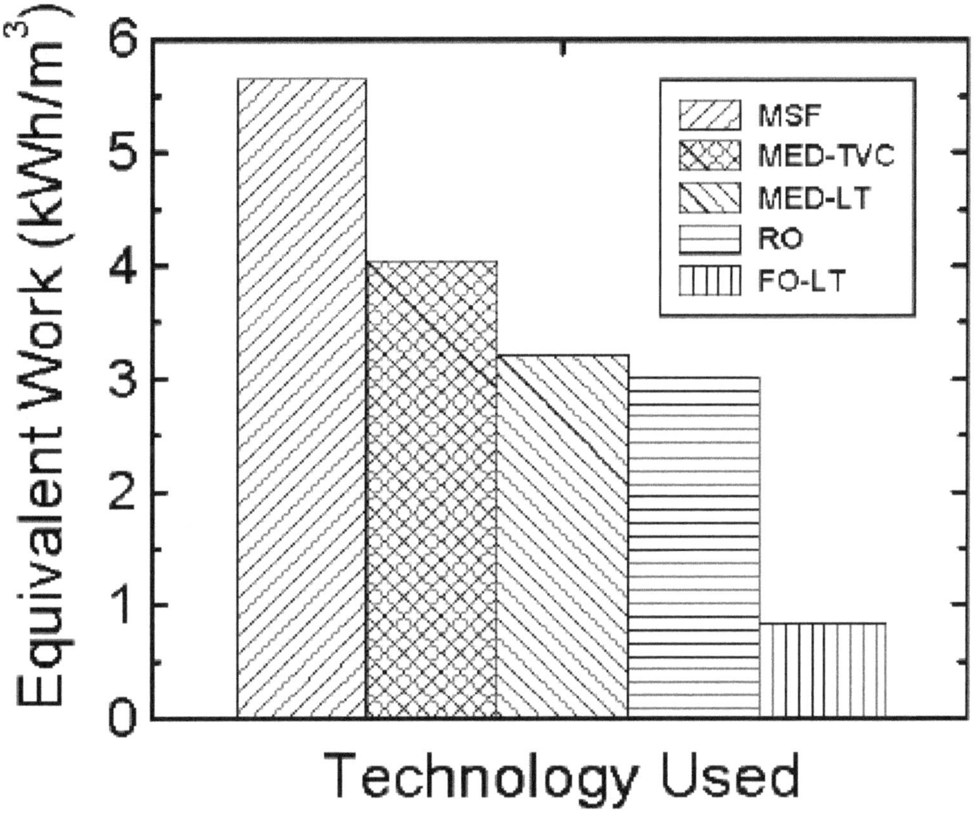

Figure 8. Estimated energy consumption for desalination processes (Source: [29]).

the existing hydraulic structures. As irrigated agriculture does not require the strict standards that apply for drinking-water requirements, opportunities appear to exist for the adoption of high-quality desalinated water, and in this way, the final cost of a cubic metre of irrigation water can be reduced [29].

3. Conclusions

Salinity arises from various natural and human-induced processes and is a major phenomenon that deteriorates soil properties, thus limiting the potentials of soils for sustainable crop production. Desalinated water is usually of high quality and can have less negative impact on soils and crops in comparison with direct use of brackish water. Thus, water desalination could have positive impacts on agriculture and the environment, such as increasing water availability and recycling poor-quality water. The use of osmotic and distillation mechanisms to recover high quality water from wastewater effluents and saline waters could be high-tech demanding especially when considering desalination of large volume of water for irrigation and other forms of utilization.

Although, the use of low-tech distillation methods could be easily adopted by peasant farmers in rural communities, the use of reverse osmosis has been said to be the most suitable for irrigated agriculture. As some of the processes involved in desalinizing saline water for sustainable crop production could be expensive, it could also be cost-effective, owing to the fact that desalination could save water for agricultural production, increase the amount and types of crops grown, the area of land cultivated and as such improve the quality of crop yield and farmers' income.

Author details

OrevaOghene Aliku

Address all correspondence to: orevaoghenealiku@gmail.com

Department of Agronomy, University of Ibadan, Ibadan, Nigeria

References

[1] Baalousha H. Desalination status in the Gaza Strip and its environmental impact. Desalination. 2006;**196**:1-12

[2] Texas Water Development Board. The Future of Desalination in Texas: Biennial Report on Seawater Desalination. Austin, Texas: Texas Water Development Board; 2010

[3] Zaide M. Drought and arid land water management: Government Focal Points. CSD-16/17 National Report, Ministry of National Infrastructure, Israel; 2016; 11pp.

[4] Yermiyahu U, Tal A, Ben-Gal A, Bar-Tal A, Tarchitzky J, Laha O. Rethinking desalinated water quality and agriculture. Environmental Science. 2007;**318**:920-921

[5] Yan N, Marschner P, Cao W, Zuo C, Qin W. Influence of salinity and water content on soil microorganisms. International Soil and Water Conservation Research. 2015;**3**:316-323

[6] El-Swaify SA. Soil and water salinity. In: Silva JA, Uchida R, editors. From: Plant Nutrient Management in Hawaii's Soils, Approaches for Tropical and Subtropical Agriculture. Manoa: College of Tropical Agriculture and Human Resources, University of Hawaii; 2000. pp. 151-158

[7] Lantin RS, Quijano CC, Reyes RY, Neue HU. Rice and problem soils in the Philippines and the humid tropics: Past development and strategies for the 21st Century. Philippine Journal of Crop Science. 1990;**15**(1):41-47

[8] Gleick PH. The World's Water 2002-2003: The Biennial Report on Freshwater Resources. Washington, DC: Island Press; 2002

[9] Postel S. Pillar of Sand: Can the Irrigation Miracle Last? Washington, DC: World Watch; 1999

[10] Krishna HJ. Introduction to Desalination Technologies. Austin, Texas: Texas Water Development Board; 2004. p. 7

[11] Qadir M, Wichelns D, Oster J, Jacobsen SE, Basra MA, Choukr-Allah R. Sustainable Management of Saline Waters and Salt-Affected Soils for Agriculture. Proceedings of the Second Bridging Workshop, 15-18 November, 2009. The International Center for Agricultural Research in Dry Areas (Aleppo, Syria), and the International Water Management Institute (Sri Lanka), 2010; 115pp.

[12] Agassi M, Shainberg I, Morin J. Effect of electrolyte concentration and soil sodicity o infiltration rate and crust formation. Soil Science Society of America Journal. 1981;**45**(5):848-851

[13] Rietz DN, Haynes RJ. Effects of irrigation-induced salinity and sodicity on soil and microbial activity. Soilless Biology and Biochemistry. 2003;**35**(6):845-854

[14] Corwin DL, Lesch SM, Shouse PJ, Soppe R, Ayars JE. Identifying soil properties that influence cotton yield using soil sampling directed by apparent soil electrical conductivity. Agronomy Journal. 2003;**95**(2):352-364

[15] Katerji N, van Hoom JW, Hamdy A, Karam F, Mastrorilli M. Effect of salinity on emergence and on water stress and early seedling growth of sunflower and maize. Agricultural Water Management. 1994;**26**(1-2):81-91

[16] Smith M, Shaw R. Desalination. WEDC Loughborough University Leicestershire LE11 3TU UK [Internet]. 1996. Available form: www.lboro.ac.uk/departments/cv/wedc/ [Assessed: 15 November 2016.]

[17] Nofal I. Desal brackish water for agriculture. The regional water knowledge sharing forum. 12-14 May, 2015, 2015; Sharm El Sheikh, Egypt

[18] Beltran JM, Koo-Oshima S. Water desalination for agricultural applications. In: Proceedings of the FAO Expert Consultation on Water Desalination for Agricultural Applications; 26-27 April 2004. Rome (Italy). FAO Land and Water Discussion Paper, 2006; no. 5, 60 p.

[19] Meerganz von Medeazza G. Water desalination as a long-term solution to alleviate global fresh water scarcity? A North-South Approach. Desalination, 2004; **169**: 287-301

[20] Mechell JK, Lesikar B. Desalination methods for producing drinking water. Texas A&M Agrilife Extension Communications. E-249, 04-10, 2010; 8pp.

[21] Buros OK. The ABC's of Desalting. Topsfield, Massachusetts: International Desalination Association; 1990

[22] Metcalf and Eddy. Wastewater Engineering: Treatment and Reuse. 4th ed. New York: McGraw-Hill; 2003

[23] Buros OK. "The ABCs of Desalting", International Desalination Association, 2000; Topsfield, Massachusetts, USA, 2nd Edition, 31pp.

[24] CH2M H. Assessment of Osmotic Mechanisms Pairing Desalination Concentrate and Wastewater Treatment. Austin, Texas: Texas Water Development Board; 2011. 159 p

[25] AWWA. Reverse Osmosis and Nanofiltration. American Water Works Association (AWWA) Manual of Water Supply Practices, Second Edition, M46, Quincy Avenue, Denver, USA, 2007; 240 pp.

[26] Loeb S, Sourirajan S. Sea water demineralization by means of an osmotic membrane. Advances in Chemistry Series. 1963;**38**:117

[27] Cath TY, Childress AE, Elimelech M. Forward osmosis: Principles, applications, and recent developments. Journal of Membrane Science. 2006;**281**:70-87

[28] McGinnis RL, Elimelech M. Energy requirements of ammonia–carbon dioxide forward osmosis desalination. Desalination. 2007;**207**:370-382

[29] FAO. Water Desalination for Agricultural Applications. Proceedings of the FAO Expert Consultation on Water Desalination for Agricultural Applications, FAO Land and Water discussion paper 5, Rome, 2006; 60 pp.

[30] FAO. La desalación de agua de mar en la agricultura: situación actual y perspectivas futuras, by J. A. Medina. Draft report. Rome, 2003. 33 pp.

6

NEREU Project: Construction of a Plasma Reactor for Reform of Greenhouse Gases for Treatment of Wastewater of the Marine Farms

André Pedral S. de Sena

Abstract

The predatory exploitation techniques used for the supply of protein resources have been systematically causing the decrease in seafood. The solution was sought in the form of marine farms, for the purpose of production of fish and seafood. Brazil created an incentive production of seafood in order to increase seafood production. In 1998, the state governments joined the project, encouraging the creation of marine farms without an assessment of the impact on the environment. In 2005, after several records of seafood production in Santa Catarina started an epidemic of white spot shrimp, decimating several fishing farms. In Bahia came the disease lethargic crab, which simply decimated almost 90% of crabs those states. In 2010, the State Government of Bahia invested in research to combat this degradation of mangroves; the project had to in essence deal with the brown mare and help save the fishing industry of the state. It was based on these principles that Nereu project emerged, which provides for the rational and intelligent use of contaminated and hypersaline water of the marine farms from the production area of fish and seafood, in order to neutralize all pathogens and produce pure water with energy cogeneration.

Keywords: colloidal carbon, desalination, hydrogen, methane, stirling engines

1. Introduction

The constant negative action of man over nature, promoting deforestation, releasing waste without treatment and chemicals in the water, grounding of the mangroves that are nurseries for many species, and overfishing causing the decrease of fish stocks and leading the fisherman to stop fishing and look for other activities. The crisis in the fisheries sector caused by the uncontrolled exploitation of the coastal zone with the consequent reduction of marine stocks makes aquaculture an important food production alternative, significantly contributing to food security and poverty alleviation on the planet [1, 2].

The aquaculture is a deliberate attempt by humans to modify and manipulate the trophic relationships controlled by nature. Thus, it can be seen that all production activities are influencing the environment and in aquaculture, such impacts can be classified into three groups: those from the environment (exogenous), those resulting from aquaculture owned (endogenous), and those caused by the aquaculture on the environment. Moreover, it is noteworthy that the environmental impacts caused by aquaculture will always be more negative than positive [2, 3].

Thus, fish farming and shellfish generate waste in the form of particles and inorganic and organic soluble material. The nature and quantity of effluent vary with the type of cultivation, crop species, and intensification level. In general, feces, excreta, and pseudo-feces are the main sources of inorganic and organic nutrients [4].

The bacteria, fungi, and protozoa participate in the degradation process of organic matter, and release of toxic gases such as sulphur become unavoidable. The ideal of sustainable aquaculture, which is consistent with the economic, social, ecological, spatial, and cultural contexts [4, 5], is at an early stage in Brazil, where mistakes are being evaluated and studied so that new management techniques may be suitable for targeting a smaller effect on the ecosystem balance.

This concern about the environmental impact was not taken into account by the Federal Government of Brazil, which despite the warnings given by CONAMA [6] decided, through the Ministries of Agriculture, Environment and Development of Industry and Trade in 1997, to create an incentive to the fishing and production of seafood through the creation of marine farms. In 1998, the state governments began to join the project, and the states of *Bahia, Santa Catarina, Ceara,* and *Espirito Santo* invested heavily in the project for installation of farms mainly shrimp, oysters, and scallops. Through tax incentives, public funding, and federal funding, there was an incentive to create these marine farms, but there was no evaluation on the impact of aquaculture in the environment [4–9].

Between 2000 and 2005, the record increase in the seafood production was achieved. A total of 75% of this production aimed at export [4, 6–9]. However, in 2005, it began a productive decline, beginning with Santa Catarina, southern Brazil, which started an epidemic of white spot syndrome in the shrimp. It is a disease that affects shrimps and is one of the four identified in Brazil contained in the OIE list of aquatic animals. The WSSV—"White Spot Syndrome Virus"—is a persistent infection, with high mortality in shrimp cultivation. This disease began to spread along the Brazilian coast, currently reaching the state of *Ceara,* with production loss of nearly 80% of shrimps. Almost concurrent appearance of this disease arose in the mangroves of *Espirito Santo* and *Bahia,* the disease lethargic crab (DCL). It was found that the causative agent of DCL is the pathogenic fungus *Exophiala cf psycrofila* that simply decimated almost 90% of crabs in these states [3, 4, 9].

In 2015, the disease proliferates reaching other states, decimating the mangrove areas and ending with the crab. After several analyses of the environmental agencies, it was found that the marine farms established in these states generate waste, which cause the problem, generating conditions for the proliferation of pathogens in the environment, often resistant to the immune system of animals due to the indiscriminate use of antibiotics in effluent ponds and crops laboratories. The DNAs of the *Exophiala cf psycrofila* and *Nimaviridae* were found in the existing mangrove in the region of *Jaguaribe*, fishing town in the state of *Bahia*. This organic material was present in a phenomenon known as brown tide and was caused by the discharge of waste from marine farms who had settled in this region [4, 6–9].

It is problematic and with the need to clean the environment of this waste, the principles of Nereu project were built, which provides for the rational and intelligent use of contaminated water and hypersaline of the marine farms, in order to neutralize all pathogens and produce pure water with energy cogeneration.

2. Farm of shrimp: the problem

Despite the changes in development models that have taken place throughout history, natural resources have always been exploited without any concern for the sustainability of the system, often exhausting these resources to the extreme. The current model of development has generated global environmental impacts, putting the biosphere at risk. Development models that revert to the situation or at least minimize impacts on the biosphere are urgently needed.

According to some data, world mariculture expanded rapidly from 16.7 million tons in 2004 to 22.8 million tons in 2014, an increase of 35.8%, whereas the fishing sector declined in this period from 83.8 to 69.9 million tons. In 2004, mariculture accounted for 16.6% of the world's fish production, which increased to 30.1% in 2014. In contrast to the increasing decline in fisheries, seafood production is seen as an alternative to maintain the world demand for aquatic products. If current development is maintained over the next 15 years, 32% of the world's total production of marine fish will come from mariculture [1, 4–7, 9].

In Brazil, mariculture has been progressively expanding, mainly since the 1990s due to government programs, which is represented by bivalve mollusks in the South and Southeast regions and by shrimps in the North and Northeast regions [4].

Brazil produced 91,405 t of mariculture products in 2014, of which 71,000 t are shrimp and 20,405 t represented by mollusks. The total value of mariculture production is US$876.8 million, of which US$256.8 million is represented by mollusks [3, 4, 8–10].

Thus, mariculture is a deliberate attempt by humans to modify and manipulate nature-controlled food relationships. It can be observed that all productive activities are impacting the environment and that in aquaculture, these impacts can be classified into three sets: those from the environment (exogenous); those resulting from aquaculture (endogenous), and those caused by aquaculture on the environment [8–10].

In addition, it is emphasized that the environmental impacts caused by aquaculture will almost always be more negative than positive.

These impacts arise mainly from the use of resources, such as space, water, raw materials, and food, reduction of biodiversity, as well as from the production of organic and inorganic waste (e.g., excreta production, introduction of microorganisms, pathogens and parasites in the environment, accumulation of residues of cultivated organisms, and release of antibiotics in the effluent of nurseries and crop laboratories) [3–5, 10].

Therefore, it can be observed that the negative impacts can be of two types: direct, for example, by the introduction of exotic genetic material into the environment and indirect, due to loss of habitat and ecological niche.

Any alteration of the physical, chemical, and biological properties of the environment caused by any form of matter or energy resulting from human activities that directly or indirectly affect the health, safety, and well-being of the population shall be considered an environmental impact, which include social and economic activities; biota, aesthetic, and sanitary conditions of the environment and the quality of environmental resources. These changes need to be quantified, since they present relative variations, which can be positive or negative, big, or small [8, 10]. The different aquaculture modalities can generate diverse environmental impacts, depending mainly on the cultivation system (closed, semi-open, and open systems), aquaculture (freshwater or marine), of the species used and especially the density and quantity of production. Due to the many variables that may influence the generation or identification of such impacts, and because it is a relatively recent activity in Brazil, few conclusive studies have been published on the possible environmental impacts caused by aquaculture, especially by mariculture. Nevertheless, in any form of production, the impact on the environment occurs through three processes: the consumption of natural resources, the process of transformation, and the generation of final products (waste) [3].

The main environmental impacts caused by aquaculture (including fish farming and shrimp farming) are conflicts with the use of water bodies, sedimentation and obstruction of water flows, over nutrition and eutrophication, discharge of pond effluents, and pollution by chemical residues used in the different stages of cultivation [2, 5].

Thus, fish, crustacean and mollusk crops generate waste in the form of particles and organic and inorganic soluble material. The nature and quantity of the effluent vary with the type of farm, cultivated species, and level of intensification. In general, feces, excreta, and pseudofeces are the major sources of inorganic and organic nutrients [5].

The water column is the container for the dissolved material, while the greater proportion of solid waste precipitates into the sediment below the growing area or in the immediate vicinity of the farm. This almost immediate deposition of daily organic matter in the sediment leads to an onset of eutrophication with the decrease of the oxygen rate dissolved in the medium. The level of water contamination of the shrimp farms in the Jaguaripe/Bahia—Brazil can be demonstrated by **Figure 1**, which contains the photo of containers of water samples, which were collected by federal agents of the Ministry of the Environment.

Bacteria, fungi, and protozoa participate in the process of degradation of organic matter, and the release of toxic gases such as sulphur become unavoidable. In this sense, the alteration

Figure 1. Sample containers of contaminated water discharged to ocean [8].

of the marine community caused by the accumulation of organic matter in the sediment has already been studied by several researchers. The ideal of sustainable aquaculture, which is consistent with the economic, social, ecological, spatial, and cultural contexts, is at a final stage in Brazil, where errors are being evaluated and studied so that new management techniques can have a lower effect on the ecosystem balance [3, 8, 10].

This concern about the environmental impact was not taken into account by the Federal Government of Brazil, which, despite the warnings given by CONAMA, decided, through the Ministries of Agriculture, Environment and Development of Industry and Commerce, to create in 1997 a project to encourage fishing and seafood production, through the creation of marine farms, in order to increase the production of seafood, mainly crustaceans and mollusks. In 1998, the state governments began to join the project, and the states of *Bahia, Santa Catarina, Ceara*, and *Espirito Santo* invested heavily in the project for the installation of farms mainly shrimp, oysters, and scallops. Through fiscal incentives, public funding, and federal funds, there was an incentive to create marine farms in these states, but there was no prior and direct evaluation of the impact of the implementation of aquaculture in these states and in the environment [4, 6–10].

Between 2000 and 2005, records of seafood production were achieved. A total of 75% of this production was for export [4, 10]. But in 2005, a productive decline began, beginning in Santa Catarina, southern Brazil, where an epidemic of shrimp of the white spot started. It is a disease that affects shrimps and is one of the four identified in Brazil that are listed in the OIE in aquatic animals. It is of viral ethology. WSSV—"White Spot Syndrome Virus" is a persistent infection throughout the animal's life and has a high mortality in shrimp cultures.

Animals that recover from the infection are persistent carriers of the virus. There are no known biological vectors. This virus has a type of flagellum and its genetic material is DNA (being genetically more stable) and is part of the *Nimaviridae* family, decimating several fishing grounds. This disease began to extend along the Brazilian coast, currently reaching the state of *Ceara*, with productive loss of almost 80% of the prawns [4, 5, 9].

Almost concomitant with the appearance of this disease, in the mangroves of *Espirito Santo* and *Bahia*, appear the disease of the lethargic crab (DCL), it was found that the agent that causes DCL is the pathogenic fungus *Exophiala cf psycrofila*, which simply decimated almost 90% of the crabs in these states [3].

By 2015, this disease spread to other states, decimating the mangrove areas and drastically reducing the population of crabs. After several analyses of environmental agencies, it was verified that marine farms, implanted in these states, generate residues that provoke the problem, generating conditions for the proliferation of pathogens in the environment, often resistant to the immune system of the animals, due to the indiscriminate use of antibiotics from nurseries and crop laboratories.

The DNAs of *Exophiala cf psycrofila* and *Nimaviridae* were found in the mangrove in the region of *Jaguaripe*, a fishing town in the State of *Bahia*—Brazil. This organic material was present in a phenomenon known as brown tide and was caused by the discharge of tailings from the fishing farms that had settled in this region (see **Figure 2**).

Thus, from the studies of the discharge effluents from tanks of shrimp farms in the *Jaguaripe—Bahia*/Brazil region, promoted by the Ministries of Fisheries and Environment, a map of the destruction and contamination of the region known as the coast of the *Dende* was created that raised the points of impact and associated the degradation of the aquatic environment to the effluents discarded by these farms, which influenced native marine life, causing the degradation and destruction of native species [8].

Figure 2. Shrimp farm: region of *Jaguaripe—Bahia*, Brazil [8].

3. Nereu project

The oceans and seas are vital to sustain life and to provide for the various human activities and are the primordial elements for the entire global economy.

In this context, it is verified that the activity of fishing and gathering of seafood stands out due to its importance in the production of food and world economy. But in order to reach the levels of production and productivity demanded today, fishing has destroyed most of the oceans' fishing grounds due to the phenomenon of overfishing.

The oceans have strategic relevance for maintaining the dignity of human life and for the economy, which converges to the challenge of integrated management of seafood production and management. Environmental resources, especially ocean resources, are currently in a vulnerable position of degradation and scarcity. Several authors argue that the main factors of the maritime crisis are the environmental pollution and the unbridled increase of the world population, without the policies on territorial planning and environment to adequately meet the new demands.

The degradation of ocean water quality is a reflection of the current pattern of consumption and lack of care with the oceans. In addition, the Industrial Revolution, in which the concern to increase fish production, without considering the externalities, resulted in the degradation and waste of marine resources. In this crisis scenario, countless losses can be raised, such as loss of ecological balance, loss of biological diversity, climatic imbalance, and profound changes in the water cycle, which affect society and the environment leading to reduced or depleted resources. Thus, today, the world population has established a situation of imbalance between the spatial pattern of availability in the oceans and the spatial pattern of demand for consumption centres' [11–13].

With the emergence of aquaculture, the first attempt was made to reduce the pressure on fishing banks, but the lack of an environmental policy to treat wastewater has increased the degradation of the oceans. To address this degradation caused by the seafood farms that affected several Brazilian states, the Ministry of the Environment analyzed the environmental issue in order to build an environmental awareness. In this sense, with the introduction of new sustainable development concepts, a new cycle began, based on the elaboration and implementation of environmental policies, in the search for negotiation and understanding between preservation of the environment and production processes.

Advances can be verified with the Green Protocol (Federal Law: institutional device of introduction of the environmental variable as a relevant criterion in economic policy decisions and project financing) and other devices of environmental dimension inserted in the decisions of public policies. Specific measures for the rational management of irrigated areas and collection for the use of water are already established by law.

From these premises, the Nereu project was conceived by associating two Chiron and Prometheus projects, which combine three basic conditions for the production of water and energy, with food production based on a clean production system. A conception was used that interrelated CH_4 reform through a systematic pyrolysis in a cold plasma reactor to obtain H_2 whose thermal cycle was used for water desalination and energy generation combining with the production of totally sterilized rock salt (see **Figure 3**).

With this conception, the Nereu[1] project has the following structure:

- Chiron[2] project: the purpose of this project is to carry out pyrolysis of CH_4 through a cold plasma reactor for the separation of colloidal carbon and hydrogen, which will feed the Prometheus reactors and produce O_2 using a modular and interchangeable structure [12, 13].
- Prometheus[3] project: it provides for the rational and intelligent use of reuse water from aquaculture systems, whether having a high content of dissolved solutes, with the purpose of sterilizing, purifying, and neutralizing pathogens present in process waste water, through a reactor of hydrogen. This process desalinates water by generating the superheated steam which is sent to a set of the stirling system, without affecting the environment and having by-products of the process the generation of clean energy, gem salt, and demineralized and deionized water (water pure; see **Figure 4**) [12, 13].

In view of these premises, the Nereu project seeks to meet the main characteristics of the Apollo Research Group projects, that is, to have a simple automation in their constructive characteristics, guaranteeing their interchangeability and maintenance, as well as their low automation costs. In addition, the Nereu Project units were conceived and developed in modular systems, proto-units, to be placed in reduced and limited physical spaces; from its constructive concept, the Nereu project can expand its productive capacity according to the needs.

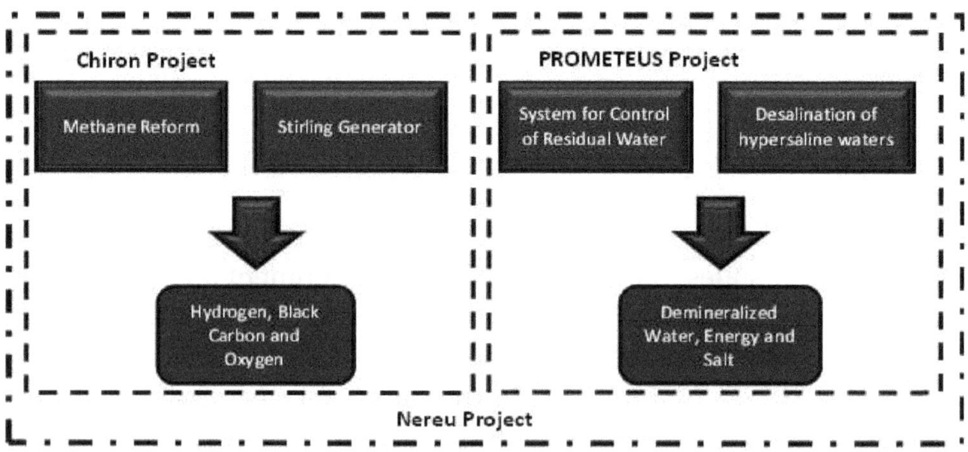

Figure 3. Diagram of the Nereu project.

[1]The Nereus project is a tribute to Nereus (Greek: swim), in Greek mythology, is a primitive sea god, depicted as an elderly character - the old man of the sea. He is known for his virtues and his wisdom. Pindar celebrates his fairness, hence his epithets "true," "benevolent," "without lying or oblivion".

[2]The Chiron project is a tribute to Chiron (Greek: hand), in Greek mythology, was a centaur, considered superior by his own peers. Chiron was intelligent, civilized and kind, and celebrated for his knowledge and skill in medicine; he was professor of Heracles and other demigods.

[3]The Prometheus project is a tribute to Prometheus (in Greek: foresight), in Greek mythology is a titan; he was a defender of humanity, known for his astute intelligence, responsible for stealing the fire of Hestia and giving the Mortals.

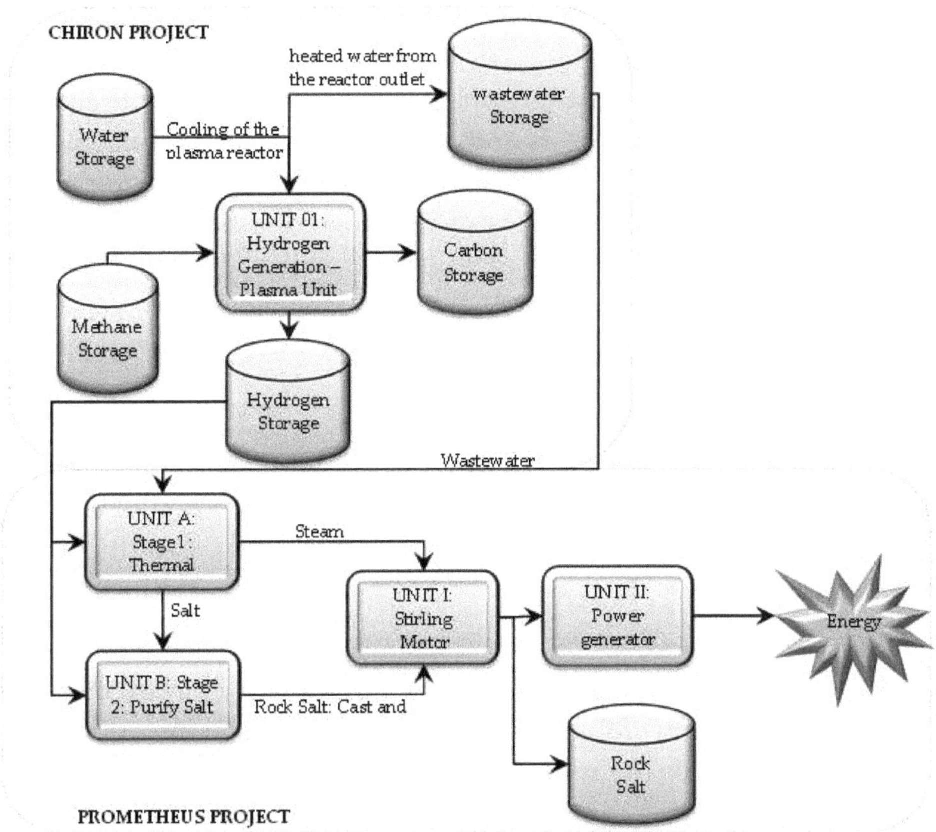

Figure 4. Process flowchart [12, 13].

3.1. Chiron project

The desalination process is a complex process because the salts are strongly bound to the water molecules, which makes conventional water treatment processes inefficient. Therefore, it is necessary to make use of other processes capable of breaking the forces of attraction between molecules [12, 14, 15].

3.1.1. Plasma generation reactor for H_2—Unit 01

In the specific case of pyrolysis of CH_4 by thermal plasma, the products formed include hydrogen, which can be used as fuel, and colloidal carbon, with many uses in the cement and rubber industry.

Conventional hydrogen and black carbon production processes are based on the incomplete combustion of hydrocarbons, subject to low yields and emission of high levels of pollutants. Due to incomplete combustion in the presence of oxygen, the process is inevitably accompanied by the emission of CO, CO_2, SO_2, NO_x, and volatile hydrocarbons. In addition, the maximum reaction temperature achievable is from 1500 to 2000 K [11–15].

The reactor of the Chiron project was designed to produce colloidal carbon and hydrogen, through thermal plasma, although using electric energy and an inert gas in the process have several advantages in its application that are compensatory, as for example: increased efficiency, reduction of the reaction time, low cost of the equipment, favoring variety of productive size, and making possible its industrial application. The use of a noble gas as a plasma gas avoids the formation of CO_2 in the process and is only necessary to initiate plasma formation and was initially replaced by H_2, and then by CH_4, another important factor is that the plasma process enabled the use of different sources of raw material, besides CH_4 gas, butane and propane, but for this project, we chose to use the gases from landfill gas. Thus, the production of hydrogen as an energy source and that of colloidal carbon as an industrial input becomes attractive and compensates the energy and operational expenditure of the system [11–13, 15, 16].

3.1.1.1. The plasma system

The Chiron project has a thermal plasma system, which is composed of a plasma arc torch, and a reactor or pyrolysis chamber designed to retain most of the heat flux from the plasma jet. The inert gas argon (Ar) was used as a plasma gas and was introduced in the torch by a controlled flow [12, 13, 15, 16].

The electric arc between the cathode and anode was initiated by a high-voltage discharge and a variable current. The formation of the plasma arc occurs between the cathode and the anode (cooled by water). The gas to be degraded, CH_4, was introduced into the pyrolysis chamber with controlled flow [12–14, 16].

The torch consisted of a cone-shaped cathode and an anode rivet, channel-shaped. The electric arc established between the electrodes (cathode and anode) was initiated by a high-voltage discharge. The gas is heated, ionized, and emerged from the torch as a plasma jet. CH_4 injected into the reactor was heated and passed then into the liner, by contacting the plasma and suffering degradation [11, 12, 16].

3.1.1.2. The process

The values of degradation of CH_4 and the efficiency of H_2 production depend directly on the CH_4 flow; these results do not depend on the argon flow. Important to note that the amount of CH_4 converted is related to the total flow of CH_4, relating to what has been transformed into solid and gaseous products. For example, when the flow of CH_4 was adjusted to 0.001 m^3/min, 59.4% of the gas was transformed into solids and the remainder 40.6% by gaseous products (see **Figure 5**) [12–16].

The proportional reduction in the conversion relates to the constant electric power applied to the torch; since the available power is controlled, the CH_4 conversion is also limited and any additional amount of CH_4 remains in the system as a nondegraded fraction. To correct this problem, argon flow was increased in the plasma torch, which caused the increase in plasma jet temperature and consequent increase in system power [11, 12].

The decomposition of CH_4 to produce hydrogen and colloidal carbon is quite promising, if aggregated with other activities that use their products. As the CH_4 conversion function is endothermic, the thermal plasma is very efficient because in these conditions of high temperatures

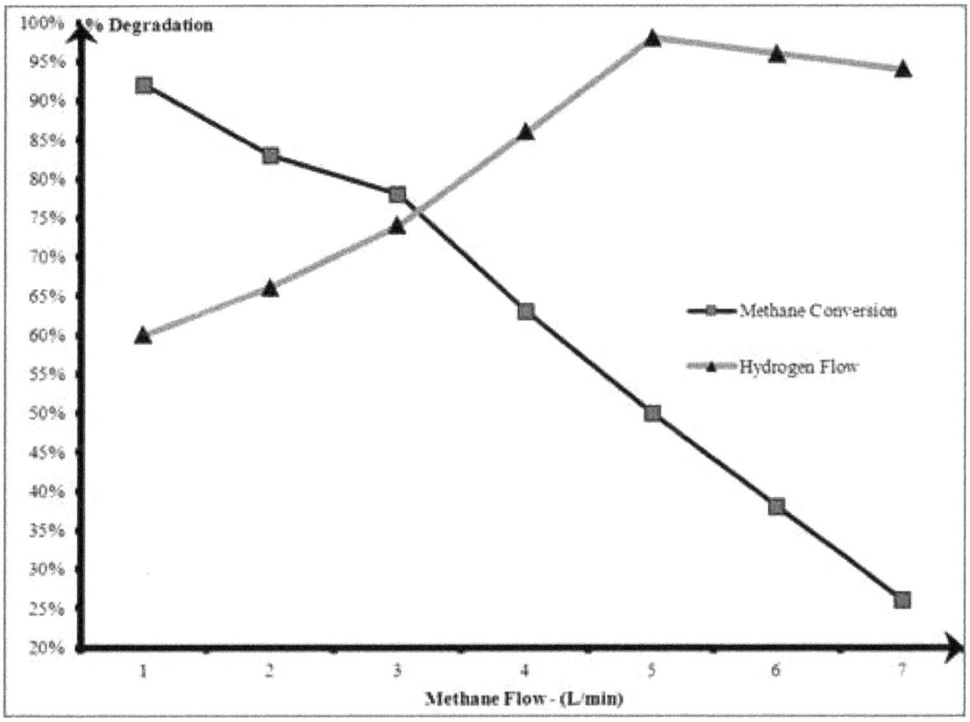

Figure 5. Percentage of degradation of CH_4 (■) and efficiency of H_2 (▲).

and high degree of ionization, chemical reactions are accelerated by the power supply, and moreover, there is no CO_2 emissions or other contaminant gases, once the system is deoxygenated [12, 13, 16].

In the Unit 01 plasma pyrolysis reactor, CH_4 is broken down mainly by the heat from the arc plasma. However, the existence of a collision process between the existing elements in the plasma catalytically assists the decomposition reaction of CH_4. Different types of crashes happen in a plasma state of this nature; dissociation of CH_4 and other molecules (CH_3, CH_2, and CH) may also occur as a result of impacts between the molecules, atoms, ions, electrons, and protons [11–15].

A CH_4 molecule can be dissociated into one or more parts to collide with another particle. Various collision processes may occur inside the plasma reactor, including the recombination of occurring particles forming a molecule [16].

The usage of Unit 01—reactor pyrolysis—has several advantages such as high energy density existing in the plasma which allows the construction of a compact reactor. Due to the plasma characteristics, a continuous flow is possible, enabling the formation of carbonaceous materials of different qualities and providing no pollution to obtain two commercial products (H_2 and C), both of high purity [12, 13].

A revolutionary constructive architecture for the treatment of methane, the reactor provides a greater CH_4 residence time in the reaction zone and also a longer residence time inside the reactor, enabling minimizing the ionization of CH_4 molecules. This feature is important for a high efficiency in the conversion of CH_4 into H_2 and C. In addition to presenting a favorable flow of the mixture of CH_4 and hydrogen flow, important for the heat exchange between the flows and good distribution of current lines coupled with good mixing between the flows, it is important to provide a homogenous treatment of particles of carbon [13, 15].

Thus, in the pyrolysis gas plasma, the energy required for reaction is supplied to the elements in the reaction chamber of the reactor by a high-energy electrical discharge. The electrical discharge at atmospheric pressure generates thermal plasma with high temperatures that can completely dissociate CH_4 molecules. The molecular decomposition in the absence of oxygen generates hydrogen and carbon, which do not bother catalysts, for there is the formation of CO and CO_2 [12, 13].

The pyrolysis of CH_4 does not produce CO or CO_2; the carbon in CH_4 molecule does not react with the oxygen because it does not exist in the process. What happens in the process is the breaking of the molecules; resulting in isolated black carbon. The process allows a simple purification of hydrogen, since the solid carbon may be extracted by physical separation of the flow [12].

3.1.2. Oxygen unit

Oxygen is the most known and widespread of the gases, being the most abundant element in the earth's crust, about 46.6% by weight. To meet the oxygen needs of the Prometheus project, a little production plant of O_2 was created in this unit, taking advantage of the project's power generation. Thus, the O_2 production system followed a PSA ("Pressure Swing Adsorption") process, which is a classical mechanical O_2 production system [16].

Pressure swing adsorption—PSA allows the viability of obtaining oxygen at various on-site scales and its low operating cost was the main drivers of this process.

In this system, there is an oxygen production plant, which is installed next to the consumer (on site). This technology allows the separation and concentration of oxygen (approximately 95% O_2) by subjecting the ambient air to a zeolite molecular sieve under low pressure (3–6 bar) for a period sufficient to adsorb carbon monoxide, steam, CO_2, and almost all of the nitrogen present in the air [13].

The process uses two metal vats containing molecular sieve (zeolite) in antiparallel, through six valves, which retains nitrogen from the air at high pressure, which will be released at low pressure, and allows the oxygen to cross the adsorbent bed as final product.

Also, there must exist in the process a filtration and drying step of the compressed air that is admitted in the sieves, protecting the zeolite from contamination so as not to compromise the efficiency of the sieve.

Thus, this is an alternative of lower cost and better energy efficiency, if compared to the cryogenic process, but produces a gas with a lower oxygen percentage, that is, a minimum purity of 92% (see **Figure 6**)

3.1.3. Power generators—using stirling engine

This type of engine operates with a thermodynamic cycle composed of four phases and executed in two-stroke piston: isothermal (= constant temperature) compression, isochronous

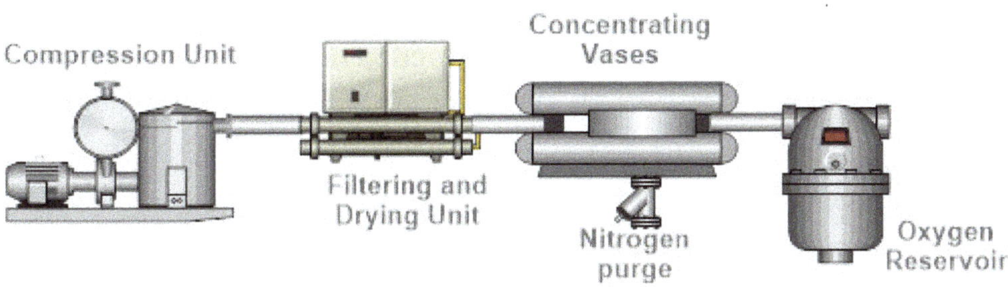

Figure 6. Oxygen unit [13].

heating (= constant volume), isothermal expansion, and isochronous cooling. This is the idealized cycle (valid for perfect gases), which diverges from the actual cycle measured by instruments. Nevertheless, it is very close to the so-called Carnot Cycle, which establishes the maximum theoretical limit of thermal machines performance.

The Stirling engine surprises for its simplicity, because they are of external combustion and are used in the process to generate energy, since they can work with more than one source of heat and thus are projected, receiving heat of the cooling water of the pyrolysis reactor, of the salt leaving the bottom of the salinization unit at a temperature of approximately 1000 K of the steam leaving the desalination unit and even of heating of solar origin (see **Figure 7**).

Because they are very simple machines with easy maintenance, easy operation, quiet operation, and low vibration, these peculiarities made possible their use in a distributed way throughout the entire production chain, reducing the cost of energy and the need for large generation units. The use of the stirling engine for power generation is the best way to reduce costs and increase energy production. Thus, the option to use the stirling engine as driver for power generation units is due to its low cost of manufacturing, the ease of working with small

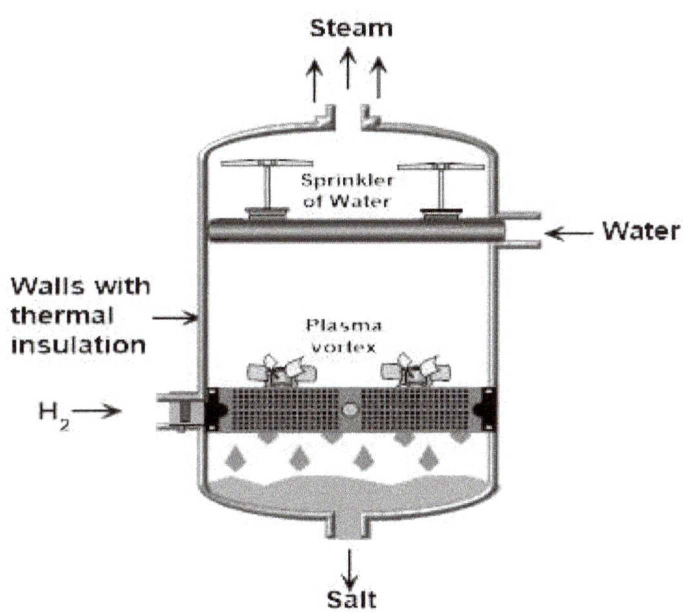

Figure 7. Prometheus unit [12, 13].

generating units, and the distribution of several generating units throughout the production unit. Thus, Stirling engines are used along the production line, using the generated steam and heat of the gem salt, which is extracted by the lower nozzle of the Prometheus unit.

3.2. Prometheus project

The need to obtain drinking water, increasing demand and neglect of much use by the population are the few factors because of which water scarcity begin to emerge in places where hitherto was not a problem. Water, unlike other natural resources, cannot be replaced, must be preserved, and distributed properly to avoid its loss.

Desalination is a technique that has been used for thousands of years in places where we cannot get enough fresh water. It is considered the future alternative to meet the needs of living beings, since 97.2% of the planet's water is salty or brackish. It is currently underutilized due to the high cost of the process, since it requires a lot of energy and sophisticated materials [12, 13].

Since the sixteenth century the desalination of sea water was used in vessels. Land desalination began in the 18th century and began to play an important role in the late 1940s and the early 1950s, especially in countries where potable water is scarce as in the Arab Gulf, USA, Caribbean and some areas of North America.

The composition of the salinized water varies according to the source, although the concentration of salts varies from place to place, the ratio between the most abundant constituents is practically constant. Among these constituents are sodium chloride, bicarbonates, calcium sulphate, and magnesium sulphate [12, 13].

Another alternative to obtain water is to reuse it, by replacing a source of drinking water with another of inferior quality where such substitution is possible, supplying the less restrictive demands made, releasing water of better quality for the noblest uses, such as domestic supplies.

The reuse of water in the urban sector can be accomplished by appropriate treatment of sewage and its reuse for potable purposes (indirect reuse) or non-potable (irrigation, fire reserve, dust control, decorative aquatic systems, etc.). In the industrial sector, the reuse of water is a frequent practice, since it reduces production costs. They are often used in cooling towers, boilers, civil construction, and maintenance of industrial facilities and within industrial processes. The recharge of aquifers is another form of water reuse [12, 13].

The Prometheus project uses a systematic rationalization in the use of water and energy, basing its process on a continuous control of what is generated and what is consumed. Thus, the project provides for the rational and intelligent use of wastewater from industrial processes, whether it has low oxygenation content and a high presence of dissolved salts, in order to desalinate and neutralize dissolved solutes in the water.

In Counterpart, there is to the production of electricity by the use of H_2 and associated thermogeneration cycles, at low cost and with high profitability, constructed in a modular and interchangeable way, without affecting the organizational structure and the environment, and having by-products of the process generation of clean energy, gem salt and demineralized and deionized water [12, 13].

Thus, the reactor receives the salinized water mixed with all the solutes of the production of mariculture, which would become a contaminant very dangerous for the environment, mainly to the mangroves; although the concentration varies according to the source, the productive stage and the type of production and the relationship between the most abundant constituents is practically constant. Among these constituents are chlorides, sulphates, iodides, bacteria, fungi, protozoans among others.

The reactor configuration used in plant is the type of rotor under pressure with an arrangement in which the saline solution is introduced into the chamber through a spray system; the spraying occurs on a vortex hydrogen plasma vaporizing it. One container captures the salts deposited in the bottom in the form of rock salt. The superheated steam is sent to Unit 03 in order to generate energy (see **Figure 7**) [12, 13].

4. The results analysis

The production of energy, water, and food in sufficient quantity and at competitive prices is important for economic development. At present, there is great concern about the environmental aspects of production in these segments, with emphasis on the use of renewable sources that do not degrade the environment and guarantee the productive continuity for future generations.

Therefore, it is no secret that fishery supplies are being affected by predatory fishing and discharges of chemical and organic waste. The disorderly growth of the world's population, coupled with the intensified demands of protein needs, has been draining the oceans. But, the solution found by mankind through aquaculture apparently solved the question of productive continuity and protein reserves. What is the problem with this activity? The vast majority of the water available for swimming pools is not suitable for return to the oceans and also is not suitable for the demands of aquaculture. These tailings are dumped in the oceans and can create areas of marine desertification in dump areas, which would lead to famine in many communities.

Over the years, the conviction that infrastructure in the world should be based on a mix of productive sources and self-sustainable systems, composing production triad—sustainability—environment. In the same way, it offers great opportunities for expansion of new business initiatives and integrated production projects; they are relatively new and suffer the discrimination of the need for productive specialization to guarantee the maximum profit of the capital employed.

As the production of energy, food, and water treatment are important elements for the economic development of societies and the maintenance of the quality of life and health of families; in recent years, emphasis has been placed on the relationship between production and the environmental issue, considering possible reductions in the availability of nonrenewable natural resources.

Thus, in order to evaluate these investments, there are indicators of analysis capable of assisting in the perception of the behavior between risk and return, the Nereu Project would have a distinct characteristic for all desalination projects. To carry out this technical-economic evaluation, it is necessary to make a hybrid analysis, taking into account four basic characteristics of this project, which are as follows:

- Technique: development of new technologies; reducing the emission of greenhouse gases into the atmosphere; producing water and energy for regions lacking these resources.
- Economic: the low cost of implementation, the low maintenance cost, and diversification of income and production.
- Environmental: carbon sequestration, eliminating greenhouse gases, and producing water and energy without generating contaminant residues.

It is worth noting that the Nereu project aimed to reduce the risks of contamination in the oceans caused by aquiculture tailings, supported by a demand from CONAMA, which provided for the eradication of pathogenic releases to coastal areas.

The project, in an experimental phase, was built in a proto-unit system in the region of the *Dende* Coast, in the city of *Jaguaripe–Bahia*, Brazil, operating for 8 months, in laboratory scale, with the support of *Funil* Farm, which invested in materials and assembly, with the purpose of neutralizing the pathogenic elements of the pool 03AD, with volume of 3.5 m^3, purge tank used to deposit filtration wastes from other pools. Thus, the Nereu project operated to make the cleaning residual water from the existing filtration system.

4.1. Techniques resulting

The Nereu project used its two reactors and seven generators with stirling engine to treat waste water from *Funil* Farm, generating energy and producing food. To perform this procedure, the Chiron Project was miniaturized to work with limited flow rates to meet the needs of the Prometheus project. Thus, the project was conceived as a pilot plant with the following technical characteristics: total flow: 100 l/h; CH_4 flow rate: 400 l/h; and average concentration of salts: 35 g/l.

It was verified that the Nereu project consumes a total of 25.4 kWh of energy to meet the energetic demands of desalination and decontamination of the waste water, and the reactor consumption, in maximum efficiency, for CH_4 pyrolysis, is 7.2 kWh, with advantages over other techniques, as the process does not release CO_2.

The percentage of degradation of CH_4 reaches the maximum value of 98.8% for an applied power of 7.2 kWh. In order to cope with the energy needs of the system, seven stirling engines of the Alpha type have been adapted which have been developed for other purposes and reused for the project. Thus, it was possible to adapt them for power generation, producing a total of 40 kWh of energy, which has its stepped drive, being connected as the energy needs are presented. The Nereu project consumed 54.4% of the energy produced, and 45.6% was made available to the electricity grid of the farm.

The C produced by the degradation of CH_4 in thermal plasma has hydrophobic characteristics, is amorphous and of high purity, presenting added value and can be commercialized, for companies producing carbon fibber and carbon nanotubes.

In spite of the Nereu project, a miniaturized version of the Chiron project and the equipment of the Prometheus project were coupled to its structure, and the average project yield was 79.6%.

The average yield in the project stems from the main constraint adopted in the operationalization and implementation of the Nereu project. In other words, due to budget constraints, it was only possible to use recycled materials, scrap, and/or products recovered from old irons.

Thus, the great problem generated by the filtration of the shrimp production ponds and dumped in the pool 03AD, without having a real destination for this residue, was solved with the Nereu project, which desalinized and sterilized this residue, producing pure water and gem salt, totally free of any pathogenic element. At this point, the Nereu project stands out in relation to the other desalination processes, because while the other systems consume energy, the Nereu project produces its own energy to maintain the system and to be commercialized in the energy market.

4.2. Economic resulting

When analyzing the feasibility of desalination projects, one should not only consider the question of return on investment, but also the impact on the entire productive and social context, especially for locations lacking these resources. It is therefore of utmost importance not to dwell on a simplistic reading of numbers and results. The Nereu project was a predetermined research funded by an investor with limited resources, with a limited value of US$35,000.00 to prove that it was technically and economically viable.

With the architecture of the project working with the thermal cycle of the H_2, the nonexistence of the generation of residues, nor greenhouse gas emanations, in the operation of the system was verified. Thus, the combination of the processes added to a system of financial control would enable the recovery of the investment in a period of 10 months. This was made possible by the generation of several marketable by-products that have added value.

Throughout the Nereu project's operating time, it was possible to operate at low production costs and it was observed that, based on the standards of taxes and input costs in Brazil, the project has a low maintenance cost, despite being fully developed with scrap. In this way, the average cost of the plant is low per hour of operation. The cost of each product, including taxes and operating costs, can be deployed as follows: pure water: US$0.12 per 1 l/h; electric power: US$0.037 kWh; Carbon: US$1.46 per 1 kg/h; hydrogen: US$1.46 per 1 kg/h; and rock salts: US$ 0.028 per 1 kg/h.

With the diversity of by-products generated, the Nereu project had a multiplicity of revenues that allowed the recovery of investments within 6 months, and an investment profit of around 10.6%, despite being a basic small plant, built from scraps. These results reflect the reality of the Brazilian market, incidental taxes and charges, labor costs, and employed automation. These values will change according to the economic, fiscal, and equipment conditions present in each country.

4.3. Environmental results

The development of projects based on environmental preservation attitudes is part of the goal plan of any research project prepared by GPAp. The goal proposed in any project of the research group is in the search for the development of actions with emphasis on sustainable development projects, the insertion of needy communities, in generation of new projects, always focused on the protection of the environment.

Contained in the thinking of sustainability is the possibility of change in the relationship between the individual and the environment. From the normative point of view, it is the necessary adoption of attitudes more coherent with the demands of environmental conservation. Thus, the development of an education strategy that allows better relations with the environment is undoubtedly one of the most strategic and effective actions to achieve the objectives proposed by the notion of sustainability.

With these bases, the Nereu project was built with a sustainable development procedure, so as not to generate any residue and to use only what one cannot take advantage of. Using natural resources such as hypersaline waters, greenhouse gases and scrap, sequestering C and generating pure water. The main characterization of sustainable development is the assurance of environmental concern. As, the plasma reactor did not generate CO_2, decomposed CH_4, with C sequestration, which can be sold or used in the production of shrimp.

The salt produced by the Nereu project was sent for testing at the university. It was verified that there were no pathogens present in the product. This had a very relevant chemical balance of chlorides, iodides, sulphides, and carbonates in the mixture, similar to the rock salt that is produced in mines in Europe and Asia. As the salt produced was processed from wastewater treatments, despite being completely free of pathogens and contaminants, under Brazilian laws, the rock salt could only be sold for animal feed.

Thus, serious environmental problems from desalination plants, such as high energy consumption, greenhouse gases, and waste generation were solved in the Nereu Project, which processed and neutralized greenhouse gases, generated its own energy for desalination, and treated the waste from the desalination process.

5. Final considerations

The Nereu project combined two different guideline projects, combining a plasma reactor for H_2 generation, a steam desalination unit and a turbo-generator of electrical energy in a totally clean system. This was created to work with desalination residues. In this way, a system of treatment of contaminated hypersaline waters of any origin has been developed. To achieve these goals, the Nereu project has adopted a sustainable development rule and, consequently, the nonproduction of waste that damages the environment.

The Nereu project, when it unified the two projects, created a totally high sustainable and closed process, ensuring that the most critical points in Prometheus, which is continuity of energy supply and combustion for thermal energy generation, were fully met by Chiron.

It should be noted that the project's application potential is directly associated with its uniqueness, simplicity, portability, modularity, and constructive speed, so the Nereu project is fully applicable to work in consortium with other desalination systems, producing energy to sustain them and treat the waste from the filtration without the environment being attacked, as it works as an alternative to take advantage of the wastewater potential without the need for large engineering projects and civil works for its implementation. In addition, the generation of energy is justified by the thermal availability of the process, compared to the conventional production of separate thermal and electromechanical energy, which, from this point of view, has led to the search for the most appropriate technology in order to provide the highest possible energy efficiency, therefore the option to use stirling engines to power the generators, due to its low cost of production and ease of maintenance.

In the process, it was possible to determine that the most relevant costs in the project come from four basic factors: energy, pure water, natural gas, and oxygen, which have a much greater influence than other costs. In this way, the Nereu project prioritized to minimize costs in three of the four items, which were energy, oxygen and water, directly impacting the final cost of the project.

As the Nereu project produces pure water, this productive cost is minimized by using the water that is provided by the Chiron subproject. With this advantage in mind, stirling engines are used for energy production, since they can be introduced at any stage of the production system in order to generate energy with any existing thermal potential, minimizing costs, and dependence on external energy to the project, stirling engines have low production and maintenance costs.

Nomenclature

ANA	National Water Agency
BR	Brazil
HDI	Human Development Index
MMA	Ministry of the Environment
Ar	Argon
CO	Carbon monoxide
CH_4	Methane
H_2	Hydrogen
NG	Natural gas
O_2	Oxygen
CONAMA	National Council for the Environment
MPA	Ministry of Fisheries and Aquaculture
IBGE	Brazilian Institute of Geography and Statistics

GPAp	Apollo Research Group
C	Carbon black
CO_2	Carbon dioxide
CH_3, CH_2, CH	Methane decomposition chains
HC's	Hydrocarbons
NO_x	Nitrous oxide
SO_2	Sulfur dioxide

Author details

André Pedral S. de Sena

Address all correspondence to: apedral@gmail.com

Universidade Federal da Bahia / UFBa, Salvador, Brazil

References

[1] FAO—Food and Agriculture Organization of the United Nations. The State of the World Fisheries and Aquaculture 2012. New York: FAO; 2012. p. 345

[2] Hall SJ, Delaporte A, Phillips MJ, Beveridge MCM. Blue Frontiers: Managing the Environmental Costs of Aquaculture. The WorldFish Center. 1st ed. Penang, Malaysia: Beveridge & M. O'Keefe; 2011

[3] Van Kouwen F, Dieprink C, Schot P, Wassen N. Applicability of decision support systems for integrated coastal zone management. Coastal Management. 2008;**36**:19-34

[4] Boscardin NR. *A produção aquícola brasileira* (Brazilian aquaculture production). In: Ostrensky A, Borghetti JR, Soto D. (Orgs.), editors. Aquicultura no Brasil: o desafio é crescer (Aquaculture in Brazil: the challenge is to grow). 1st ed. Brasilia: UNB; 2014. pp. 27-72

[5] Vita R, Marin A, Jimenez-Brinquis B, Cesar A, Marin-Guirao L, Borredat M. Aquaculture of Bluefin tuna in the Mediterranean: Evaluation of organic particulate wastes. Aquaculture Environment Interact. 2010;**69**:57-70

[6] Plano Nacional de Recursos Hídricos - quinquênio 2014/2019 (National Water Resources Plan - 2014/2019). 1st ed. Brasilia: ANA; 2014. p. 476

[7] ANA (Water National Agency). Relatório Avaliativo dos Recursos Hídricos do Brazil - 2014 (Evaluation Report of the Brazilian Water Resources - 2014). 1st ed. Brasilia: ANA; 2015. p. 274

[8] Conama – Conselho Nacional De Meio Ambiente (National Environmental Council). Relatório de Maricultura e Regras de Implantação (Mariculture Report and Deployment Rules) [Internet]. 2015 [Updated: 2015]. Available from: http://www.mma.gov.br/port/conama/relatórios/relatorioanual.pdf [Accessed: May 23, 2016]

[9] MPA – Ministério de Pesca e Aquicultura (Ministry of Fisheries and Aquaculture). Guia de Orientação para A Regularização da Aquicultura em Águas da União (Guidance for the Regularization of Aquaculture in the Waters of the Union). 1st ed. Brasília: MPA; 2015

[10] Dubois S, Marin-Leal JC, Ropert M, Lefebvre S. Effects of oyster farming on macrofaunal assemblages associated with Lanice conchilega tubeworm populations: A trophic analysis using natural stable isotopes. Aquaculture. 2007; **271**:336-349

[11] Rogdakis ED, Antonakos GD, Koronaki IP. Thermodynamic analysis and experimental investigation of a Solo V161 Stirling cogeneration unit. (Original Research Article). Energy. 2012;**45**(1):503-511

[12] de Sena APS, Coutinho OA; Lima JR, Ailton de SS. Neptune project: A reactor of cold plasma to treat seawater for use in the food production. In: Terrell J, editor. Water Filtration System. 1st ed. New York: Nova Publishers; 2016. pp. 27-70

[13] Chiron project: Reform of methane from a cold plasma reactor with hydrogen production for desalination and energy cogeneration. Energy Conversion and Management. 2014;**85**:933-943

[14] Kim KS, Seo JH, Nam JS, Ju WT, Hong SH. Production of hydrogen and black carbon by methane decomposition using DC-RF hybrid thermal plasmas. IEEE Transactions on Plasma Science. 2005;**33**:813-823

[15] Zhanghua WU, Guoyao YU, Limin Z, Wei D, Ercang L. Development of a 3 kW double-acting thermoacoustic stirling electric generator. Applied Energy. 2014;**136**(1):866-872

[16] Gallego JS, Barrault J, Batiot-Dupeyrat C, Mondragón F. Production of hydrogen and MWCNTs by methane decomposition over catalysts originated from $LaNiO_3$ perovskite. Catalysis Today. 2010;**149**:365-371

7

Solar Thermal-Driven Desalination Pursuing Products of Pure Water and Salts and Leaving Minimum Impact to Environment

Ben Xu, Peiwen Li and Penghua Guo

Abstract

Desalination, removal of salt and other minerals from seawater, brackish water, and wastewater, is becoming a promising solution for providing the increasing need of freshwater. It is highly desirable that environmentally friendly renewable energy resources be utilized for water treatment to minimize the consumption of fossil fuels. Given that most desalination systems can directly use thermal energy, concentrated solar thermal energy is very suitable for application to the water treatment. To avoid the potential negative impacts from disposing the concentrates, recovery of important minerals from concentrates to achieve zero discharge is a promising option. The recent technology development on solar thermal energy storages has shown that sea salts are very promising materials for large-scale thermal energy storage. Hence, a full separation of salts and water in desalination process becomes a necessity in advanced water treatment technologies, which should be achieved in an economically feasible way. Literature review and studies about innovative concept of full separation desalination system will be presented in this study. A full separation device integrated with conventional multieffect distillation or multistage flashing water treatment system will be introduced into the system design to enhance the water productivity and thermal efficiency.

Keywords: desalination, full separation multieffect distillation (FSMED), concentrating solar power (CSP), thermal energy storage (TES)

1. Introduction to desalination process

The population growth, urbanization, and climate change lead to increase in household and industrial uses of more freshwater, while natural freshwater reserves cannot meet the demands. Over one-third of the population lives in water-stressed countries, and by 2025 this figure is projected to rise to two-thirds [1]. Water with sufficient quantity and good quality for household uses and industrial applications is critical to health and well-being, as well as the opportunities to achieve human and economic development. Desalination, removal of salt and other minerals from seawater, brackish water, and wastewater, is a promising solution to grow the supply for fresh water. Presently, many arid areas have to rely on desalination to provide major quantities of safe water [2].

According to the International Desalination Association (IDA), there are 18,426 desalination plants in operation in more than 150 countries in June 2015, producing about 86.8 million cubic meters of water per day [3]. This number is continuously growing as the need for fresh water supply grows. **Figure 1** shows the world desalination plants per geographical area, and over half of these plants are in the Middle East.

A simplified process sequence of a typical desalination plant is provided in **Figure 2**. A brief introduction of the source intake, pretreatment, and post-treatment is given below, whereas the detailed description of the desalination process will be given in the next section.

Basically, there are two types of intake facilities to access reliable water sources: subsurface and open ocean intakes [6]. Proper design of the intake facility would not only protect downstream equipment and reduce adverse effects on aquatic life but also be beneficial to system productivity and capital/operating costs. The choice of the intake facility depends on the geographical conditions and desalination technology employed. For example, the raw feed water obtained from the intake well (belong to subsurface intake) is pretreated via slow filtration which is beneficial to the pretreatment process. However, the intake well is usually used for small-scale desalination plant due to their small productivity [5].

In addition to dissolved solids, the raw feed water generally contains other impurities such as silt, algae, bacteria, and even small aquatic life. The pretreatment process, involving some filtration and physical-chemical processes, is mainly used to improve the quality of the raw feed water to meet the requirement of different desalination technologies. Generally, the membrane desalination requires a higher degree of pretreatment than distillation technologies. To ensure that the product water from the desalination process meets statutory water quality standards for public health and protection of the water distribution system, post-treatment of the product water is necessary. This process may involve pH adjustment, disinfection, boron removal, and addition of minerals and corrosion inhibitors.

Despite the tremendous improvements in conventional water treatment technology, desalination industry is still facing several practical challenges on high-energy consumption, using either membrane-based technologies or thermal-based phase change approaches. Therefore, it is highly desirable that environmentally friendly renewable energy resources should be utilized for water treatment to minimize the consumption of fossil fuels. More details on this issue can be found in Section 3.

Another key concern associated with the desalination is the potential negative environmental impacts caused by the concentrated discharges to the environment, such as the adverse effects on water and sediment quality, aquatic lives, and the functioning of ecosystems. The concentrate is generated as a side product of the desalination process, which contains most of the

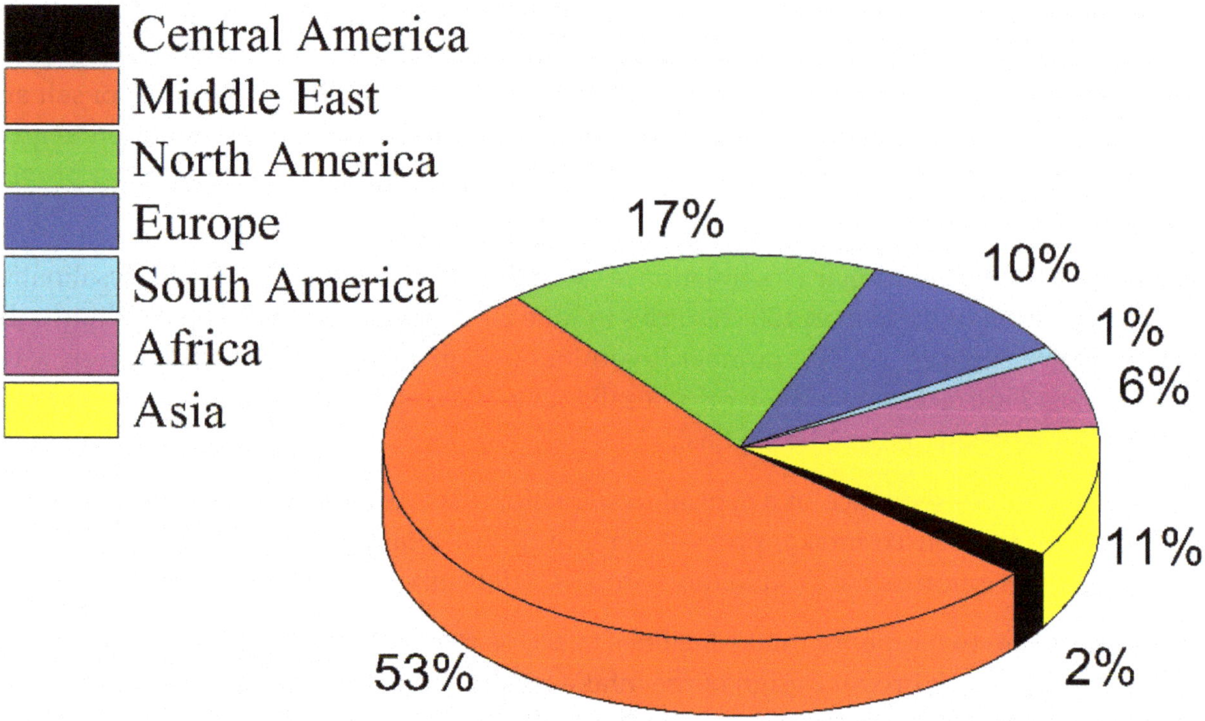

Figure 1. World desalination plants per geographical area [4].

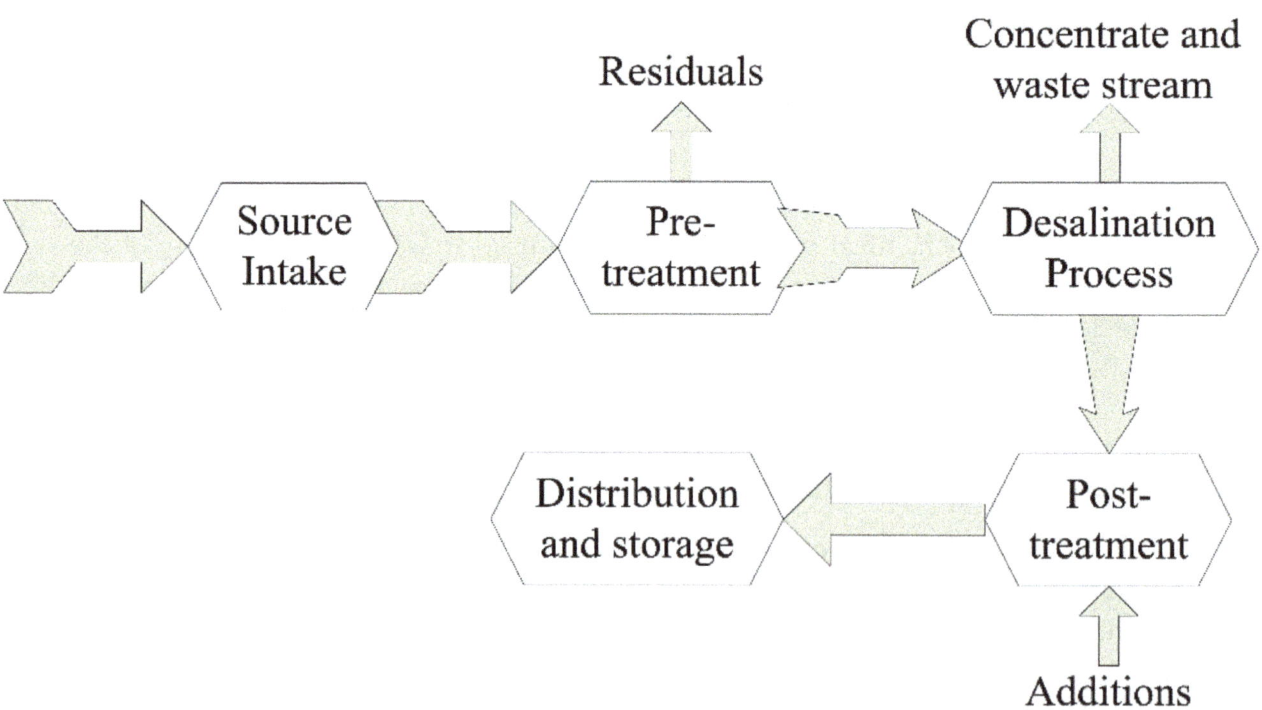

Figure 2. Typical sequence of desalination treatment and distribution processes [5].

minerals and contaminants of the source water and pretreatment additives in concentrated form. Based on a survey of concentrated disposal methods in 203 desalination plants in USA [7], it turns out that 87% of all 203 desalination plants are using surface water discharge and sanitary sewer discharge methods, while the rest part includes deep-well injection, evaporation ponds, and spray irrigation methods, unfortunately none of them has any zero-liquid discharge technology applied.

Obviously, most of the desalination brines are disposed to the sea or the sewer lines, and this will lead to adverse effects to the environment. To avoid the potential negative impacts from disposing the concentrates, exploring the options of recovering the dissolved salts from the desalination effluent to achieve zero-liquid discharge is necessary. The existing method to recover salts from desalination is solar pond approach which requires extensive areas of land. To better understand all the pros and cons, a brief review of various desalination technologies will be provided in the next section, then a novel water and solute full separation process using solar thermal energy will be introduced. Unlike the existing process which uses crystallizer, the proposed approach is a once-through process involving zero recirculation and zero-liquid discharge. It is expected that the proposed full separation thermal-driven desalination process in this work can address problems of concentrated discharge disposal, solute recovery, and high-energy consumption.

2. Classification of desalination technologies

Currently, most of the desalination plants have been implemented in large scales; there are more than 15,000 desalination plants installed by 2010 in the world, with a production capacity of 65 million m^3/day for both domestic consumption and industrial water production. There are a lot of different desalination technologies, some of them have already been fully developed at large scales, whereas others are still in pilot scales for demonstration or laboratory scales for research and development. **Figure 3** provides a list of common desalination technologies.

Basically, desalination process can be categorized as two major ones: thermal desalination and mechanical desalination. Thermal desalination utilizes the heat from combustion, power block, or even renewable energy to evaporate seawater. Vapor compression can be combined with thermal and mechanical desalination, which has the capability of increasing volumes and efficiency of the whole process. Thermal desalination has three classifications, which include filtration, evaporation, and crystallization. Mechanical desalination is discussed in this section and thermal desalination will be introduced in the next section.

Reverse osmosis (RO) is a membrane separation process that recovers water from a saline solution pressurized to a point greater than the osmotic pressure of the solution. The saline water is fed to the porous membranes at high pressures. The hydrophilic membranes allow only water to pass through it, as shown in **Figure 4**. This technique requires high-pressure pumps and costly membranes; also the membrane is susceptible to fouling and needs frequent replacement, which results in high cost of maintenance. Incorporation of energy recovery system reduces specific energy consumption and product cost but increases the capital cost. The product cost is significantly affected by the price of electricity in this technique.

Pressurizing the saline water accounts for most of the energy consumed in RO. The osmotic pressure required to separate water from the brine is related to the salt concentration; therefore,

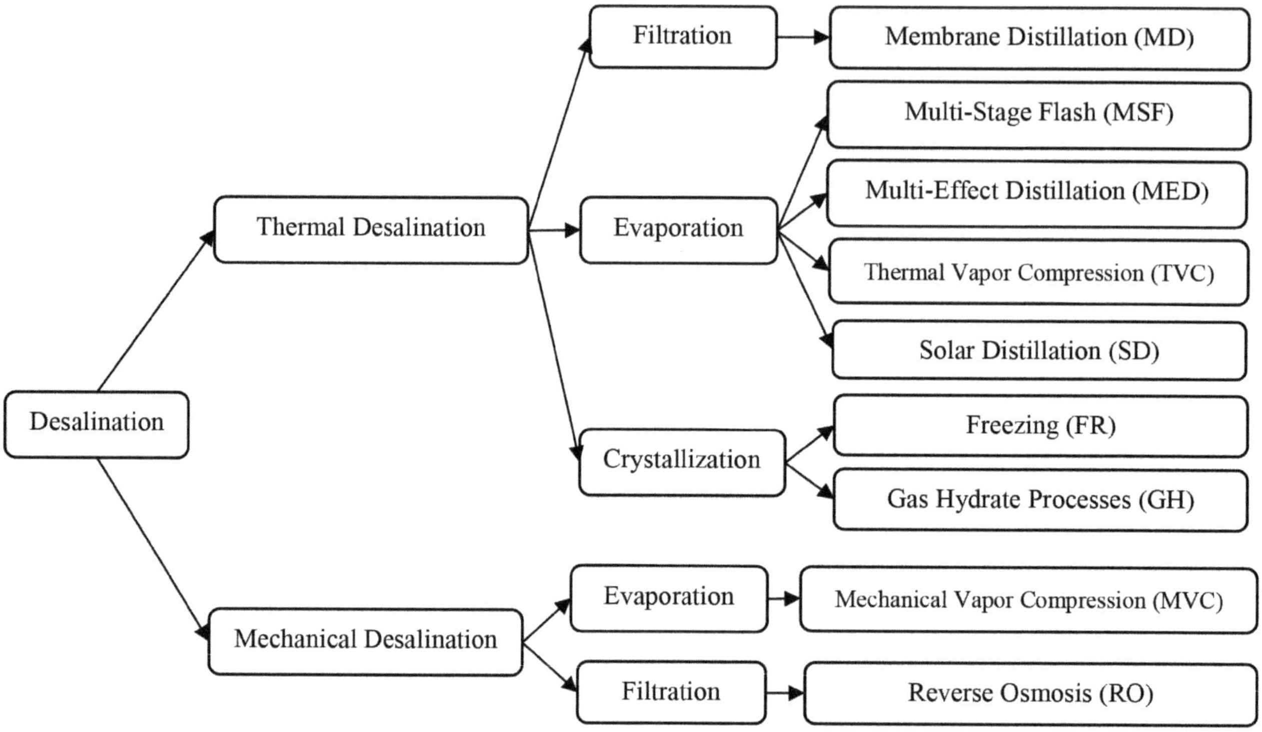

Figure 3. A list of contemporary desalination technologies.

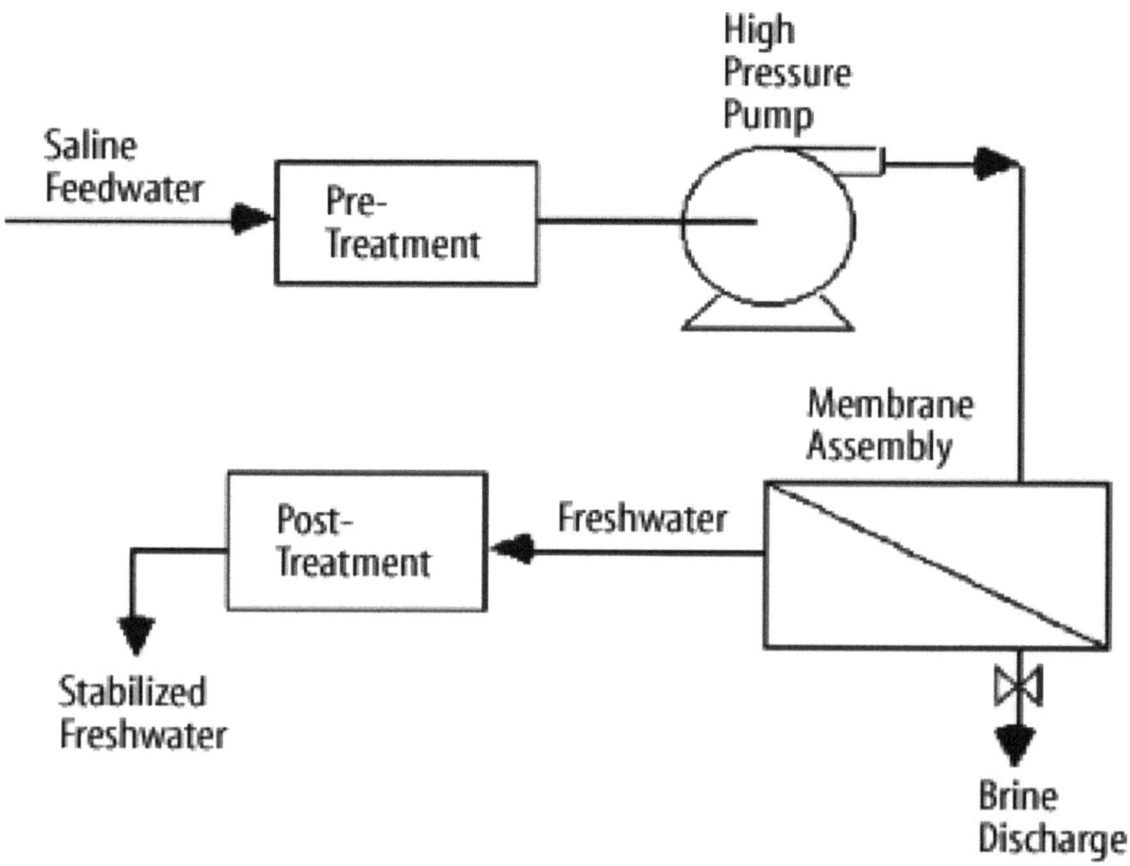

Figure 4. Schematic of RO [8].

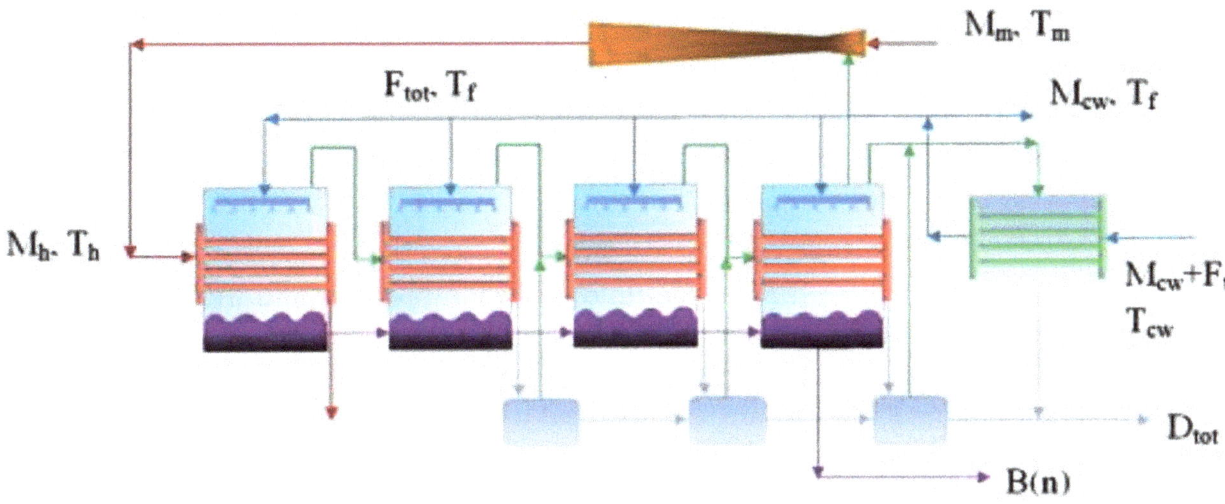

Figure 5. Principle of vapor compression [9].

RO has been considered as a practical approach for brackish water, due to the fact that brackish water only requires low to intermediate pressure range (10–15 bar) for the RO process, which is only one-fifth of the pressure for seawater RO process.

To improve the system efficiency, vapor compression can be added to a multieffect distillation (MED) process, as shown in **Figure 5**. The reuse of vapor is the key for vapor compression process, where the vapor is generated from the distiller after recompression. The first module in **Figure 5** can be heated up by utilizing the recovered heat from the partially recompressed vapor on stage. The vapor can be compressed either by a mechanical compressor or by a steam ejector, which can be categorized as mechanical vapor compression (MVC) and thermal vapor compression (TVC), respectively,

For TVC, motive steam (in ejector) at higher pressure is withdrawn from another process, for example, a steam power cycle or an industrial process steam. MVC is useful for small- or medium-scale desalination plants. MVC units typically can generate fresh water up to about 3000 m^3/day, while TVC units have much more capacity, with daily fresh water generation of 36,000 m^3. Most of MVC systems have only one stage, while TVC systems usually have several stages. This difference arises from the fact that the pressure and temperature increased by the compressor and its capacity are limited.

3. Thermal-driven desalination systems and the application of molten salts as heat transfer fluid in concentrating solar power plants

3.1. Thermal-driven desalination systems

Conventional thermal-driven desalination technologies are broadly classified into two major categories: filtration and evaporation. The thermal distillation systems will heat saline water and separate the relatively pure vapor for subsequent condensation and use. Membrane separation systems usually drive high-pressure pumps that overcome osmotic pressure differentials or create electric fields that drive electro-migration of ions in solution. **Figure 6** summarizes all these methods.

3.1.1. Multistage flash (MSF)

MSF is based on the principle of heating the fluid at certain pressure and then flashing it at lower pressure to form vapor, as shown in **Figure 7**. This vapor is collected and condensed which gives purified water. Here, the brine water is fed through a series of feed water heaters, to recover the energy from flashed steam, and then fed to an unfired boiler or a heat exchanger. The brine water gains maximum heat at unfired boiler and then flashed in several stages with decreasing pressure, each stage giving out some amount of steam. The difference of pressure between subsequent stages is the main factor influencing steam production in each stage. Highly concentrated brine is discharged from the last stage. The main problem of MSF process is the low-performance ratio which causes lower efficiency; however, it has lower scaling problems than that of MED due to relatively simpler design [10]. The number of MSF plants grows since its conception. And the cost of MSF equipment has been reduced by 50% in the last 20 years; however, this is accompanied with an increasing unit size [11]. Therefore, MSF units are economical with large capacities. In case of solar-coupled MSF system, the low-pressure steam can be formed by circulating water through the field of parabolic trough collectors.

An optimization study indicates that there is significant declining trend of product water cost with increasing top brine temperature (TBT). For a 30-stage MSF plant product, water cost is 1.15 $\$/m^3$ for 347 K TBT, and that for 377 K TBT is 0.95 $\$/m^3$ [13]. However, the TBT corresponding to the minimum product cost cannot be achieved due to the problem of scaling at high TBT. The area of technical optimization for MSF can be new corrosion-resistant alloys and corrosion and scale-inhibiting techniques.

3.1.2. Multieffect distillation (MED)

The governing principle for MED is to boil inlet seawater or brine in different evaporation effects, as shown in **Figure 8**. In the first effect, heat given by steam from waste heat source or solar collector is used to vaporize seawater. The generated vapor passes to the next effect. This vapor loses its latent heat to evaporate a part of seawater fed in the next effect, and so on. Flow schemes for MED systems include forward feeding, backward feeding, parallel feeding, and parallel feeding with cross flow. In forward feed, the direction of brine flow and steam flow is same and all the feed seawater is sent to first effect; in backward feeding, the direction of brine flow is opposite to steam flow and the seawater is firstly introduced into the last effect. Backward feed scheme makes more sense thermodynamically, but the first effect receives highest concentration brine at a high temperature. This escalates the scaling problems; therefore, this scheme is avoided. In parallel feed scheme, feed is equally divided and distributed to different effects. MED units typically operate below 120°C TBT, to avoid the problem of scaling.

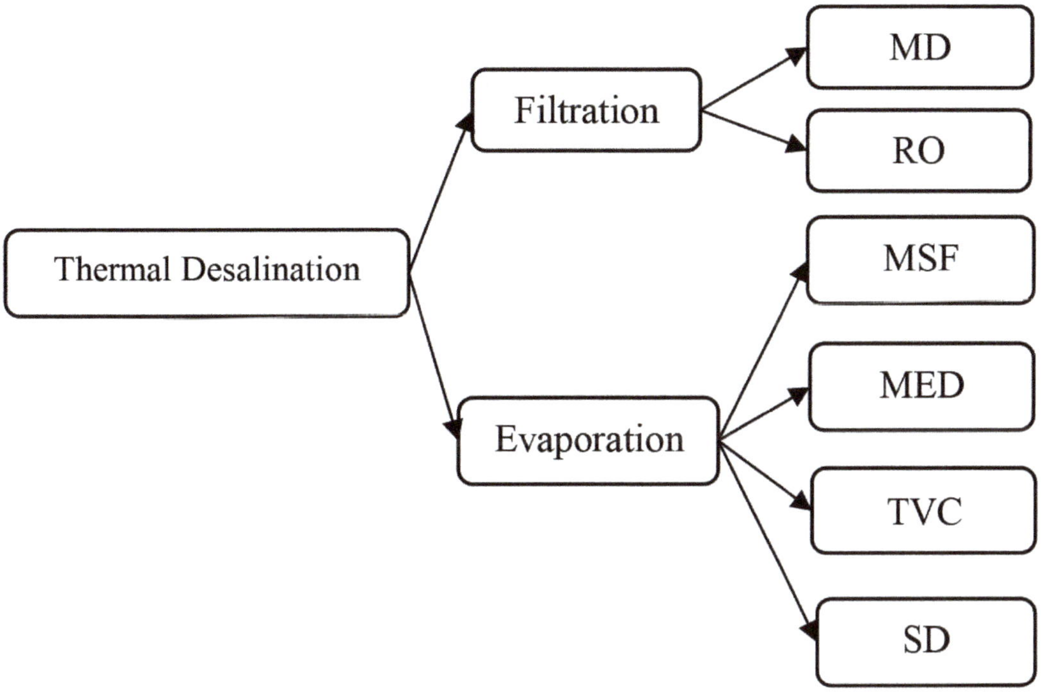

Figure 6. Thermal desalination technologies.

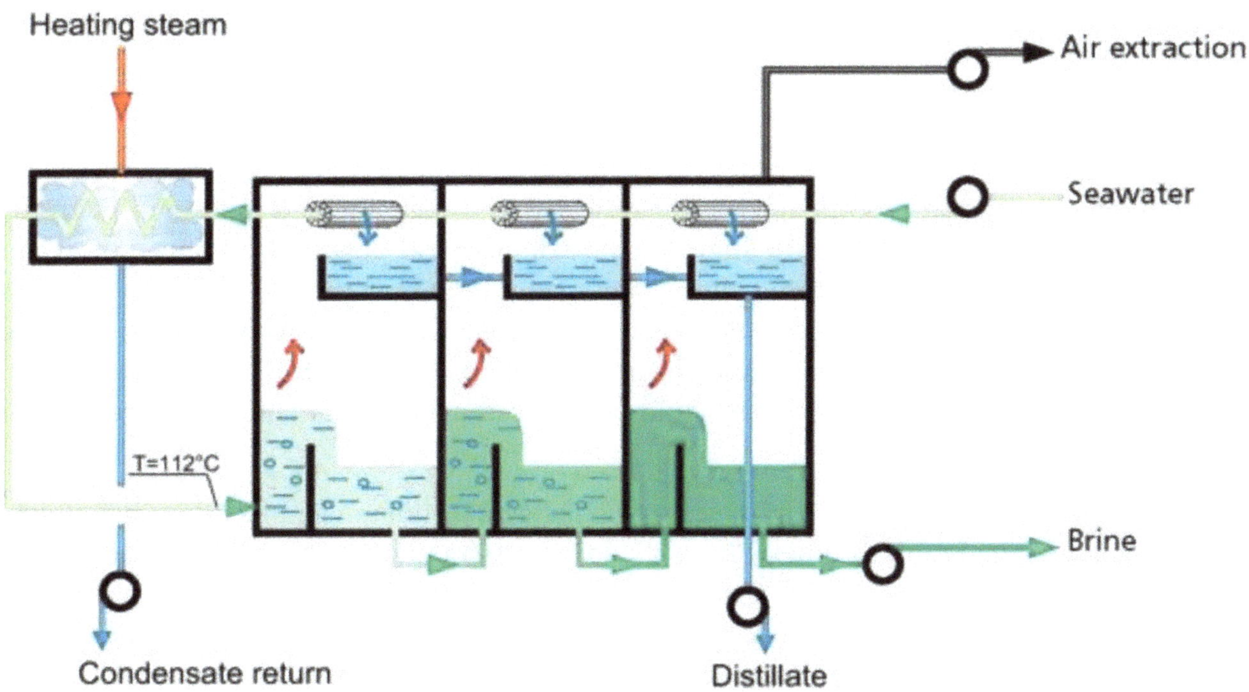

Figure 7. Schematic of MSF [12].

And the minimum brine temperature is determined by temperature difference between the last effect and ambient temperature necessary for heat transfer. MED system is usually combined with mechanical vapor compression or thermal vapor compression to use a part of steam from any of the effects and increase its pressure. MED systems can operate at low TBT, thus enabling

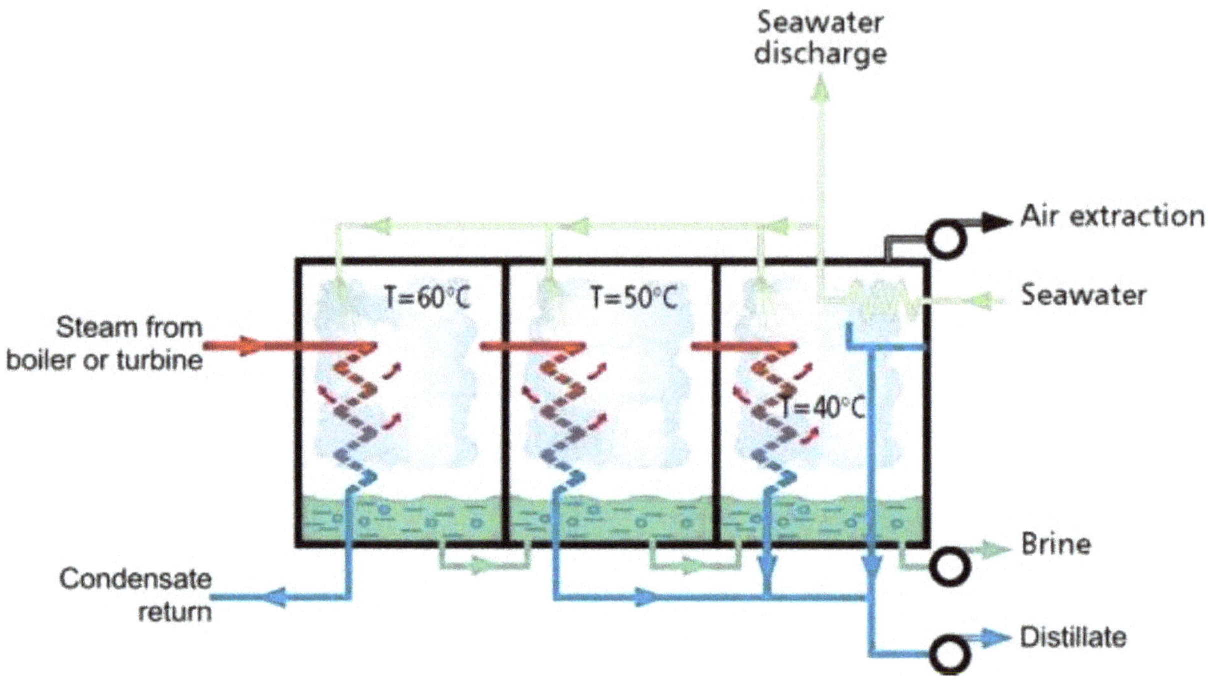

Figure 8. Schematic of triple-effect MED [14].

them to use cheap heat sources such as waste heat or solar collectors. For example, an existing diesel waste heat-coupled MED plant's energy consumption is 2 kWh/Ton for water pumps [15].

3.1.3. Membrane distillation (MD)

In membrane distillation process, feed water is heated and then allowed to flow through the porous hydrophobic membrane. The high pressure or electrical potential is applied on water vapor to produce fresh water from saline water [16]. The vapor pressure difference across membrane causes water vapor molecules to flow and it is condensed on the other side of the membrane. MD has the following characteristics: high porosity, hydrophobic and low thermal conductivity; the membrane thickness should be reduced and to maintain high pore size for the increase of flux.

The following are the operating parameters which effects distillate yield like flow rate, thickness of air gap, membrane thickness, porosity, long-term operation, thermal conductivity of membranes, and porosity. It is still a big challenge to develop high and efficient membrane technology, since MD has basic challenges to meet to compete with MSF, MED, and RO. In addition, MD system consumes a lot of energy [17, 18].

3.2. Application of molten salts as heat transfer fluid in concentrating solar power plants

Despite the tremendous improvements in conventional desalination technology, the desalination process is very energy consuming. For instance, to produce 1000 m^3/day of desalinated water in a year, nearly 10,000 tons of oil is required [16]. More detailed energy consumption data for different desalination processes are listed in **Table 1**.

Desalination process	Max seawater temperature (°C)	Energy consumption	
		Thermal energy (kJ/kg)	Electric power (kWh/m^3)
MSF	120	184–222	2.5–4
	90	252–327	2.5–4
	70	462–567	2.5–4
MED	80	176–231	1.2–1.8
	70	235–294	1.2–1.8
	60	294–394	1.2–1.8
RO (without energy recover)	–	–	7–8
RO (with energy recovery)	–	–	5–6

Table 1. Energy consumption of seawater desalination process [19].

Due to high expenses of the conventional energy resources, renewable energy sources can provide alternatives. Since most desalination systems can directly use thermal energy, concentrated solar thermal energy is very suitable for water treatment. Furthermore, the potential negative environmental impact from the desalination process is the discharge of high concentrated brine to the environment [20]. To avoid the potential negative impacts from disposing of the concentrates, the recovery of important minerals from concentrates to achieve zero discharge is a promising option. Recent development on concentrating solar thermal power generation has shown that the sea salts are very promising materials for large-scale thermal energy storage in concentrating solar power (CSP) plants. This may dramatically change the technology for desalination and water treatment because the salts may be collected for better value, rather than disposed.

A typical CSP plant consists of a solar collection system and a traditional power block. Solar energy is concentrated in the collection system, and the heat from direct normal insulation (DNI) is collected by a heat transfer fluid (HTF). The HTF circulates to the power block and transfers the collected heat to generate steam. The steam drives a turbine to produce electricity in a steam cycle [21–23]. Four most commonly used solar-concentrating methods are parabolic trough collectors, linear Fresnel reflectors, solar power tower collectors, and parabolic dish collectors. Parabolic trough and solar power tower are the two solar-concentrating methods mostly used in current CSP plants. HTF is the most important part in CSP plant beside energy storage media, because it can transfer the heat from the solar field to the power block. Currently, several typical HTFs are used in commercialized CSP plants, including air, water/steam, synthetic oils, organics, and molten salts [24, 25]. A good candidate of HTF in CSP should have the following characteristics: low freezing point, good thermal stability, low viscosity, high specific heat capacity, as well as acceptably low chemical corrosion rate, while low cost is another key criterion for industrial applications [26, 27].

Concentrates from desalting process contain inorganic salts and other compounds that may be purified for commercial value. For example, the recent technology development on solar thermal energy storages has shown that sea salts are very promising materials for large-scale thermal energy storage [28–31]. At present, salt mixtures have been considered as promising HTF candidates due to their thermal stability at high temperatures [32]. So far, alkali nitrate and nitrite mixtures are the two most successful HTF candidates. Solar salt (KNO_3 40 wt%-$NaNO_3$ 60 wt%), Hitec ($NaNO_3$ 7 wt%-KNO_3 53 wt%-$NaNO_2$ 40 wt%), and HitecXL ($NaNO_3$ 7 wt%-KNO_3 45 wt%-$Ca(NO_3)_2$ 48 wt%) are three commercialized alkali nitrate and nitrite mixtures. Other nitric salt-based HTFs include Sandia Mix ($NaNO_3$ 9–18 wt%-KNO_3 40–52 wt %-$LiNO_3$ 13–21 wt.%-$Ca(NO_3)_2$ 20–27 wt%) developed by Bradshaw and Brosseau [33] at the Sandia National Laboratories and SS-500 ($NaNO_3$ 6 wt%-KNO_3 23 wt%-$LiNO_3$ 8 wt%-$CsNO_3$ 44 wt%-$Ca(NO_3)_2$ 19 wt%) developed by Halotechnics Inc. [34]. Reddy et al. [35] developed a mixture of alkali-fluoride and carbonate salt (LiF-NaF-K_2CO_3) as the HTF with a working temperature range of 400–900°C. Li et al. [31, 36–38] recently developed a chloride salt mixture (NaCl 7.5 wt%-KCl 23.9 wt%-$ZnCl_2$ 68.6 wt%), its melting temperature is about 850°C, while its viscosity is only 0.325 Pa s at 300°C. The cost of this ternary chloride salt mixture is expected to be below 1 $/kg.

Consequently, a full separation of salts and water in desalination process needs to be developed, the byproducts of which can be utilized as the HTF in CSP plants. Furthermore, such a full separation desalination system is expected to become a necessity in advanced water treatment technologies in an economically feasible way.

4. Heat and mass transfer analysis of a novel full separation system using solar thermal energy from concentrated solar power (CSP)

4.1. Motivations

As discussed in Section 1, the disposal of high-temperature-concentrated brine is a big concern to the environment. It has been clearly documented that the discharged high-temperature concentrates can cause a lasting change in species composition and abundance in the discharge site and affect aquatic life [21]. To avoid the potential adverse effects from disposing of the concentrates, the recovery of important minerals from concentrates to achieve zero discharge is a promising option. Concentrates from desalting process contain inorganic salts and other compounds that may be purified for commercial applications. Section 3.2 introduced a ternary halide salts as a potential candidate of thermal storage material; this may dramatically change the technology for desalination and water treatment. Therefore, a full separation of salts and water for desalination becomes necessary in new water technologies. A proposed full separation multieffect distillation (FSMED) desalination system can accommodate the demands of using high-temperature thermal energy to effectively separate salts and water at 100%, resulting in simultaneous collection of pure water and dry salts.

4.2. Concepts

To achieve 100% water extraction, an innovative concept of FSMED system using solar thermal energy is proposed. In the proposed FSMED desalination system, a full separation tank (FST) is integrated to a conventional forward feed MED water treatment system to enhance the water productivity and thermal efficiency, as shown in **Figure 9**. The main advantage of forward feed system is to be able to operate at high top brine temperature. Although the mean temperature of heat addition in forward feed is lower as compared to backward feed, it is better from the operational point of view. In backward feed arrangement, the highest brine temperature occurs at the highest brine salinity, thereby causing severe scaling problems [39].

The heat source for the FST is from concentrated solar thermal system. The air feed to the FST is preheated by the exhaust air/steam mixture to achieve heat recovery. The FST receives concentrated brine from the last effect of the MED. The concentrated brine from the MED is atomized into tiny droplets and fully evaporated in the FST due to the effective convective heat transfer between water droplets and hot airflow. The brine vaporizes into steam leaving behind salt down to the bottom of the FST, resulting in the simultaneous collection of pure water and dry salts. The hot air/steam mixture leaves from the top of the FST and passes through the first effect of MED to heat the feed seawater. There is no need to retain the brine for an internal convective heat transfer process, which eliminates the scaling problem. The only section subject to a hot environment is the pipe introducing brine to the nozzles and sprayers, which can be insulated and maintained at low temperature. Therefore, this novel system can operate effectively at either very high temperatures using primary heat or at medium to low temperatures using exhaust heat.

4.3. Heat and mass transfer analysis

The proposed process is a closed-loop system which increases the nonlinearity and complexity of resulting system of equations, unlike conventional systems which are open loop. The mass and energy balance analysis for the FSMED system in steady-state operation will be introduced in this section. **Figure 10** shows the energy balance of the FSMED system.

According to **Figure 10**, the input energy involves the energy from the heat source and the feed seawater, as shown in Eq. (1)

$$E_{\text{in}} = m_{\text{air}} \cdot C_{p,\,\text{air}} \cdot (T_{\text{hs}} - T_{n,\,\text{out}}) + \left[\beta \cdot m_{s,\,n}\left(h_{\text{gs,\,hs}} - h_{\text{gs},\,n}\right)\right] + m_{\text{csw}} \cdot h_{\text{sw,amb}} \tag{1}$$

where m_{air} is the mass flow rate of the dry air and m_{csw} is the mass flow rate of the cooling seawater. The subscript n denotes the number of the effects, then $T_{n,\,\text{out}}$ and $h_{\text{gs},\,n}$ represent the parameters in the last effect. The subscript hs means the parameters are obtained under the temperature of the heat source. β is the bleed steam fraction, which indicates the ratio of mass flow of bleed steam to the mass flow of steam from which it is extracted. To reduce the thermal energy consumption, a small fraction of steam (steam 21 in **Figure 9**) from the last effect is bled and mixed with the air steam (steam 13 in **Figure 9**), and then this mixture is heated up and blown into the FST. The bleed fraction is a very important parameter which influences other crucial parameter like salinity of brine and energy requirement.

The output energy is carried out by the saturated water from each effect, collected salts and the seawater rejected from condenser, as listed in Eq. (2)

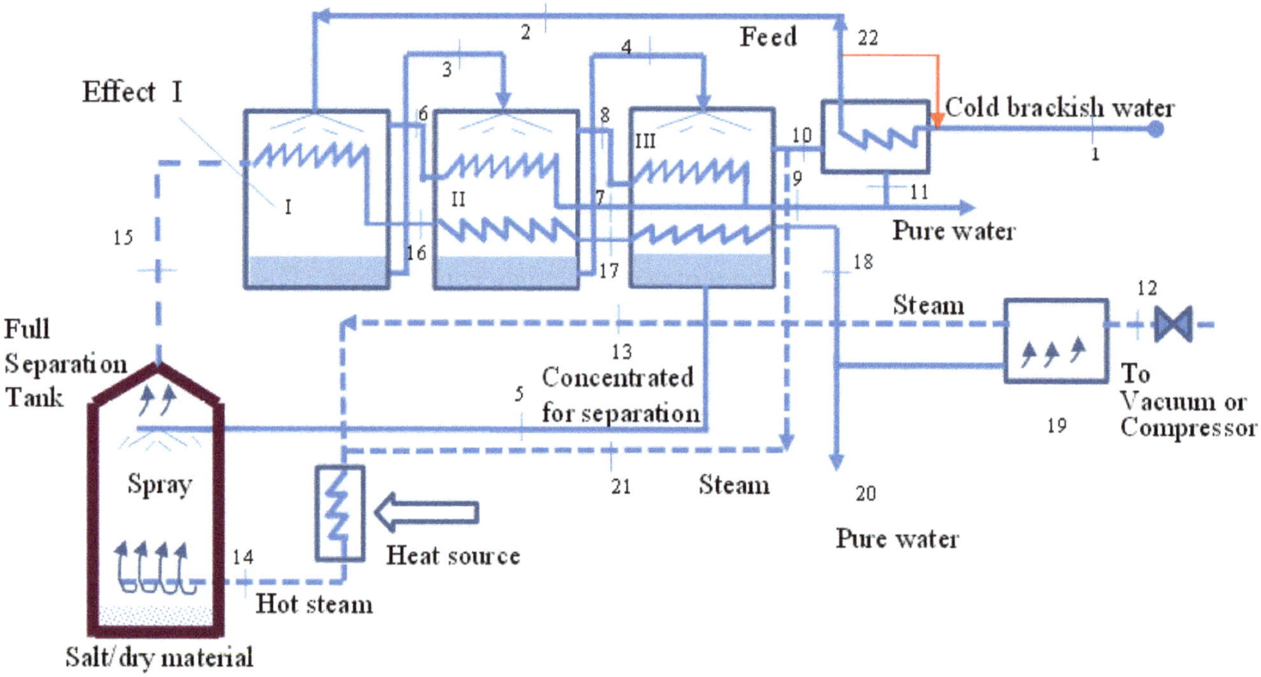

Figure 9. Concept of the FSMED desalination system.

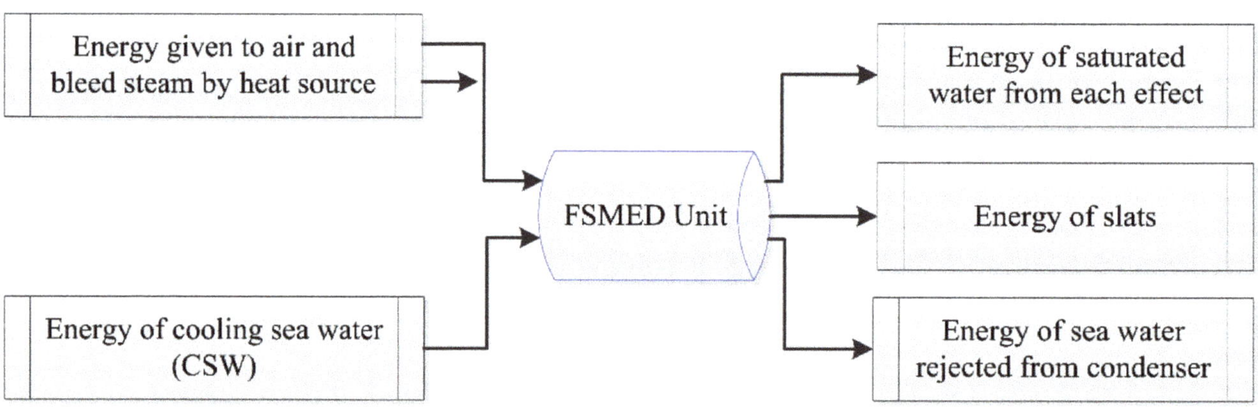

Figure 10. Energy balance of the FSMED system [40].

$$E_{\text{out}} = (1-\beta)\cdot m_{s,n}\cdot h_{fs,n} + \sum_{i=1}^{n-1} m_{s,i}\cdot h_{fs,i} + m_{\text{salt}}\cdot C_{p,\text{salt}}\cdot (T_{\text{FST}} - T_{\text{amb}}) + (m_{\text{csw}} - m_{\text{sw}})\cdot h_{\text{fsw,WH}} \quad (2)$$

where the subscript WH means the thermal properties are obtained under the temperature of the feed seawater heater.

The mass flow into the system is the cooling seawater, and the total mass flow out involves the rejected seawater, collected salts, and the pure water. The mass flow rates, salinity of different streams, and total energy consumption under different temperature of hot air going into the

FST and varying bleed steam fraction can be calculated through detailed energy and mass balance analysis of each component of the FSMED system, which is not shown here for the sake of brevity.

An example of the calculated energy consumption for FSMED system with 3, 4, and 5 as the total number of effects is shown in **Figure 11**, in which the heat source temperature varies from 150 to 400°C. The bleed steam fractions for 3, 4, and 5 effects are 0.6, 0.57, and 0.55, respectively. It is apparent that a five-effect system is the least energy consuming. The minimum energy consumption for five-effect forward feed FSMED desalination system is estimated to be about 350 kJ/kg of seawater, which compares very well to thermal energy requirements of existing MED plant which require 394 kJ/kg of seawater [19]. More detailed operating data for the five-effect FSMED system are listed in **Table 2**.

4.4. Preliminary assessment of water droplet evaporation path in FST

To achieve 100% water extraction, detailed knowledge of the water droplet behaviors in the FST is essential to the design optimization of the FSMED desalination system. It is thus necessary to determine the lifetime and trajectory of tiny water drops as a function of the drop size, the ambient temperature, and the spray-injection parameters. Given that the analysis of vaporization of saline droplet swarm is complex (involving crystallization process), the present

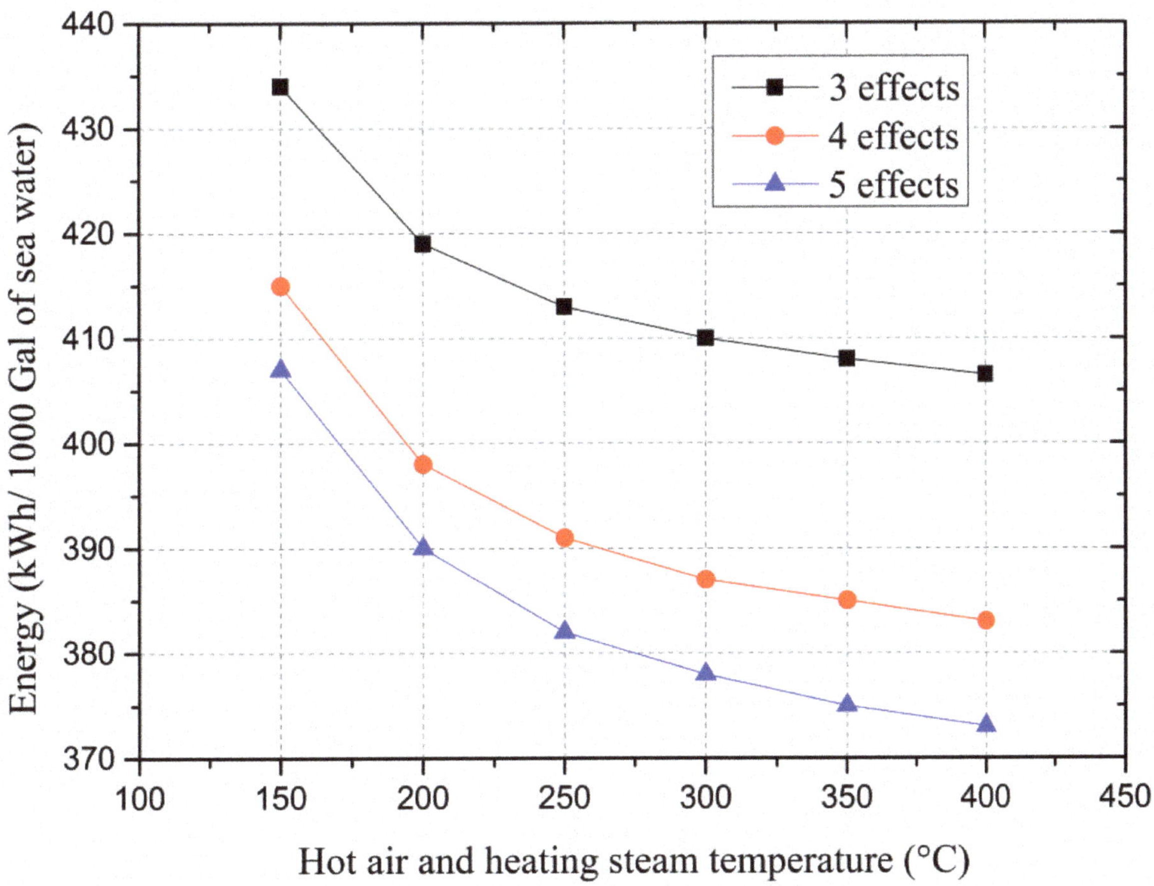

Figure 11. Thermal energy requirement of FSMED desalination system versus hot air and heating steam temperature.

Parameters	Value					
Heat source temperature (°C)	150	200	250	300	350	400
Mass flow rate of dry air (kg/s)	3.844	2.291	1.600	1.209	0.957	0.781
Mass flow rate of seawater (kg/s)	6.160	5.716	5.515	5.401	5.327	5.276
Mass flow rate of pure water (kg/s)	0.965	0.965	0.965	0.965	0.965	0.965
Energy input by heat source (kWh/1000 Gal of seawater)	407.6	390.4	382.5	378.1	375.2	373.2

Table 2. Calculated results for a FSMED system with five effects under different heat source temperature [40].

study mainly focuses on the evaporation of a single water droplet in hot air to make a preliminary assessment.

A simplified non-equilibrium vaporization model is adopted to describe the movement and evaporation behavior of a single water droplet in the FST. The model is described based on the following assumptions:

1. No temperature gradient in the droplet;
2. Thermal properties are uniform in the droplet;
3. Droplet maintains a spherical shape during falling;
4. Vapor mass fraction in the hot airflow is 0.

The droplet size change is related to the heat/mass transfer that determines the evaporation of the droplet. The conservation equations and mass transfer equation are described in the following.

Mass conservation equation:

$$\frac{d}{dt}\left(\frac{4}{3}\pi r_s^3 \rho_l\right) = -G \tag{3}$$

Energy conservation equation:

$$2\pi r_p \overline{\lambda}(T_\infty - T_s)Nu = \frac{4}{3}\pi r_s^3 \rho_l C_{p,l}\frac{dT_s}{dt} + GL \tag{4}$$

Mass transfer equation:

$$G = 2\pi r_s \overline{\rho} \overline{D} \cdot Sh \cdot Ln(1 + B_m) \tag{5}$$

where B_m is the Spalding mass transfer number, and is given by

$$B_m = \frac{Y_{Fs} - Y_{F\infty}}{1 - Y_{Fs}} \tag{6}$$

The droplet velocity and position are evaluated using the following equations:

$$-\frac{du_x}{dt} = \frac{3C_D}{8r_s}\left(\frac{\overline{\rho}}{\rho_l}\right)u_r \cdot u_x \tag{7}$$

$$-\frac{du_y}{dt} = \frac{3C_D}{8r_s}\left(\frac{\overline{\rho}}{\rho_l}\right)u_r \cdot (u_y - u_\infty) - g\frac{\rho_l - \overline{\rho}}{\rho_l} \tag{8}$$

The lifetime of the water droplet in the FST under different diameters and temperatures is compared in **Figure 12**. The droplet lifetime increases with an increase in the droplet diameter but decreases with an increase in the ambient temperate. In addition, the lifetime decreases with the ambient temperature increase in a nonlinear way, with a gradual decrease in the slope.

To confine the vaporization within the FST with certain size, the variations of the falling distance in the vertical direction and stopping distance in the horizontal direction were calculated for a water droplet with 400 μm in diameter under different injection parameters, as shown in **Figure 13**.

For the case of $\theta = 0°$, the falling distance decreases with the increase in the initial injection velocity. This is because the lifetime of the droplet decreases with the increase of the injection velocity. For the case of $\theta = 10°$, when the injection velocity is below 15 m/s, the influence of the lifetime decrease on the falling distance is larger than that of vertical component of the injection velocity. Hence, the falling distance decreased slightly with the initial injection velocity. When the injection velocity exceeds 15 m/s, the effect of vertical component of the injection

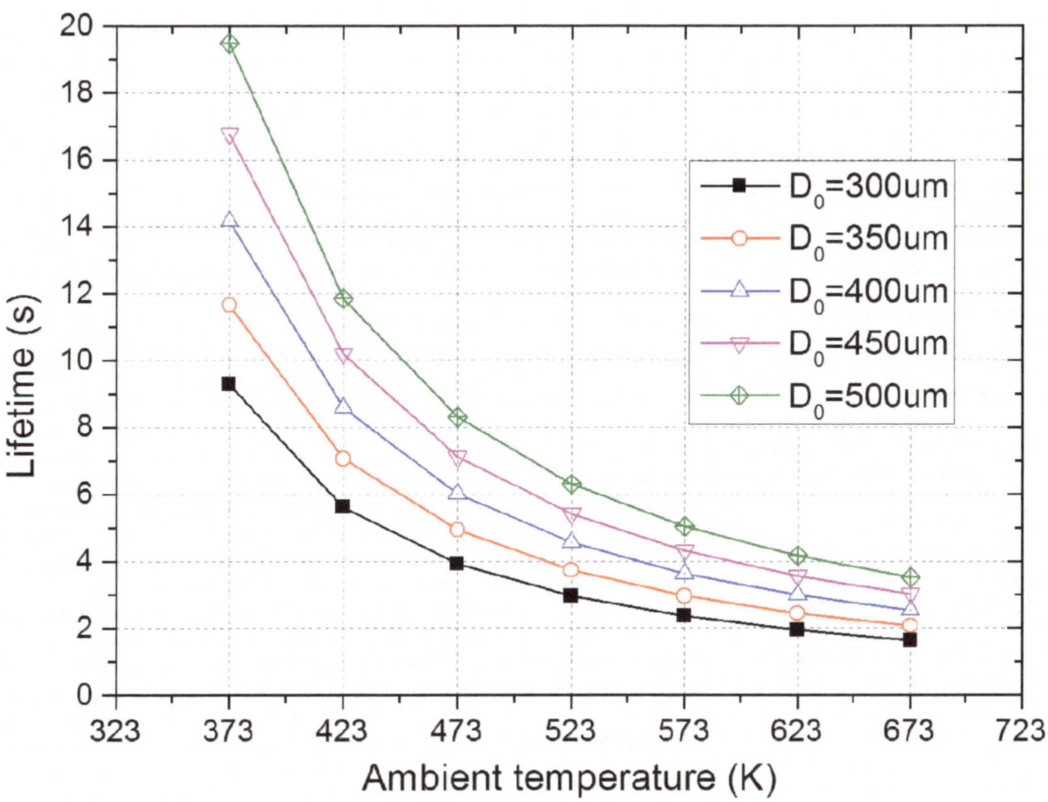

Figure 12. Comparison of the lifetime under different droplet diameters and ambient temperatures.

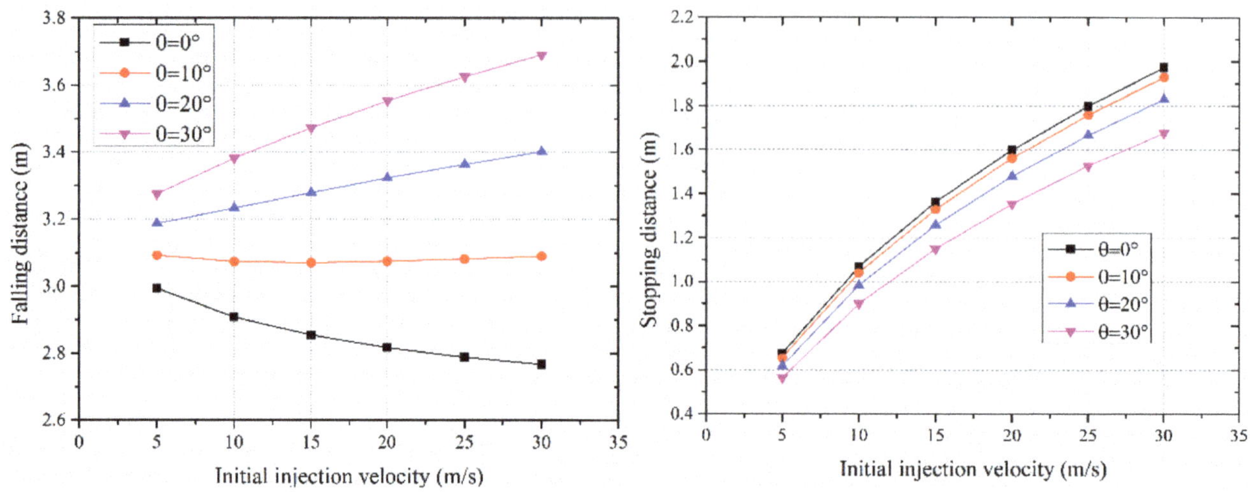

Figure 13. Variation of the falling and stopping distance with the injection velocity and angle.

velocity overshadows that of the lifetime decrease, so the falling distance increased slightly with the initial injection velocity. For the cases of $\theta = 20°$ and $30°$, because the increase of vertical component of the injection velocity always has greater influence than the lifetime decrease, the falling distance increases with the initial injection velocity. The stopping distance decreases with the injection angle due to the decrease of the horizontal component of the initial injection velocity.

5. Concluding remarks

Desalination is a promising solution for providing the increasing need of freshwater, due to the increasing growth of population and climate change. Despite the tremendous improvements in conventional water treatment technology, the desalination process is very energy consuming. Therefore, it is highly desirable that environmentally friendly renewable energy resources should be utilized for water treatment. Since most desalination systems can directly use thermal energy, concentrated solar thermal energy is very suitable for water treatment. Furthermore, recent developments on CSP have shown that the sea salts are very promising materials as HTF in thermal energy storage system, which in fact provides an alternative to desalination and water treatment by collecting the salts through full separation from seawater and brackish water, instead of disposing the high concentrate to the environment.

To achieve 100% water extraction, an innovative concept of FSMED desalination system using solar thermal energy was proposed in the present study. Compared with conventional MED, the FSMED technology will overcome the limitation of operating temperature for water desalination. The concentrated brine from the MED is atomized into tiny droplets and fully evaporated in the FST by using the high-temperature heat from concentrated solar thermal energy system, resulting in simultaneous collection of pure water and dry salts. A simplified non-equilibrium vaporization model is used to investigate the lifetime and trajectory of a tiny water droplet in the FST as a function of the droplet size, the ambient temperature, and the spray-injection parameters. The relationship of the water droplet size and falling distance with the hot air-steam temperature, and initial injection/spray parameters is investigated and presented. Results from the study provide important guidance to the design of such a water treatment system. It is expected that the proposed full separation thermal-driven desalination process can address problems of concentrated discharge disposal, solute recovery, and high-energy consumption.

Acknowledgements

The authors are grateful to the support of the New Faculty Supporting Program (NFSP) from the University of Texas Rio Grande Valley and the University of Arizona under its program of Water, Environmental and Energy Solutions (WEES).

Author details

Ben Xu[1], Peiwen Li[2*] and Penghua Guo[3]

*Address all correspondence to: peiwen@email.arizona.edu

1 Department of Mechanical Engineering, The University of Texas Rio Grande Valley, Edinburg, TX, USA

2 Department of Aerospace and Mechanical Engineering, The University of Arizona, Tucson, AZ, USA

3 School of Energy and Power Engineering, Xi'an Jiaotong University, Xi'an, Shanxi, China

References

[1] Elimelech M, Phillip WA. The future of seawater desalination: Energy, technology, and the environment. Science. 2011;**333**:712–717

[2] Liyanaarachchi S, Shu L, Muthukumaran S, Jegatheesan V, Baskaran K. Problems in seawater industrial desalination processes and potential sustainable solutions: A review. Reviews in Environmental Science and Bio/Technology. 2014;**13**:203–214

[3] IDA. Available from: http://idadesal.org/desalination-101/desalination-by-the-numbers. June 30, 2015. [Accessed: July 12, 2016]

[4] IDA. 2014. The IDA Desalination Yearbook 2014–2015

[5] WHO. Desalination for safe water supply: Guidance for the health and environmental aspects applicable to desalination. In: World Health Organization. 2007

[6] Goosen MA, Mahmoudi H, Ghaffour N. Today's and future challenges in applications of renewable energy technologies for desalination. Critical Reviews in Environmental Science and Technology. 2014;**44**:929–999

[7] Mickley MC. Membrane Concentrate Disposal: Practices and Regulation. Desalination and Water Purification Research Program Report, US Bureau of Reclamation. 2001

[8] Available from: http://resource2.rockyview.ab.ca/science8/Sci8%20U1%20MixFlow/Sci8U1%20Lessons/Sci8U1L4.html

[9] Available from: http://www.arabwatersource.com/water-blog/category/desalination, July 9, 2014

[10] Van der Bruggen B, Vandecasteele C. Distillation vs. membrane filtration: Overview of process evolutions in seawater desalination. Desalination. 2002;**143**(3):207–218

[11] Borsani R, Rebagliati S. Fundamentals and costing of MSF desalination plants and comparison with other technologies. Desalination. 2005;**182**(1):29–37

[12] Multiple Stage Flash. Available from: http://www.sidem-desalination.com/en/Process/MSF/. [Accessed: March 12, 2016]

[13] Fiorini P, Sciubba E. Thermoeconomic analysis of a MSF desalination plant. Desalination. 2005;**182**(1):39–51

[14] Multiple Effect Distillation. Available from: http://www.sidem-desalination.com/en/Process/MED/. [Accessed March 12, 2016]

[15] Ophir A, Lokiec F. Advanced MED process for most economical sea water desalination. Desalination. 2005;**182**(1):187–198

[16] Sharon H, Reddy KS. A review of solar energy driven desalination technologies. Renewable and Sustainable Energy Reviews. 2015;**41**:1080–1118

[17] Kesieme UK, Milne N, Aral H, Cheng CY, Duke M. Economic analysis of desalination technologies in the context of carbon pricing, and opportunities for membrane distillation. Desalination. 2013;**323**:66–74

[18] Ali MT, Fath HE, Armstrong PR. A comprehensive techno-economical review of indirect solar desalination. Renewable and Sustainable Energy Reviews. 2011;**15**(8):4187–4199

[19] Deng RY, Xie L, Lin H, Liu J, Han W. Integration of thermal energy and seawater desalination. Energy. 2010;**35**:4368–4374

[20] Laspidou C, Hadjibiros K, Gialis S. Minimizing the environmental impact of sea brine disposal by coupling desalination plants with solar saltworks: A case study for Greece. Water. 2010;**2**(1):75–84

[21] Zhang HL, Baeyens J, Degrève J, Cacères G. Concentrated solar power plants: Review and design methodology. Renewable and Sustainable Energy Reviews. 2013;**22**:466–481

[22] Barlev D, Vidu R, Stroeve P. Innovation in concentrated solar power. Solar Energy Materials and Solar Cells. 2011;**95**:2703–2725

[23] Vignarooban K, Xu X, Arvay A, Hsu K, Kannan AM. Heat transfer fluids for concentrating solar power systems—a review. Applied Energy. 2015;**146**:383–396

[24] acio J, Wetzel T. Assessment of liquid metal technology status and research paths for their use as efficient heat transfer fluids in solar central receiver systems. Solar Energy. 2013;**93**:11–22

[25] Tian Y, Zhao CY. A review of solar collectors and thermal energy storage in solar thermal applications. Applied Energy. 2013;**104**:538–553

[26] Cordaro JG, Rubin NC, Bradshaw RW. Multicomponent molten salt mixtures based on nitrate/nitrite anions. ASME Journal of Solar Energy Engineering. 2011;**133**:011014

[27] Bauer T, Laing D, Tamme R. Recent progress in alkali nitrate/nitrite developments for solar thermal power applications. Molten Salts Chemistry and Technology. Trondheim, Norway. June 2011

[28] Wang K, Molina E, Dehghani G, Xu B, Li P, Hao Q, Lucas P, Kassaee MH, Jeter SM, Teja AS. Experimental investigation to the properties of eutectic salts by NaCl-KCl-$ZnCl_2$ for application as high temperature heat transfer fluids. In: ASME 2014 8th International Conference on Energy Sustainability, Paper ES-FuelCell 2014-6578, American Society of Mechanical Engineers; Boston, MA, June 29-July 2, 2014

[29] Vignarooban K, Pugazhendhi P, Tucker C, Gervasio D, Kannan A. Corrosion resistance of Hastelloys in molten metal-chloride heat-transfer fluids for concentrating solar power applications. Solar Energy. 2014;**103**:62–69

[30] Li PW, Chan CL, Hao Q, Deymier PA, Muralidharan K, Gervasio DF, Momayez M, Jeter SM, Teja AS, Kannan AM. Halide and oxy-halide eutectic systems for high performance high temperature heat transfer fluids. SunShot Concentrating Solar Power Program Review. 2013:85–86

[31] Xu B, Li P, Chan C. Application of phase change materials for thermal energy storage in concentrated solar thermal power plants: A review to recent developments. Applied Energy. 2015;**160**:286–307

[32] Peng Q, Wei X, Ding J, Yang J, Yang X. High-temperature thermal stability of molten salt materials. International Journal of Energy Research. 2008;**32**(12):1164–1174

[33] Bradshaw RW, Brosseau DA. U.S. Patent No. 7,588,694. Washington, DC: U.S. Patent and Trademark Office; 2009

[34] Raade JW, Padowitz D. Development of molten salt heat transfer fluid with low melting point and high thermal stability. Journal of Solar Energy Engineering. 2011;**133**(3):031013

[35] Wang T, Mantha D, Reddy RG. Novel high thermal stability LiF-Na_2CO_3-K_2CO_3 eutectic ternary system for thermal energy storage applications. Solar Energy Materials and Solar Cells. 2015;**140**:366–375

[36] Li P, Molina E, Wang K, Xu X, Dehghani G, Kohli A, Hao Q, Teja AS. Thermal and transport properties of NaCl-KCl-$ZnCl_2$ eutectic salts for new generation high temperature heat transfer fluids. Journal of Solar Energy Engineering. 2016;**138**(5):054501

[37] Vignarooban K, Xu X, Wang K, Molina EE, Li P, Gervasio D, Kannan AM. Vapor pressure and corrosivity of ternary metal-chloride molten-salt based heat transfer fluids for use in concentrating solar power systems. Applied Energy. 2015;**159**:206–213

[38] Xu B, Li PW, Chan C. Application of phase change materials for thermal energy storage in concentrated solar thermal power plants: A review to recent developments. Applied Energy. 2015;**160**:286–307

[39] Al-Mutaz IS, Wazeer I. Development of a steady-state mathematical model for MEE-TVC desalination plants. Desalination. 2014;**351**:9–18

[40] Peri A. Development and analysis of a novel thermal driven water-solute separation process [Master thesis]. University of Arizona; 2015

8

Responding to Water Challenges through Desalination: Energy Considerations

George Arampatzis, Avraam Kartalidis and Dionysis Assimacopoulos

Abstract

Desalination technology and reverse osmosis in particular, is used by several island authorities in Greece to address water scarcity. However, this is a highly energy-intensive technique, requiring the consumption of significant quantities of fossil fuels. The case of Syros island is presented to demonstrate the strong water-energy link in the operation of desalination plants. The use of renewable energy sources as a means for reducing water cost from desalination is also discussed. A simple algorithm to calculate estimating water costs with renewable energy sources (RES) is presented and is applied in the island of Patmos and in Hermoupolis, Syros island.

Keywords: desalination, reverse osmosis, RES-powered desalination, hybrid energy systems, water scarcity, water-energy nexus

1. Introduction

Economic advancement and prosperity depend on the significant resources of energy and water, the links between which are equally complex and important: energy production necessitates significant quantities of water, and water supply requires great amounts of energy. The study of the "water-energy nexus" demands a holistic view of the production and consumption chains of each resource [1] presents a view of the complexity of the water–energy nexus), focusing on the core operational components, i.e., infrastructure and technologies [2].

Energy requirements along the stages of the water chain depend on the quantity and quality of water and usually increase over time [3]. Energy intensity will increase with growth in water demand, as abstraction of water is carried out from deeper boreholes with lengthy interbasin water transfers, strict water-quality standards, and limitations on wastewater discharge. Despite all this, there are opportunities for improving energy efficiency. According to the European Innovation Partnership for water [4], there is an important potential for increasing the efficiency of energy use in water supply systems by using low-energy technologies in water treatment and wastewater cleaning processes. Also, the use of renewable resources is beneficial for the total energy efficiency of these systems.

The sources of water (in terms of type, quality, and seasonal availability) and the end-use characteristics (such as consumption patterns and infrastructure) play an important role on the energy intensity of the water supply chain. An increase of 35% in energy consumption is expected by 2035 due to the impacts of climate change on water resources [5], the increased competition among water users, and the "loop effect" of the energy-water interlinks. This might cause 85% increase in the water consumption and thus extra energy requirements are needed for water distribution and treatment [6, 7]. As these interlinkages are established, the quantification of the energy-water relationship is a priority issue, particularly in water-scarce regions.

In water-scarce areas, the main supply management methods are (1) water transports from areas with abundant water sources to areas with water scarcity; (2) intensive use of the available resources; and (3) use of nonconventional water resources (e.g., wastewater reuse, desalination). All these supply methods cause an increase in the energy consumption for water supply and as a consequence, the cost of water is higher either for consumers or governments. Thus, any improvements in the energy use and the efficiency of water supply may lead to profits for the environment as well as important reductions in the cost of water.

In this chapter, the water–energy nexus is presented with emphasis in remote areas. The island of Syros, located in the Cyclades island complex, is used as a case study to demonstrate the links between energy consumption and water production and the need to increase energy-use efficiency in the isolated water and electrical energy systems of the Greek islands. The desalination option with RES is explored through modeling, and as a case study, Hermoupolis in the island of Syros and Patmos island is selected. The rich history of all the desalination methods that have been applied in the Greek islands is also described in brief.

2. The water system of the Greek islands

Greece has the longest coastline in Europe, approximately 14,000 km. The country has about 2500 islands (**Figure 1**), with a total area of 21,580 km^2. The islanders are estimated to be 1,633,433 [8].

The climate on the Greek islands is typical Mediterranean (dry summers and wet winters), with relatively low precipitation (less than 400 mm per year, especially for the Aegean islands [9]). The storage, surface, or groundwater of sufficient quality and quantity is impossible on some islands due to the low precipitation and the geological formations. Furthermore, there is a peak in water demand on the summer period (due to tourism and irrigation), the water losses are high due to the leakages in the old distribution networks, the aquifers are degraded due to over-pumping, and above all, an integrated water management system is

Figure 1. Map of Greek islands and complexes.

absent in most of the islands. These facts intensify the problems of water scarcity and create water deficiencies in the local areas (**Table 1**).

In each island, the specific conditions shape the approach used to address water deficiency problems. The current practices for domestic water supply according to Ref. [10] are: (a) desalination of sea or brackish water, (b) water transfer from the mainland at high cost (about 10 €/m^3), (c) dams, and (d) boreholes. The state officials regard desalination as an appropriate long-term settlement to the water-shortage problem.

Complex	Number of islands	Population	Population water balance [11]
Dodecanese	24	191,084	Negative
Ionian	17	211,954	Positive
Cyclades	28	118,000	Negative
North Aegean	14	202,360	Negative
Sporades	7	16,792	Negative
Saronic	7	63,467	Negative

Table 1. Island complexes in Greece [8].

3. Energy-intensive water desalination

Desalination was first introduced into Greek islands during 1960s, based on solar still collectors. Five systems of various sizes were installed until 1973, without great success, mainly due to operational breakdowns, poor maintenance [12], and conflicts of interest with the water tanker owners. The success in operation of these desalination units would result in a significant decline in the quantities of water they transferred to the islands.

In 1969, a multistage-flash system was installed in Syros, powered by fuel oil. The high cost of fuel in addition to mechanical problems led to the shift toward RO [13]. The next desalination attempt took place in the island of Corfu in 1977. A reverse electrodialysis (ED) plant with a daily capacity of 15,000 m^3 was installed for the treatment of low-salinity brackish water (up to 2000 ppm). The operation of this desalination plant ceased due to functional problems happened after few years [14]. Its specific energy consumption was 1.7 kWh/m^3 [15].

Before the Syros RO installations, the first RO units for the public community water supply were installed in 1981–1982 in the islands of Ithaca and Mykonos. The reasons for the domination of RO are: (1) compact design, (2) relatively low water needs, (3) modular operation to meet seasonal and diurnal variations in demand, (4) easy operation, (5) relatively low-energy consumption comparing to other desalination methods, and (6) fast installation (2–3 months).

Currently, the total capacity of the installed desalination units (for public use in the islands) is about 60,000 m^3/day spread over 39 islands (**Figure 2**) with 9000 m^3/day of brackish water feed, and 51,000 m^3/day of seawater feed. Very small islands of Pserimos and Rho are not included in the map of **Figure 2**. Most desalination units are located on islands, which are not connected to the mainland electricity grid (31 out of 39 islands), and the rest are installed on islands that are interconnected to the main grid, through underwater cables. The rate of development of new desalination units was quite low until the early 2000s, but rose very fast prior to the Athens Olympic Games in 2004 (**Figure 3**). In the coming years, almost five more islands are about to install and operate desalination plants.

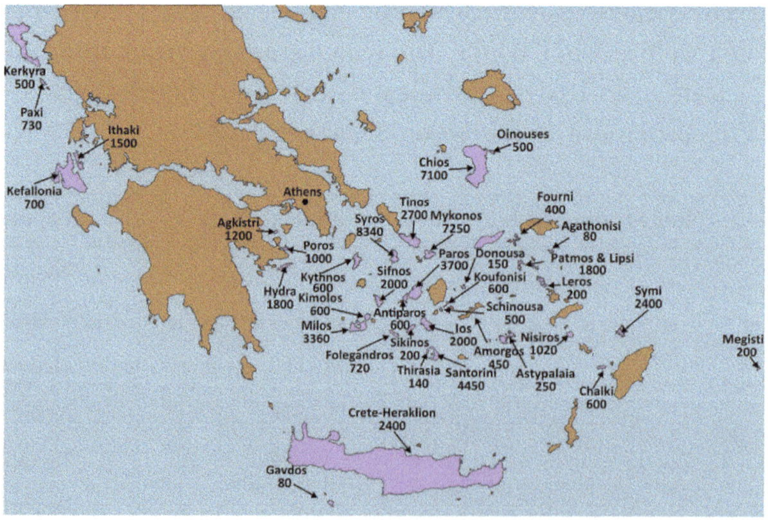

Figure 2. Desalination units in the Greek islands ([13, 16–22]; TEMAK SA, personal communication).

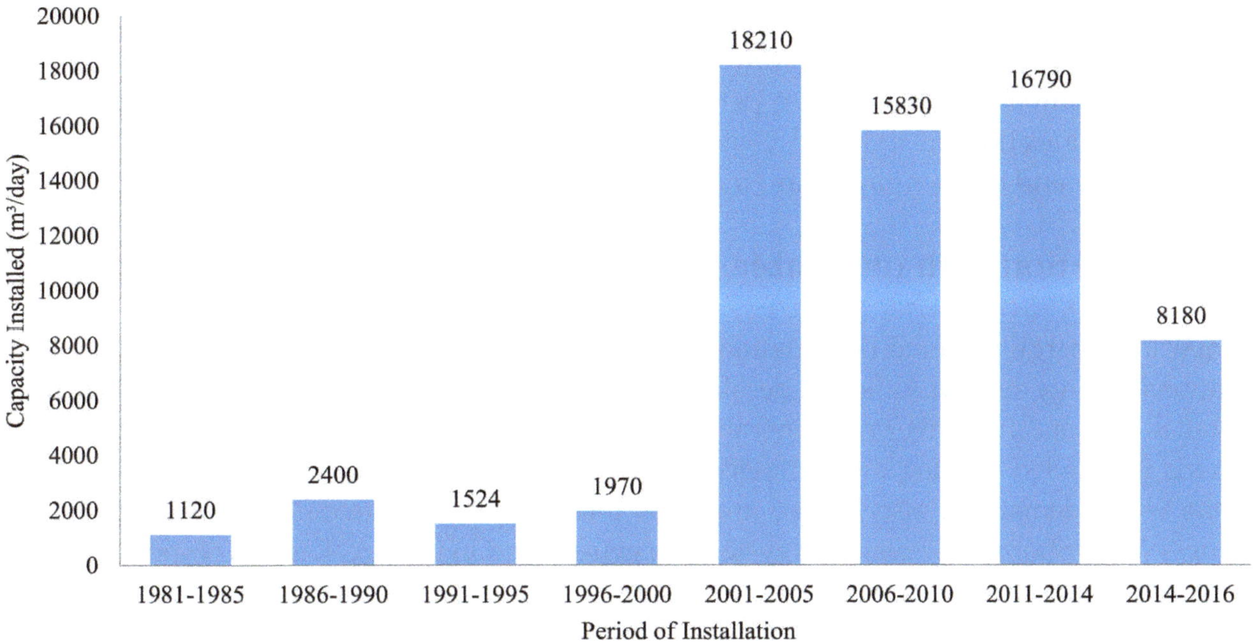

Figure 3. Desalination facilities in Greece, 1981-2014 ([13, 16–22]; TEMAK SA, personal communication).

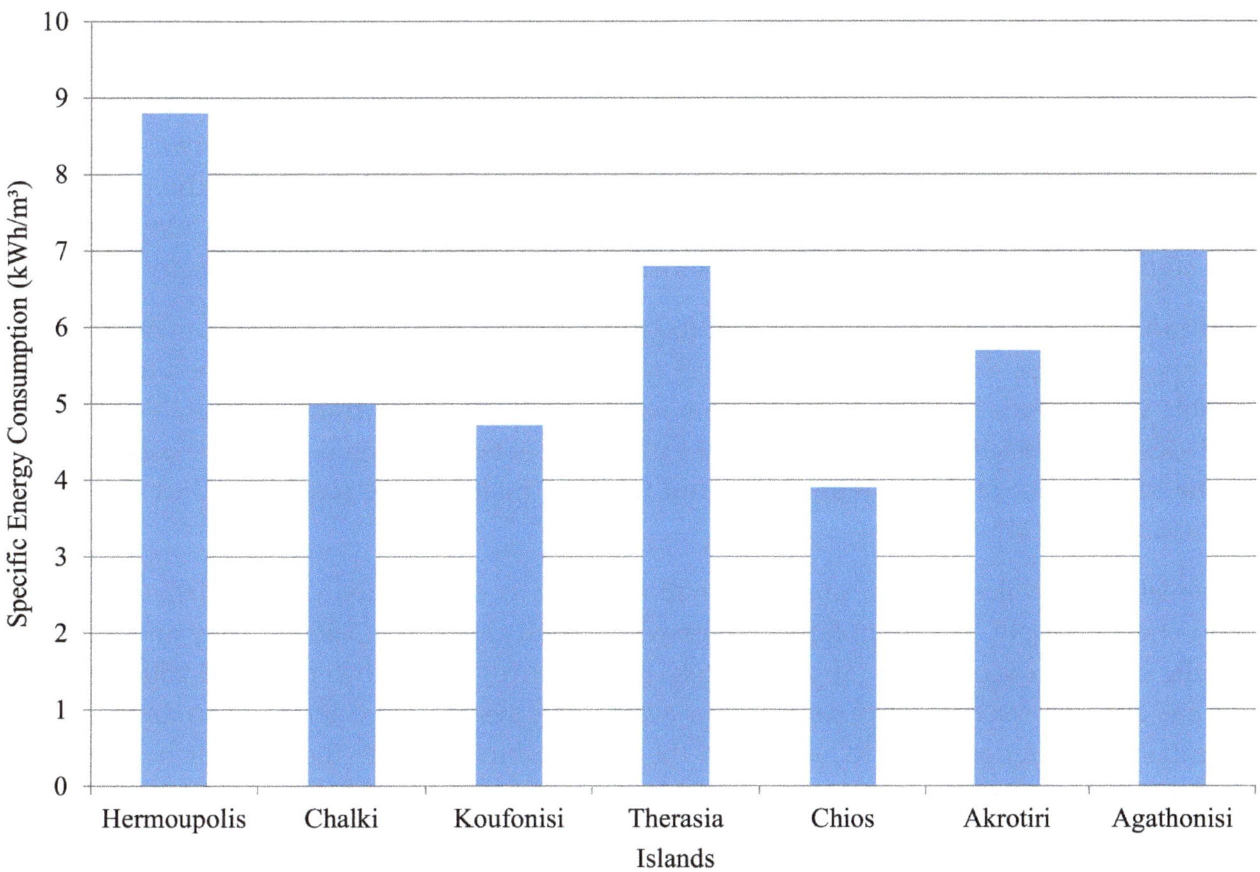

Figure 4. Specific energy consumption of Greek desalination plants (Syros WSSC, personal communication, 2014; TEMAK SA, personal communication, 2014).

The specific energy consumption for a number of desalination units installed on the islands is presented in **Figure 4**. These data were acquired through the desalination manufactures and local water companies (TEMAK SA, personal communication, 2014; Culligan Hellas SA, personal communication on technical characteristics of existing desalination units, 2014). Both high pressure and booster pumps are included in these data.

4. Desalination with renewable energy sources

Renewable energy powered desalination offers noteworthy benefits in comparison to use conventional energy supply. Although desalination systems with renewable energy have increased installation costs, they have lower operating costs due to the fact that costly fuel oil used in most islands is avoided. During the summer period, higher wind speed and solar radiation occur while water demand is also higher, and thus these have been identified as possible energy sources for supporting desalination systems in general and in the Greek islands especially.

Annual mean wind speed in most areas varies from 5 to 7 m/s, but it can be much higher in some areas. Annual solar radiation varies from 1400 to 1700 kWh/m^2, and in the islands of Milos, Nisyros, Kimolos, and Thira have high-enthalpy geothermal fields. Since the installation of solar stills, several attempts have been made to use renewable energy for desalination, both on pilot and commercial systems. The Greek state has promoted desalination systems powered by renewable energy by prioritizing the licensing procedure for the installation of the renewable systems.

A desalination unit with geothermal energy as an energy source with capacity of 80 m^3/day using multi-effect distillation (MED) technology was installed in the framework of a European project and operated effectively for demonstrations in the island of Kimolos in 1997–1998 [23]. The water production cost was estimated at € 1.7/m^3. However, at the end of the project, the desalination unit was abandoned. A very small RO unit with a capacity of 4.8 m^3/day, powered by a stand-alone 15 kW wind turbine, was installed as a pilot system in Therasia in 1997.

Hydriada, a floating RO desalination unit (80 m^3/day) with power supplied by a wind turbine and photovoltaics, was launched in 2007. It was a promising prototype designed to meet the potable water needs of 300 inhabitants in a small and arid island. Iraklia, a suitable island in Cyclades was selected as a pilot area. Early technical problems emerged and were compounded by the indifference of the local authority and the high maintenance costs, and thus the project was abandoned [24].

In the island of Milos, a successful desalination unit was constructed during 2007–2009 using wind energy supply. Three similar RO units with a total capacity of 3360 m^3/day and very low specific energy consumption (approximately 3.5 kWh/m^3) were powered by a 850 kW wind turbine. A supervisory control and a data acquisition system were installed to assist and to optimize the operation of the RO units and the wind turbines as both the RO unit and the wind turbine are connected into the island's autonomous power grid. Optimization is accomplished by the use of operation modes depending on climate conditions, the potable water demand forecasting, and the level of the water tanks. One more important innovation of this project was its ownership status: the wind turbine and the desalination unit are private investments, and they work under contract agreement with the island municipality for supplying water.

A promising mechanical vapor compression (MVC) desalination system with the exploitation of wind energy (330 kW) was built in the island of Symi in 2009. The installed system was not successful due to important technical problems that caused very low water production. This system was stand-alone, with specific energy consumption of about 14.5 kWh/m^3 [25]. Its operation ceased in 2011 due to a fire and it was never repaired. An autonomous RO unit of 8 m^3/day was installed in 2013 on the very small island of Strogyli with the use of 20 kWp photovoltaics as power source. The unit provides potable water to a small army camp.

5. The case of Syros island: comparing the operation and energy costs of desalination

The urban water supply in Syros is covered mainly by desalinated water. The island power grid is autonomous and the electrical energy supply relies on diesel fuel transferred from the mainland. Currently, there are 13 RO units distributed in five regions (**Figure 5**). According to the Syros Water Company (Syros Water Supply and Sewerage Company [26] personal communication) and the Hellenic Electricity Distribution Network Operator [27], more than 11% of the electricity produced in the island is used for potable water production and the total desalination power demand, when plants are running at full capacity is 5.2% of the conventional installed power production units.

These plants require 2.08 MW of power and the cost of energy is the primary operating cost for the Water Supply Company. The total operational cost is estimated at about €1.2–1.6/m^3. The contribution of energy to the water cost is €0.7 m^3 (about 45%), with an average energy cost of €0.086/kWh [12].

The following indicators will be used to assess the energy-for-water nexus: (a) energy consumption per volume of water sold, (b) energy cost per volume of water sold, and (c) energy

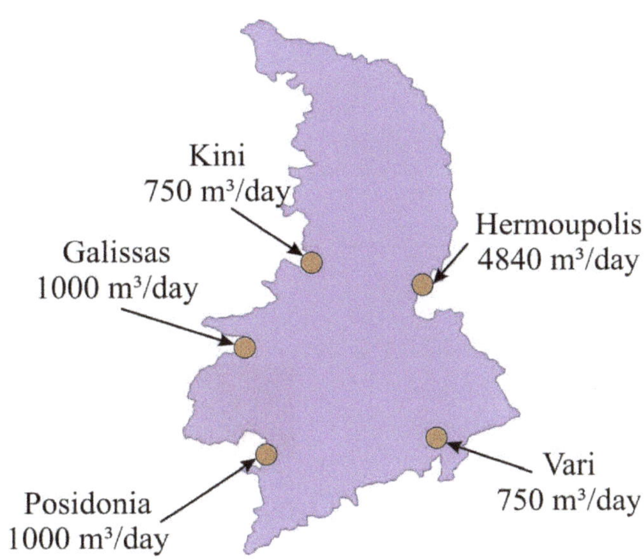

Figure 5. Desalination units in Syros.

consumption per capita. These analyses take into account only the desalination plant of Syros capital, Hermoupolis, for which the required input data are available. For comparison purposes, similar results are presented for the two major water supply systems in Greece (**Table 2**), the Athens, and the Thessaloniki water supply systems.

5.1. Assessment indicators

The energy consumption per volume of water sold is presented in **Figure 6** for three water energy systems. The Athens water system has a value 0.75 kWh/m^3, Thessaloniki's water system has twice that, and Hermoupolis about 15 times. The cost of energy also follows such

	Hermoupolis DEYA Syrou	Athens EYDAP	Thessaloniki EYATH
Population served	13,400	4,500,000	1,000,000
Water production (hm^3)	1.05	385.5	89.7
Water billed (hm^3)	0.89	337	66.3
Annual per capita consumption (m^3/year)	66.6	74.9	66.3
Energy consumption (GWh)	10	251	119
Reference year or period	2010–2013	2009	2008
Energy cost (€/kWh)	0.086	0.089	No data (assumed same as Athens)
Data sources	[27]	[28, 29]	[29, 30]

Table 2. Data used for the calculation of the assessment indicators.

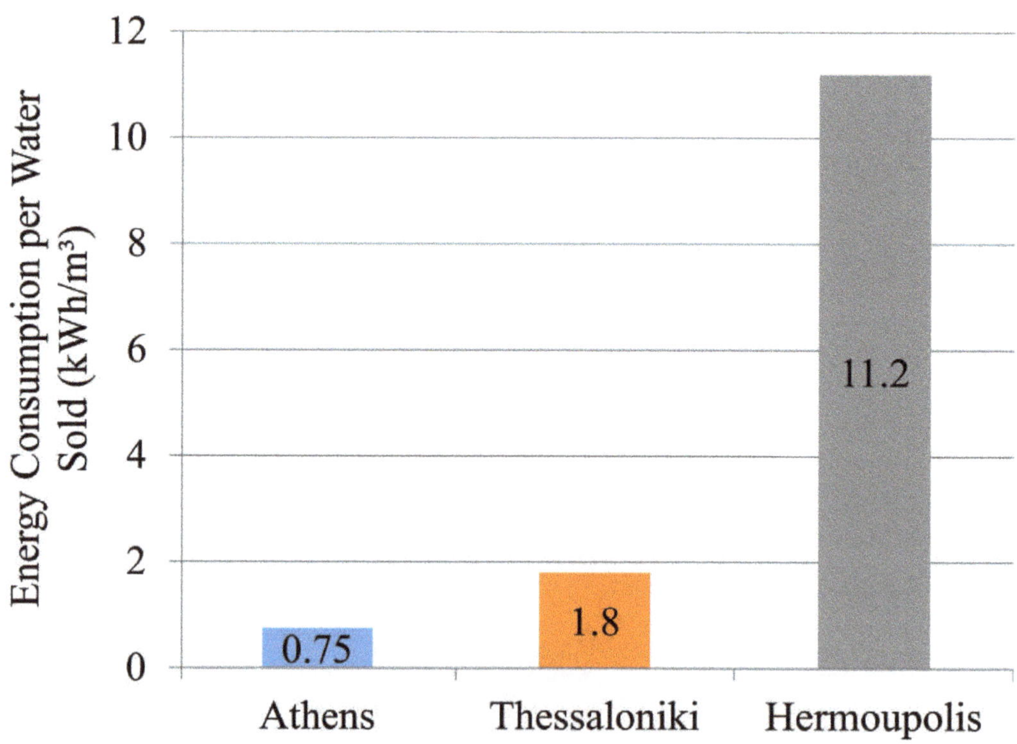

Figure 6. Energy consumption per volume of water sold in the water supply systems of Hermoupolis, Athens, and Thessaloniki [27–30].

a pattern, as seen in **Figure 7**. Similar ratios between water systems also apply for the case of the per capita energy demand for water supply. Syros has a value 14 times than that of Athens and more than twice that of Thessaloniki (**Figure 8**).

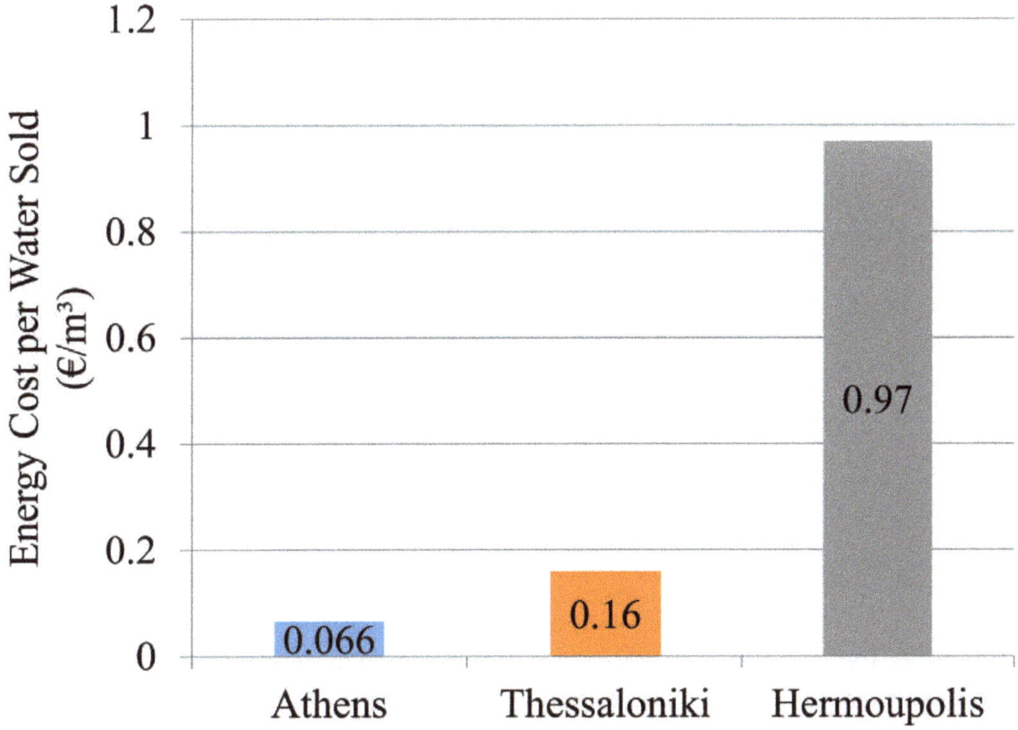

Figure 7. Energy cost per volume of water sold in the water supply systems of Hermoupolis, Athens, and Thessaloniki [27–30].

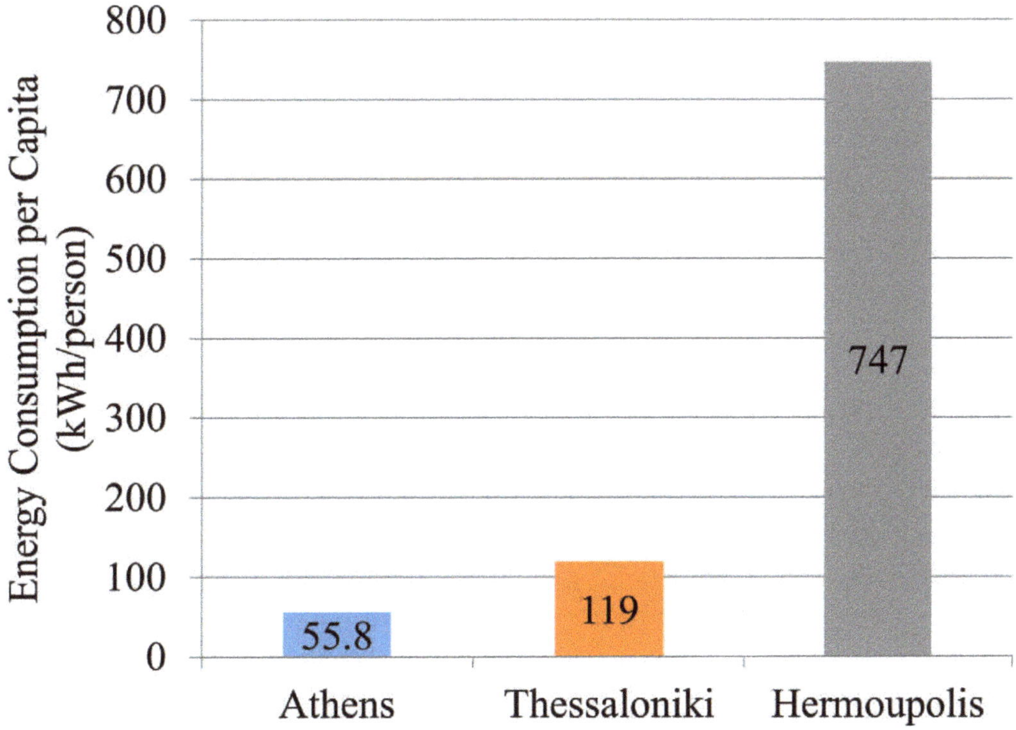

Figure 8. Energy consumption per capita in the water supply systems of Hermoupolis, Athens, and Thessaloniki [27–30].

5.2. Monthly energy consumption

Water consumption in Syros is high during summer, due to the tourism. For instance, consumption during August is around 85% higher than in February (**Figure 9**). The same

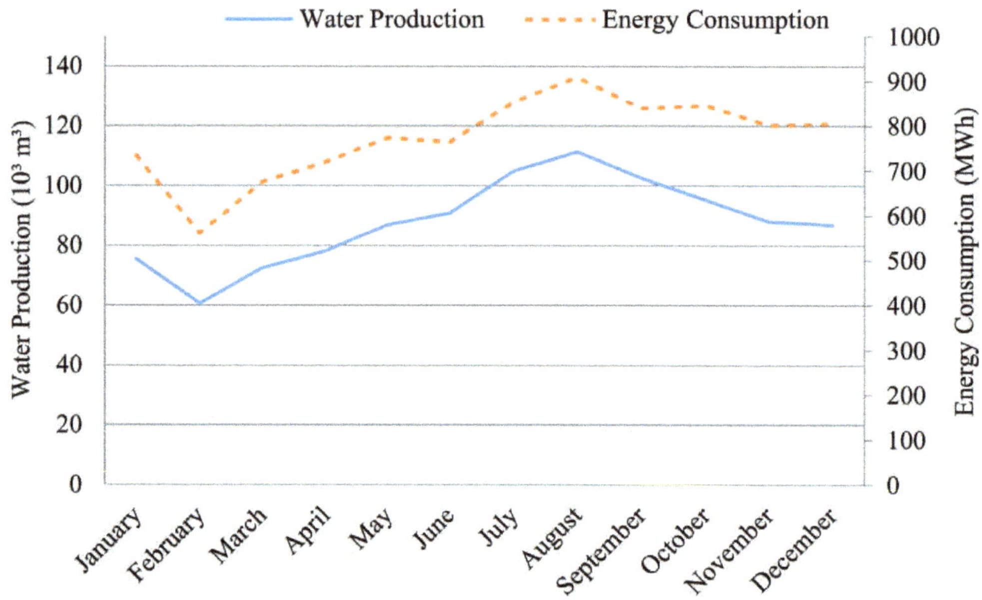

Figure 9. Monthly water production and energy consumption for the Hermoupolis desalination plant (average values for 2010–2013) [27].

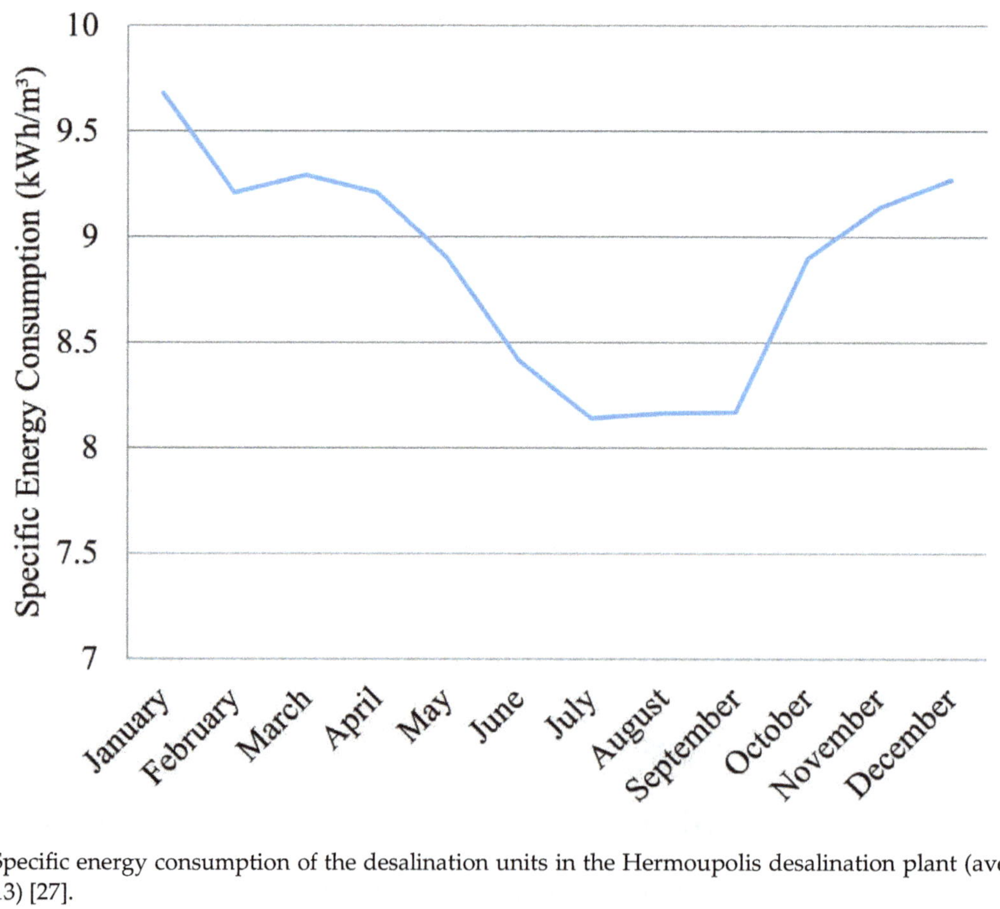

Figure 10. Specific energy consumption of the desalination units in the Hermoupolis desalination plant (average values for 2010–2013) [27].

pattern applies to the energy consumption of the desalination. However, the desalination process is significantly more efficient in the summer period (**Figure 10**). Specific energy for the Hermoupolis plant is 8.1 kWh/m^3 in August and 9.7 kWh/m^3 in February, a reduction of 16.5%. This reduction can be credited to three main factors which are as follows: (i) the units operate continuously during summer, with less start-and-stop cycles, (ii) the temperature of the seawater during summer period is high, affecting properties such as viscosity and lowering the energy demand of the high-pressure pump [31], and (iii) as the Hermoupolis desalination plant comprises several desalination units, each one with different capacities and specific energies, making extensive use of the unit with the lower specific energy also lowers the specific energy of the desalination plant as a whole.

6. Satisfying energy needs by using RES: a design methodology

A method for assessing the desalination system with RES is presented in the following section. In most cases, the desalination systems are interconnected with the electrical grid of the island, where the energy mix is dominated by diesel or heavy oil and a small percentage is covered by RES or other sources. Some islands are interconnected with the mainland and thus the energy mix contribution comes from lignite, natural gas, and RES.

The renewable sources considered in the present study are solar and wind energy. Energy demand for desalination is covered by the produced renewable energy and the excess is supplied to the grid. If the renewable energy is not sufficient to cover the demand, then energy is obtained through the grid.

The optimal configuration is the one that minimizes the water cost from the investor point of view. A configuration is defined by a set of design parameters such as the desalination capacity, the photovoltaic installed power, etc. A methodology, using hourly simulation, was developed to identify the optimum configuration. The islands to be examined are Syros and Patmos.

6.1. Design methodology

6.1.1. Water demand

Water demand can be derived from real historical data, from average monthly values or from other sources. The developed methodology requires daily water demand values throughout the year. However, if only monthly average data are available, then it can be assumed that water demand is the same for every day (Q_{daily}) of the month.

6.1.2. Desalination unit

Desalination capacity (Q_{cap} in m^3/day) is one of the design parameters. The specific energy demand (Sp in kWh/m^3) is a specification parameter of a given desalination unit and water salinity. Thus, the average power that is needed during normal operation is as follows:

$$P_{des} = Sp \cdot \frac{Q_{cap}}{24\ \text{hr}} (\text{kW}) \tag{1}$$

The hours of operation of the desalination unit per day can be derived as follows:

$$T_{op} = \min\left(\frac{Q_{daily}}{Q_{cap}/24\ \text{hr}}, 24\ \text{hr}\right) (\text{hr}) \tag{2}$$

The min operator denotes that hours of operation needed for each day cannot be more than 24 hr. The desalination might not be able to fully cover the demand and in this case, there is a water demand deficit. The operating hours of the desalination unit is a critical factor for the exploitation of the renewable sources. In case of solar energy, the best strategy corresponds to the operation of the unit in the period around noon (**Figure 11**). In case of wind energy, then the optimum period depends on the micro scale of the location of the power plant (in most cases, the afternoon is the windiest time).

The hourly distribution of energy requirements for desalination is given by Eq. (3):

$$E_{DES}(t) = OP(t) \cdot P_{DES}(\text{kWh}) \tag{3}$$

where $OP(t)$ is the desalination production distribution, representing the percentage of operation for each hour t of the year ($OP(t)$ is 100% when the desalination unit is in operation for an hour without breaks, 0% when the desalination is not operating at all, and a value between 0 and 100% in other cases).

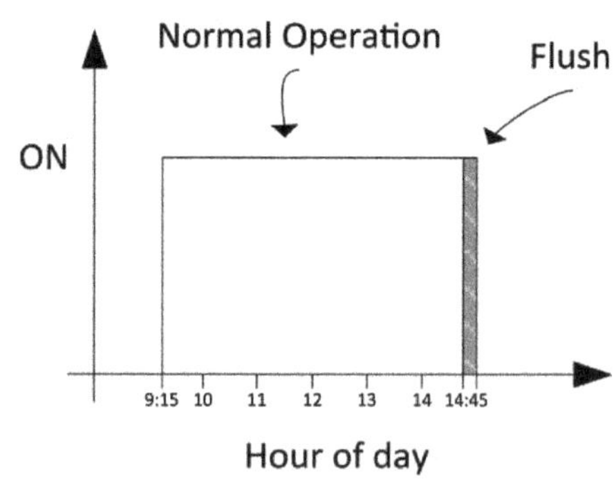

Figure 11. Reverse osmosis operation schedule (example).

6.1.3. Photovoltaic model

The installed power of the photovoltaic (PV) system is a design parameter, denoted by P_{PV} (kWp) and the energy produced by a PV array, in a given hour t, is given by:

$$E_{PV}(t) = \eta_{inv} \cdot \eta_{oeff} \cdot P_{PV} \cdot \frac{\overline{G}_T(t)}{G_{STC}} \cdot 1\ \text{hr} \quad (\text{kWh}) \tag{4}$$

where $\overline{G}_T(t)$ is the mean hour solar radiation for every hour in (kWh/m^2), G_{STC} is the standard test condition reference radiation (1000 W/m^2), η_{inv}, and η_{oeff} are the inverter efficiency and overall efficiency respectively.

6.1.4. Wind energy conversion system model

The total installed power of the wind turbines is also a design parameter (P_{WEC} in kW). The energy produced by the wind energy conversion (WEC) system in a given hour t, is given by:

$$E_{WEC}(t) = \eta_{oeff} \cdot P_{WEC}(V(t)) \cdot 1\text{ hr} \quad (\text{kWh}) \tag{5}$$

where $P_{WEC}(V(t))$ is the power produced by the wind energy system for wind speed $V(t)$. In the case of a single wind turbine, the power from the WEC system is given by the power curve of the wind turbine.

6.1.5. Energy management scheme

The scheme concerning the energy flows is quite simple. There is no sale of extra electrical energy to the grid, but net metering method is applied with annual clearance.

The renewable energy consumed by the RO unit is the sum of the wind turbine energy and photovoltaic energy produced during the operating time of the unit:

$$E_{RESav}(t) = (E_{WEC}(t) + E_{PV}(t)) \cdot OP(t) \tag{6}$$

The energy produced from renewable sources when the RO unit is not operating is as follows:

$$E_{rest}(t) = E_{WEC}(t) + E_{PV}(t) - E_{RESav}(t) \tag{7}$$

The energy difference among the available renewable energy and energy needed for desalination is as follows:

$$E_{NET}(t) = E_{RESav}(t) - E_{DES}(t) \tag{8}$$

If $E_{NET}(t) \geq 0$, then the renewable energy is not fully exploited by the RO unit. Therefore, the total amount of renewable energy that can be infused into the grid is.

$$E_{Togridav}(t) = E_{NET}(t) + E_{rest}(t) \tag{9}$$

If $E_{NET}(t) < 0$, then there are additional energy needs and they are going to be covered from the grid.

$$E_{Fromgrid}(t) = |E_{NET}(t)| \tag{10}$$

Due to technical limitations or legislation, there is an upper limit to the amount of energy that can be infused into the grid, depending on the relation between the conventional and RES energy production in the island. Thus, the energy that can be infused to the grid will be

$$E_{Togrid}(t) = \min\left(E_{Togridav}(t), E_{gridS}(t)\right) \tag{11}$$

E_{gridS} is the RES margin of the autonomous grid system.

The annual energy demand for the desalination is

$$E_{DESan} = \sum_{t=1}^{8760} E_{DES}(t) \tag{12}$$

and the renewable energy collected in one year is

$$E_{RESan} = \sum_{t=1}^{8760} (E_{WEC}(t) + E_{PV}(t)) \tag{13}$$

6.1.6. Economic evaluation

Economic evaluation will allow the identification of the most profitable and optimal configurations with the use of water cost as the main decision parameter. The levelized water cost (WC) per cubic meter can be calculated as:

$$WC = \frac{(IC_{EN} + IC_{RO}) \cdot R + OM + EC}{WP} \tag{14}$$

where: IC_{EN} is the installation cost of the power system.

IC_{RO} is the reverse osmosis installation cost.

R is the annuity factor (n is the life time of the investment and i the interest rate):

$$R = \frac{i}{1 - (1+i)^{-n}} \tag{15}$$

OM is the annual operation and maintenance cost that contains consumables (CN), (filters, spare parts), chemicals for posttreatment and pretreatment (CHM), membrane replacements (MR) and labor (LB):

$$OM = CN + CHM + MR + LB \tag{16}$$

EC is the annual cost of energy and EC_{SP} is the specific cost of energy (€/kWh):

$$EC = EC_{SP} \cdot \max\left(\sum_{t=1}^{8760} E_{Fromgrid}(t) - \sum_{t=1}^{8760} E_{Togrid}(t), 0\right) \tag{17}$$

WP is the annual water production:

$$WP = \sum_{t=1}^{8760} \frac{Q_{cap}}{24\ \text{hr}} OP(t) \tag{18}$$

6.2. Case study

The installation of renewable energy systems for covering water demand will be examined for the city of Hermoupolis in the island of Syros and for the island of Patmos. As detailed above, Hermoupolis has high energy needs for water production; therefore, the introduction of RES technologies could be beneficial in minimizing the cost of the water. The island of Patmos still covers its water needs by water transports and local drills. However, two units of 600 m^3 each have already been installed and are planned to be operational in the near future.

6.2.1. Model inputs

6.2.1.1. Water demand

Monthly water production for Hermoupolis is given in **Figure 9**. Maximum water demand for Patmos, according to a recent public tender, was estimated at 236,000 m^3 per year, but not all of the demand will be covered by the desalination. Thus, an estimation can be derived from the amount of water that was transferred to the island by tanker ships, according to Special Secretariat for Water. For 2014, this value was 68,654 m^3. Due to the lack of monthly distribution data, we can assume that the profile is the same as the previous years (monthly data exists). Also, due to the rebound effect on desalination, the water produced by desalination can be up to 20% more than water transferred. However, the case of covering all water needs through desalination will be examined.

6.2.1.2. Meteorological data

The data needed for the simulation are the wind speed and the solar radiation as hourly time series for 1 year. For Hermoupolis, the average wind speed at wind turbine height is 6.5 m/s and the total solar radiation at the photovoltaic inclination is 1847 kWh/m^2. As the island of Patmos is an important site of cultural heritage, the wind energy option is not examined. The annual solar radiation is 1935 kWh/m^2.

6.2.1.3. Installation cost

Total cost has been estimated at 2.082 M€ using data from the Syros Water Service [32]. The installation cost of the reverse osmosis unit in Patmos installation was estimated at 0.739 M€. The specific cost of the photovoltaic system was calculated for a fixed price of 1200 €/kWp and the installation cost of the wind turbine system is estimated at 1500 €/kW.

6.2.1.4. Operation and maintenance cost

These costs are either fixed or dependent on the water production:

- *Labor* is a fixed cost, in a small desalination plant estimated at $LB = 25000$ € per year and per person. When the annual water production is low, the labor cost is an important factor in the water cost. One person was employed for the RO unit of Patmos and four persons for the unit of Hermoupolis.

- *Chemicals cost* is variable and depends on the water production. The specific cost of the chemicals is estimated from 0.02 to 0.05 $\$/m^3$ [33], but in some cases it can be as high as 0.23 $\$/m^3$ [34]. For the Greek islands a value of 0.065 €/m^3 is proposed [35]. Thus, the chemical costs are $CH = WP \cdot 0.065€/\mathrm{m}^3$.
- *Membrane* cost is the most difficult to estimate, due to the fact that the life span of a membrane depends on many parameters. Proper use of the reverse osmosis system can produce water of potable quality for 5 years or more. In literature and in feasibility studies, the life span of the membrane varies from 3 to 5 years. Water production in the islands is limited and the periodical operation of the reverse osmosis can incur a specific replacement cost that is three times higher. For this reason, membrane cost varies from 0.04 to 0.34 €/m^3[36]. In this study, membrane replacement cost will be an average 0.15 €/m^3, which agrees with data from real plants in similar areas $MR = WP \cdot 0.15€/\mathrm{m}^3$[17].
- *Consumables* and *other* costs: will be taken equal to $CN = WP \cdot 0.04€/\mathrm{m}^3$ [16].

6.2.1.5. Other inputs

The specific energy for Patmos is assumed to be$Sp = 5.5\ \mathrm{kWh/m^3}$ and for Hermoupolis, a monthly specific energy is used (**Figure 10**). The flushing duration will be 15 min ($fm = 0.25$ hr). The cost of electrical energy for Hermoupolis is 0.086 €/kWh according to local data, and for Patmos a value of 0.1 €/kWh was assumed.

6.2.2. Results

The water cost in the island of Patmos, using conventional energy sources, is estimated at $WC = 1.62€/\mathrm{m}^3$. **Figure 12** presents the cost breakdown graph, where it is obvious that all cost components have approximately the same participation in water cost. If the RO unit is sized to produce water to cover all the demand of Patmos, then the water cost drops to $WC = 1.154€/\mathrm{m}^3$.

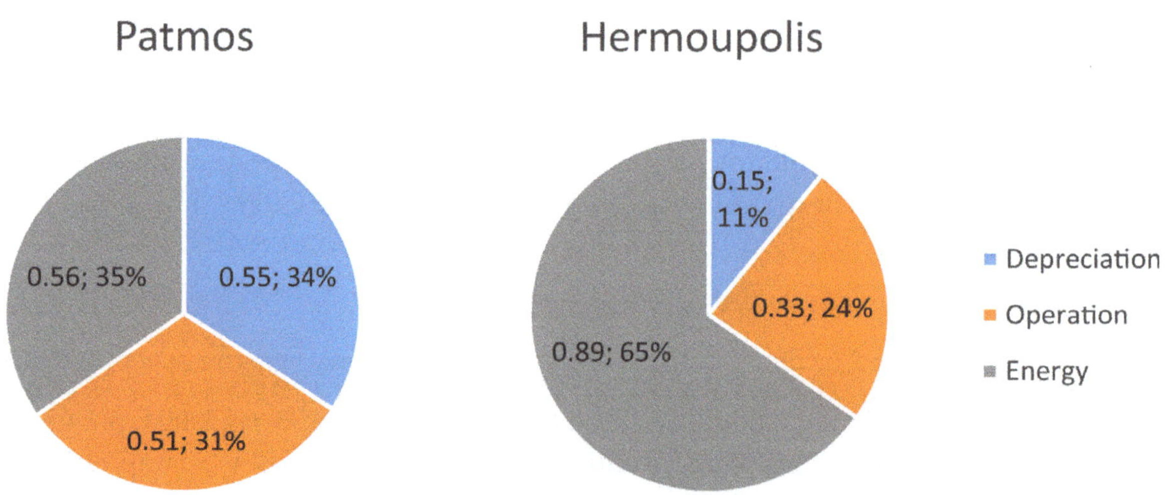

Figure 12. Water cost breakdown into the basic components for Patmos and Hermoupolis.

When installing photovoltaic units in the island of Patmos, water cost reaches a minimum of about 1.52 €/m^3, which can be achieved when the installed power of the photovoltaic system is about 200 kWp (**Figure 13**).

Water cost in the city of Hermoupolis is estimated at 1.26 €/m^3, but the contribution of each cost component is different. Energy contributes 61% of the total water cost (**Figure 12**).

The installation of photovoltaics minimizes water cost to the value of 1.16 €/m^{3}for a photovoltaic power plant of 2.3 MW The reduction in water cost is 0.1€/m^3 (the same as is in Patmos). On the other hand, when installing wind turbines, the water cost can be as low as 1.12 €/m^{3}for total wind turbine installed power of 1750 kW. In **Figure 14**, the water cost is presented versus the installed power of each technology.

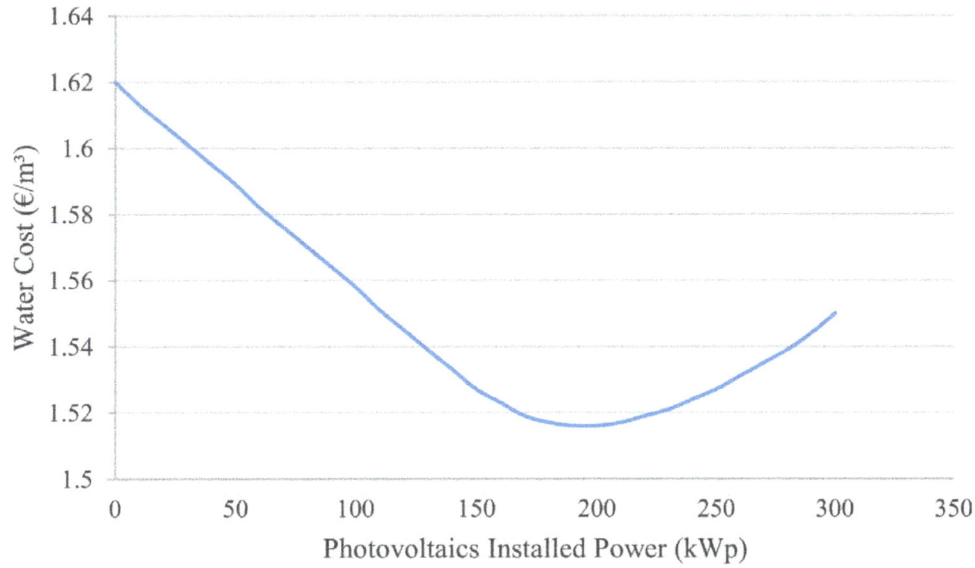

Figure 13. Water cost versus photovoltaic install power for Patmos.

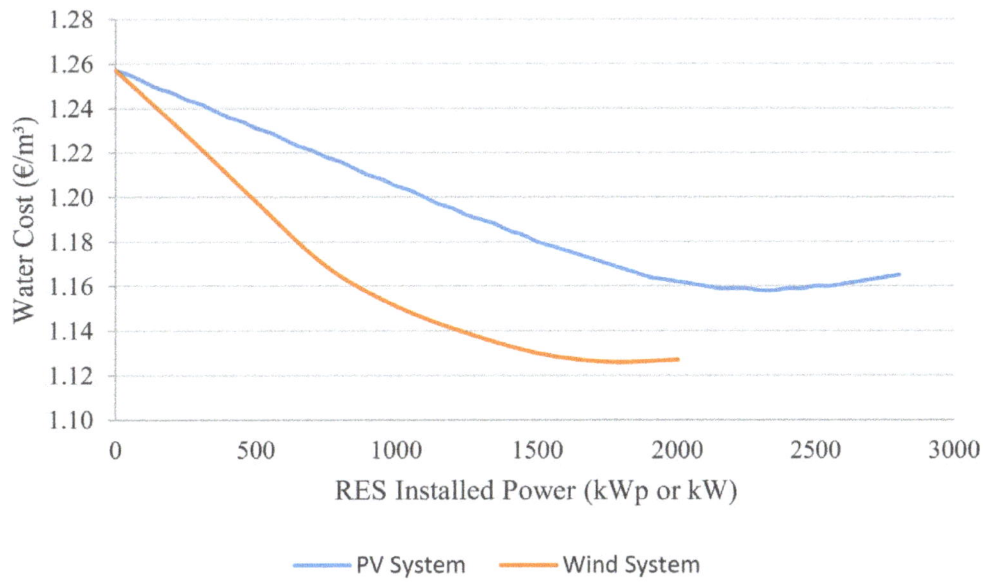

Figure 14. Water cost versus wind and photovoltaic install power for Hermoupolis. One technology each time.

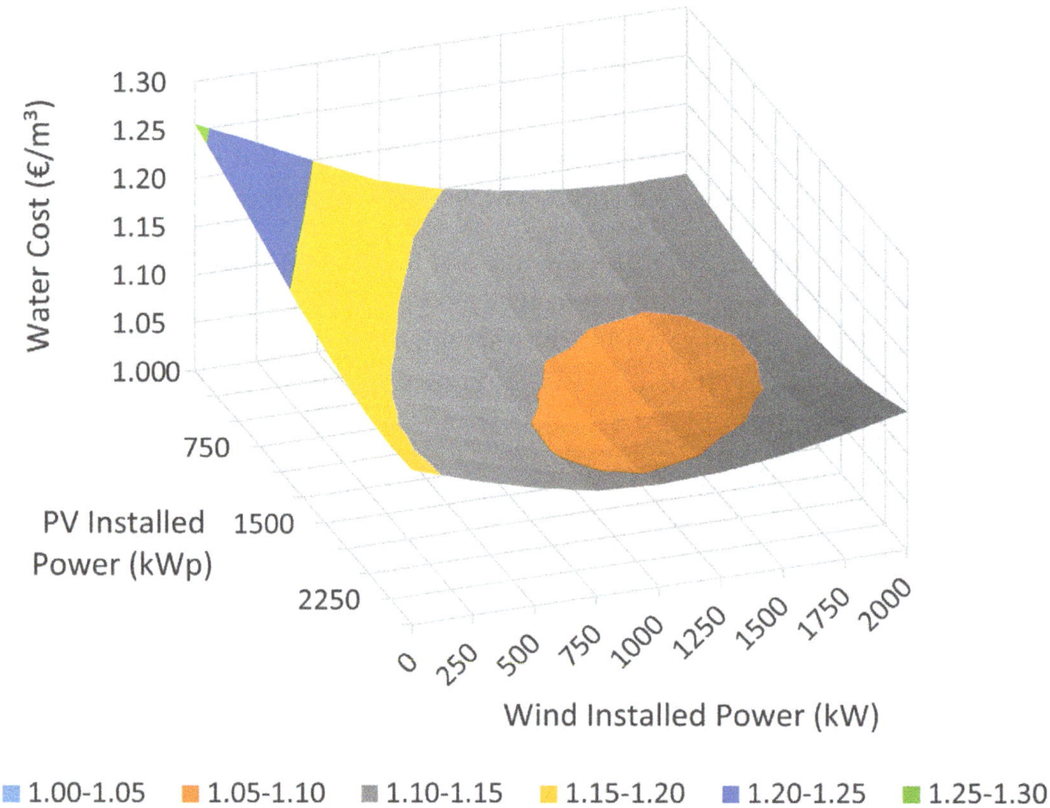

Figure 15. Water cost versus wind and photovoltaic installed power for Hermoupolis when both technologies exists simultaneously.

When both photovoltaics and wind turbines are installed simultaneously, water cost has an absolute minimum value at 1.09 €/m^3, which is 0.17 €/m^3 less than the case with conventional energy. This value can be achieved for various installed power combinations of each technology. The lower values for both the photovoltaics and wind turbine-installed power (which consequently gives the lower installation costs) are for 1250kWp and 1250 kW, respectively (**Figure 15**).

In both cases, an increase in water demand during the next years has not been taken into account; even so, the installation of renewable energy technologies will be beneficial and will make water cheaper due to higher utilization of the desalination units and due to cheap energy sources.

7. Discussion and conclusions

The practice of desalination has been identified as a promising solution to the water scarcity problem of the Greek islands (Cyclades and Dodecanese). The early efforts during the 1960s with the use of solar distillation technology were not successful, as the technical know-how was very limited and maintenance was inadequate. Also, the operational costs (labor) were very high due to the lack of automation technologies. The advances in desalination technology, combined with the need to meet water demand in the rapidly developing tourist areas, enhanced the prospects of desalination as a supplementary or even primary source of potable water. Both at the pilot and commercial levels, efforts have been made to improve the performance of desalination technology. Economic considerations and technical limitations are the most important reasons that such efforts are not always successful.

The primary challenge in the desalination practice and specifically in the reverse osmosis technology is the reduction of the energy consumption of the units. As energy is a very high cost factor for desalination, various methods for lowering the specific energy consumption can help decrease the cost of water. Another important option is the use of cheaper energy sources. Renewable energy technologies can offer a feasible way to reduce the pressure on energy systems for water production, and thus developing renewable-energy-powered desalination plants has become a priority in the Greek energy and water sectors. Energy storage is not a practical solution due to its high investment cost and the low lifetime of battery systems, but the net metering method can lower water cost by up to 13.5%. As a state of the art technology, hybrid systems, combined with operation optimization methods, provide appropriate solutions for cheaper energy; however, wind farms and photovoltaic arrays often face disapproval, especially in the tourism-dependent islands, as local inhabitants find that the local esthetic value is reduced. Environmental reasons and land-use conflicts may also prevent the integration of renewable-energy-powered RO systems. To construct efficiently operated, optimal RO desalination powered by renewable energy systems, the state should promote pilot systems through the relevant studies.

Water deficiency in most of the arid islands is treated with water transfers at the expense of the central government. The operation and maintenance cost of the desalination units is subsidized. In an era of economic crisis, this condition may not be feasible and there is a pressing need to move on toward sustainable solutions. Private sector participation, though not always socially acceptable by the locals, has to be examined. Moreover, as water and energy resources are due to face increasing pressure over the next few decades, the evaluation of trade-offs and the encouragement of cross-sector planning will be critical for further development and for their sustainable management.

The Greek experience indicates that constructing and operating RO units is made possible by state support (in the form of subsidies). The energy intensity of RO units remains high. However, RO technology, although with high cost, is reliable and offers potable water of adequate quality and quantity. Therefore, technical and operation policy challenges (including securing adequate funding) should be addressed so that the implementation of RO technology will further enhance energy efficiency and will be a part of an integrated water management system.

Author details

George Arampatzis*, Avraam Kartalidis and Dionysis Assimacopoulos

*Address all correspondence to: arab@chemeng.ntua.gr

Environmental and Energy Management Research Unit, School of Chemical Engineering, National Technical University of Athens, Athens, Greece

References

[1] Marsh DM. The water-energy nexus: A comprehensive analysis in the context of New South Wales [dissertation]. Sydney: University of Technology Sydney; 2008

[2] Scott CA, Pierce AS, Pasqualtti JM, Jones LA, Montz EB, Hoover HJ. Policy and institutional dimensions of the water–energy nexus. Energy Policy. 2011;**39**(10):6622–6630. DOI: 10.1016/j.enpol.2011.08.013

[3] Schumacher L. Water-Energy Nexus [Internet]. 2007. Available from: www.naruc.org/Committees/CommitteePresentations.cfm?c=10

[4] EIP-Water. Water-Energy Nexus [Internet]. 2015. Available from: www.eip-water.eu/priorities/water-energy-nexus

[5] IPCC, editor. Climate change 2014: Impacts, adaptation, and vulnerability. Part A: Global and sectoral aspects. Contribution of working Group II to the fifth assessment report of the intergovernmental panel on climate change (Report No. 978-1-107-05807-1). United Kingdom and New York, NY, USA: C. U. Press; 2014

[6] International Energy Agency, editor. World Energy Outlook 2012. Paris, France: IEA; 2012

[7] Rodriguez DJ, Delgado A, DeLaquil P, Sohns A. Thirsty Energy. Washington, DC: World Bank; 2013

[8] Hellenic Statistical Authority. Announcement of the Results of the 2011 Population Census for the Resident Population [Internet]. 2012. Available from: www.statistics.gr/portal/page/portal/ESYE/BUCKET/General/A1602_SAM01_DT_DC_00_2011_02_F_EN.pdf

[9] Hellenic National Meteorological Service. 2015. Available from: www.hnms.gr/hnms/english/climatology/climatology_html

[10] Nokas H, editors. Estimation and comparative analysis of water production cost in the Cyclades Islands. Syros Island: Water Directorate, Prefecture of Cyclades; 2002

[11] Special Secretariat for Water. Integrated River Basin Management Plans [Internet]. 2013. Available from: wfd.ypeka.gr

[12] Deligiannis E-E, Belesiotis V. Solar application in desalination: The Greek Islands experiment. Desalination. 1995;**100**(8)27-34. DOI: 10.1016/0011-9164(96)00006-9

[13] Vakondios G. Potable Water Production With Desalination [Internet]. 2013. Available from: www.edeya.gr/2013-09-23-10-58-06/2013-09-23-11-11-03/uliko/hmerides-1/texnologies-afalatwsis-anaduomenes-taseis-1

[14] Dafnis I. Local reverse osmosis units of Corfu municipality. In: Potable Water: Quality Control, Legislation, Problems of Quality and Water Management in Corfu Prefecture; Corfu. 2005

[15] Arnold J. Operating experiences of a 15,000 cubic meter per day municipal desalting plant at Corfu, Greece. Desalination. 1979;**30**(9):145–153. DOI: 10.1016/S0011-9164(00)88442-8

[16] Cuenca J. Report on water desalination status in the Mediterranean Countries. Murcia: IMIDA. Murcia: IMIDA; 2002

[17] Dagkalidis A. Greek sectorial studies—study on desalination. Athens: Piraeus Bank; 2009

[18] Dalmira A. Desalination Unit in Schoinousa [Internet]. [Updated: 2013]. Available from: www.koinignomi.gr/news/politiki/politiki-kyklades/2013/12/27/monada-afalatosis-gia-ti-shoinoysa.html

[19] Kampourakis N. Deyath Project Report 2011-12 [Internet]. 2013. Available from: www.thira.gr/nomika-proswpa/deya-thiras/16-dimos/nomika-proswpa-synthesi/deya-thiras.html?layout=blog

[20] Imerisia. New Desalination Unit from Temak [Internet]. Available from: www.imerisia.gr/article.asp?catid=26519&subid=2&pubid=7421123&tag=9464#

[21] Sychem. Desalination of Seawater in Paros Island [Internet]. 2011. Available from: www.sychem.gr/el/erga/afalatosi-yperdiithisi-erga/item/112-afalatwsh-thalassinoy-neroy-ston-dhmo-paroy

[22] Ventouris E. Kimolos Desalination Plant [Internet]. 2013. Available from: www.gokimolos.gr/el/component/content/article/15-news/68-dimos-kimolou.html

[23] Karytsas C, Alexandrou V, Boukis. The Kimolos Geothermal Desalination Project [Internet]. 2002 . Available from: pangea.stanford.edu/ERE/db/IGAstandard/record_detail.php?id=5365

[24] Tsamopoulos M. Cyclades: Desalination System is Rotten in Iraklia [Internet]. 2014. Available from: www.protothema.gr/greece/article/400961/kuklades-katarrei-sustima-afalatosis-stin-irakleia-

[25] Bundschuh J, Hoinkis J, editors. Renewable Energy Applications for Freshwater Production. Leiden: CRC Press/Balkema; 2012

[26] Syros Water Supply and Sewerage Company. DEYA Syrou Energy Requirements [Internet]. 2014

[27] Hellenic Electricity Distribution Network Operator S.A. Informative Report of Energy Production on Non-Interconnected Islands [Internet]. June 2012. Available from: www.deddie.gr/en/miniaia-deltia-ape-kai-thermikis-paragwgis-sta-mi-diasundedemena-nisia/2012

[28] Athens Water Supply and Sewerage Company. Annual Report & Annual Bulletin [Internet]. [Updated: 2011]. Available from: www.eydap.gr/userfiles/Presentations/ependutes/enimerotika-deltia/etisio_deltio_2011.pdf

[29] Georgalas L. Approach of Water-Energy Relation in the Water Supply-Sewage System [Internet]. 2010. Available from: www.nomosphysis.org.gr/articles.php?artid=4237&lang=1&catpid=2

[30] Thessaloniki Water Supply and Sewerage Company. Annual Bulletin [Internet]. 2011. Available from: www.eyath.gr/misc/ETISIO_DELTIO_2009.zip

[31] Water Reuse Association – Desalination Committee, editor. Seawater Desalination Power Consumption. USA: Water Reuse Association; 2011

[32] Wilf M. Fundamentals of RO-NF technology. In: International Conference on Desalination Costing; Limassol, Cyprus. December 2004

[33] Vince F, Marechal F, Aoustin E, Breant P. Multi-objective optimization of RO desalination plants. Desalination . 2008;**222**(1-3):96–118. DOI: 10.1016/j.desal.2007.02.064

[34] Lamei A, Van der Zaag P, Von Munch E. Basic cost equations to estimate unit production costs for RO desalination and long-distance piping to supply water to tourism-dominated arid coastal regions of Egypt. Desalination. 2008;**225**(1-3):1–12. DOI: 10.1016/j.desal.2007.08.003

[35] Karagiannis IC, Soldatos PG. Current status of water desalination in the Aegean Islands. Desalination. 2007;**203**(6). DOI: 10.1016/j.desal.2006.04.006

[36] Avlonitis S, Kouroumbas K, Vlachakis N. Energy consumption and membrane replacement cost for seawater RO desalination plants. Desalination. 2003;**157**(1–3, 1 August 2003):151–158. DOI: 10.1016/S0011-9164(03)00395-3

9

Pulsating Flow Effects on Hydrodynamics in a Desalination Membrane Filled with Spacers

Armando A. Soares, João Silva, Eliseu Monteiro and Abel Rouboa

Abstract

A previously developed and validated two-dimensional computational fluid dynamics (CFD) model to study the hydrodynamics in a desalination membrane filled with spacers in zig-zag arrangements has been further developed to include the effects of a pulsating flow with the profile of a heartbeat. Numerical solutions were obtained with *Fluent* for pulsating laminar flows in channels filled with four different spacers and four lengths of cells. Hydrodynamics was investigated for unsteady state, using a characteristic function of a heartbeat, in order to study the influence of temporal variation in the hydrodynamic behavior. The results show the velocities distribution, streamlines, pressure drop and the wall shear stress on the impermeable wall of the membrane, for Reynolds numbers up to 100. The reduction in the distance between the filaments of the spacers, leads to the appearance of more active recirculation zones that can promote mass transfer and decreasing concentrations layers. On the other hand, this reduction increases the pressure drop and consequently the energy expended in the process. Further, the characteristic function of heartbeat demonstrates promising results, with regard to the energy consumption in the process and optimization of the recirculation zones.

Keywords: desalination, membrane, reverse osmosis, zig-zag spacers, heartbeat flow profile

1. Introduction

The planet has not only freshwater, but from the enormous volume of water available on our planet, only 3% is not salty. This resource is limited and finite, vital for the existence of life on earth and for the economic and social development [1].

The exploitation of natural water resources combined with the increase in the world population, the changes in the life style, the inefficient use of water and its contamination, among other factors, leads to scarcity of natural freshwater to respond the necessities of the population [2]. Due to these factors, the demand for potable water, has led to an increasing need to find new alternative sources of drinking water.

Desalination provides a good alternative to reduce the problem of the scarcity of drinking water, once it provides clean water which otherwise would not be accessible for agricultural, industrial and services supply [3, 4].

Membrane systems are widely used in water treatment processes. Depending on the pore size, membrane processes can be classified as microfiltration (MF), ultrafiltration (UF), nanofiltration (NF) and reverse osmosis (RO). The MF and UF are used to remove small colloidal particles and bacteria. For removal of viruses and large molecules such as proteins, the UF is only used. The NF and RO are capable of removing the smallest particles and components, like salts dissolved in water. In the case of RO, it is possible to remove almost all the components dissolved in the water [4].

The RO is the most used membrane separation method. This is applied at chemical, textile, petrochemical, electrochemical and paper industry, as well as in the treatment of municipal wastewaters [5]. But, the major application of RO process is the desalination of brackish and salty water to obtain potable water [6]. This is due to its versatility to operate over a wide range of feed water salinities, lower intake and pumping costs, minimal space requirements and operation at ambient temperatures [7]. Spiral wound modules (SWMs) have been largely used in industrial applications to obtain potable water, due to their low cost and high membrane area to volume ratio [8]. The performance of a SWM is affected by many factors such as leaf geometry (number and dimensions), spacers, fouling propensity, cleaning ability and operating conditions [9]. Despite several benefits of membrane technology over other separation techniques, the phenomena inherent to any membrane separation method are concentration polarization and fouling. As the filtration process occurs though the semipermeable membrane, the rejected impurities begin to accumulate on the membrane surface. After a certain period of time, a concentration layer is formed on the membrane surface due to the solute. This phenomenon, in which, the concentration of solute is higher in the vicinity of the membrane as compared with the observed in the fluid, is known as concentration polarization [10, 11]. This phenomenon can significantly affect the performance of the membrane, for example, the permeate water quality and productively are affected by fouling deposited on the membrane surface. On the other hand, from the literature it is well known that spacers are essential to separate the sheets of membranes in both spiral wound modules based thin rectangular channels (slits). Spacers are also important in the optimization of the mass transfer enhancement and pressure loss minimization [12, 13].

One important way by which the efficiency of module can be enhanced is to modify the hydrodynamics in membrane channel, which can be done by improving feed channel spacers design and using the unsteady state to induce a temporal variation in the feed velocity. This can prevent the accumulation of solutes near the membrane surface, thereby lowering the concentration gradient between the bulk fluid and the membrane surface, and reducing fouling phenomenon [7].

The hydrodynamics inside the feed channel using computational fluid dynamics (CFD) has been investigated awhile. Cao et al. [14] investigated, trough CFD, the effect of three different cylindrical spacers arrangements: spacers touching the same channel wall, spacers touching opposite channel walls and spacers suspended in the channel. The flow simulations, in the spacer-filled channels, showed the distribution of velocity and turbulent kinetic energy as well as the detailed flow patterns. It was found that the presence of the spacer in the membrane is responsible for the local shear stress distribution on the membrane surface and improves the production of the eddy activities. It is also found that spacer configuration, geometry and position in the channel are crucial in the distribution of eddies and high shear stress on the membrane surface. The investigation also suggests that transverse spacer cylinders suspended in the channel are more desirable. Furthermore, decrease in the distance between transverse spacer cylinders can produce more active eddies and reduce the distance between shear stress peaks and consequently improve the mass transfer at the membrane surfaces. Besides that, it was demonstrated that the pressure drop is significantly increased by reduction in the distance between transverse cylinders which increase the operating costs. Schwinger et al. [15] perform CFD studies on how the use of spacers in membrane systems can affect its hydrodynamics. The flow fluid was studied in three different spacer configurations, the cavity, zig-zag, and submerged spacers. Were considered, in a cell, a single filament adjacent to the wall and centered in the channel for Reynolds numbers from 90 to 768 for different values of the length of the mesh and filament diameter. The results obtained show that the flow around the filament increases the wall shear stress and promotes the formation of large recirculation zones behind the filaments. Furthermore, under the same conditions, identical Reynolds number and filament diameter, a single filament adjacent to a membrane wall produced a larger recirculation zone than a single filament in the center of the channel. According to the authors, the utilization of the spacer in a SWM should enhance the mass transfer and reduce the accumulation of solute in the membrane surface while maintaining a low pressure loss along the channel. Saeed et al. [16] used a CFD tool to investigate the flow patterns associated with fluids within the membrane module. The effects on flow patterns through a spacer filled RO membrane with the secondary structure of the membranes (feed spacer filaments) at various angles with the inlet flow are analyzed. The results revealed that the alignment of the filaments in the direction of flow has a great influence on the generation of recirculation streams in channels filled with spacers. The optimization of the orientation of the filaments can lead to desirable recirculation patterns inside the modulus, and therefore, may increase the performance of the membrane. Sousa et al. [17] studied the hydrodynamics in a desalination membrane. CFD techniques were used to study the hydrodynamics of feed channels of a desalination membrane filled with elliptical spacers in zig-zag arrangements and transverse in relation to the flow, for four spacers and four cell lengths. They used *Re* values between 10 and 300 and the inlet velocity profiles are considered fully developed. The results showed that for *Re* values above 100, are formed recirculation zones downstream of the spacers. When the spacers are closer, there is a formation of more active recirculation zones. The authors state that these can increase the mass transfer and together with the pressure drop, there may be an increase in process efficiency [17]. Amokrane et al. [18] used a model based on CFD which combines the fluid dynamics and the mass transfer in SWMs with zig-zag spacers, to investigate the impact of new spacers design. They affirm that zig-zag arrangement appears to have more advantages in the performance of RO membranes, in comparison, with submerged spacers. The results highlight that the incorporation of new spacer designs, such as elliptic and oval shapes, generates lower pressure drop (Δp) compared to conventional geometries. However, concentration polarization layer thickness, induced by elliptic and oval spacers, is higher compared to that of a circular spacer, while the mass transfer coefficient obtained with circular spacer is higher [18]. Al-Bastaki and Abbas [19] used a cyclic feed flow to improve the performance of membrane modules. The operation in cyclic mode generates flow instabilities, which in turn disturbs the fouling layer and increases the permeation rates. The cyclic operation can, however, need additional devices such as pulsation generator, oscillating pistons, and pneumatic valves. Lau et al. [20] performed a tri-dimensional study of the unsteady flow in order to optimize the spacer mesh angle for the SWM feed spacer. The results show that the presence of unsteady flow can significantly interrupt the development of the concentration of solute in the membrane surface. Filaments with different geometries and different angles may lead to different degrees of disruption in the concentration of the membrane layer. The filaments can be optimized in terms of concentration polarization and power consumption.

In the present investigation, a previously developed and validated two-dimensional CFD model to study the hydrodynamics in a desalination membrane filled with spacers in zig-zag arrangements [17], is further developed to include a characteristic function of a heartbeat within a cell of a semipermeable desalination membrane, in order to study the global and local impact of hydrodynamic behavior on the velocity, pressure drop, and wall shear stress for different spacer arrangements. The numerical approach used allows uncoupling of several phenomena and contributes to a better understanding of the mechanisms through which the shape of the flow profiles can affect membrane performance.

2. Governing equations and boundary conditions

An incompressible Newtonian fluid was used to simulate the bi-dimensional unsteady state flow in a feed channel of a SWM with zig-zag spacers (**Figure 1**). The governing equations are the continuity and Navier-Stokes equations. For a bi-dimensional flow, the governing equations can be written as:

Continuity equation

$$\frac{\partial v_x}{\partial x} + \frac{\partial v_y}{\partial y} = 0, \tag{1}$$

xx—Momentum equation

$$\rho\left(\frac{\partial v_x}{\partial t} + \frac{\partial}{\partial x}(v_x^2) + \frac{\partial}{\partial y}(v_x v_y)\right) = -\frac{\partial p}{\partial x} + 2\mu\frac{\partial}{\partial x}\left(\frac{\partial v_x}{\partial x}\right) + \mu\frac{\partial}{\partial y}\left(\frac{\partial v_x}{\partial y} + \frac{\partial v_y}{\partial x}\right) + \rho g_x \tag{2a}$$

and

yy—Momentum equation

$$\rho\left(\frac{\partial v_y}{\partial t}+\frac{\partial}{\partial x}(v_x v_y)+\frac{\partial}{\partial y}(v_y^2)\right)=-\frac{\partial p}{\partial y}+2\mu\frac{\partial}{\partial y}\left(\frac{\partial v_y}{\partial y}\right)+\mu\frac{\partial}{\partial x}\left(\frac{\partial v_x}{\partial y}+\frac{\partial v_y}{\partial x}\right)+\rho g_y \tag{2b}$$

where x and y are the spatial coordinates, ρ is the fluid density, μ is the dynamic viscosity, p is the static pressure, and g is the gravitational acceleration. The velocity components in x and y directions are v_x and v_y, respectively.

In the present study, the Reynolds number is defined as

$$Re=\frac{\rho\,\overline{v}\,h}{\mu} \tag{3}$$

where h is the height of the cell and $\overline{v}$ is the average inlet velocity in a cell, see **Figure 1**.

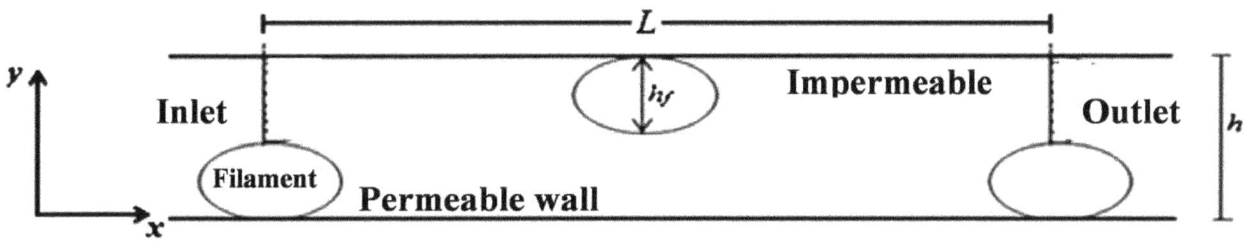

Figure 1. Schematic representation of a membrane cell with elliptical zig-zag filaments.

2.1. Inlet velocity profiles

In order to obtain a better numerical approach taking into account the flow periodicity, it was necessary to determine the velocity profile at the inlet and outlet of the cell [21]. As far as our knowledge in previous investigations, for example Refs. [8, 17, 22], it was considered a fully developed velocity profile at inlet. However, the velocity profiles between the filaments are not fully developed because there is no enough distance between filaments to allow the full development of the fluid flow. Therefore, for a given average velocity, the inlet velocity profile was obtained using the following procedure: for the value of Re studied, the corresponding average velocity is imposed at the inlet boundary. On the outlet boundary was considered the outflow option present in *Fluent*. With this condition, it is assumed that there are no diffusive fluxes of all variables at the exit in the flow direction, or a zero diffusion flux at the exit is expected to have a small impact on your flow solution.

For the case of impermeable walls, when the solution is achieved, the boundary condition at the inlet is replaced by the solution obtained for the velocity profile at the outlet boundary. This process was repeated until the differences between the inlet and outlet profiles were less than 0.15%. Similarly, in order to verify the permeability influence on the flow, a study was also done for a semipermeable membrane. To this, the bottom wall was considered permeable, with a constant flow rate of permeate. The value of permeate velocity (v_p) was 0.1 mm/s, value taken from the literature [7, 8]. Since there is permeability, part of the flow is lost through the permeable wall and, thus for the conservation of mass principle be satisfied, the permeate flow rate has to be added to the outflow to obtain the inlet velocity profile. That is,

$$v_y(0,y) = v_y(L,y) \tag{4a}$$

$$v_x(0,y) = v_x(L,y) + \frac{L}{h - h_f} v_p(L,y) \tag{4b}$$

where the second term on the right-hand side of Eq. (4b) is the velocity profile in the outlet boundary obtained for the case of impermeable walls, when the imposed inlet flow rate is equal to the permeate flow rate.

The time-dependent velocity at the inlet was obtained from Doppler ultrasound images for a heartbeat, in the abdominal aorta, with a period of $T = 1.53$ s. **Figure 2** shows the normalized inlet velocity as function of time, $v_h(t)$, obtained from a polynomial fit of the Doppler data. Once given, the normalized velocity of the heartbeat was necessary to make an adjustment thereof, in order to encompass the typical velocities of a desalination membrane. Thus, the inlet velocity profile is given by

$$v_x(0,y,t) = v_x(0,y) \cdot v_h(t) \tag{5}$$

Then the normalized velocity $v_h(t)$ was multiplied by the velocity profiles, already determined for steady-state flow and so we obtained the inlet velocity profiles used in the present study, Eq. (5). For the impermeable membrane $v_x(0,y) = v_x(L,y)$.

The no-slip condition was applied at the wall boundaries.

Note that in the presence of permeability, it is needed to differentiate inlet and outlet profiles because some fluid is permeated through the permeable wall, see Eq. (4b).

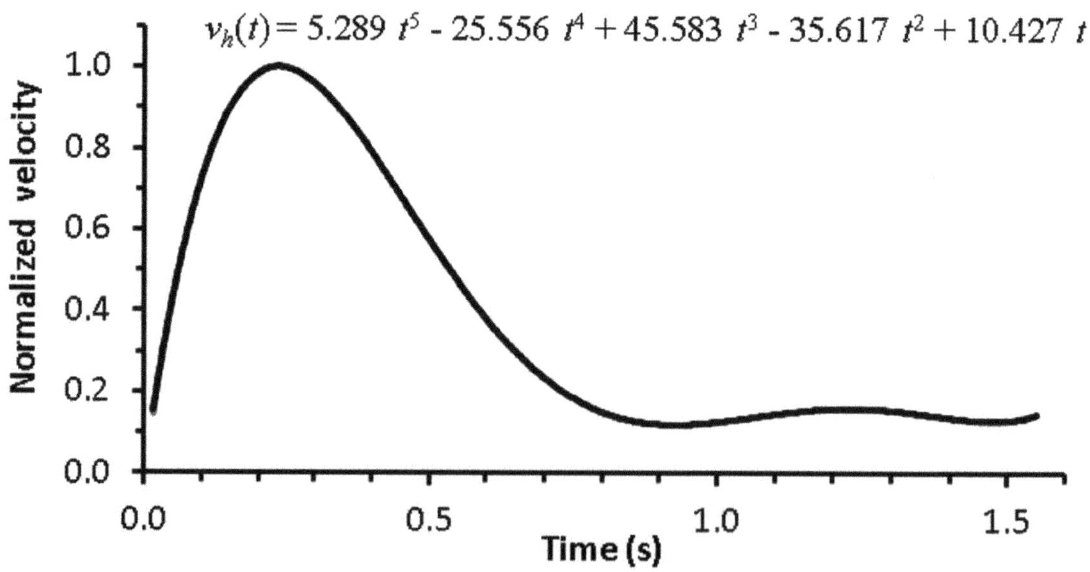

Figure 2. Normalized velocity for a heartbeat with a period $T = 1.53$ s.

3. Results and discussion

In this study, two-dimensional simulations using different feed channel geometries filled with spacers (different cells lengths and different shape of the elliptical filaments) were performed. We studied the impact on the velocity field, streamlines, average wall shear stress (*WSS*), and pressure drop for the cells lengths, $L = 4$, 6, 8, and 10 mm (**Figure 3**) and height of 0.7 mm. Four elliptical filaments configurations (*F*1, *F*2, *F*3, and *F*4) were also tested.

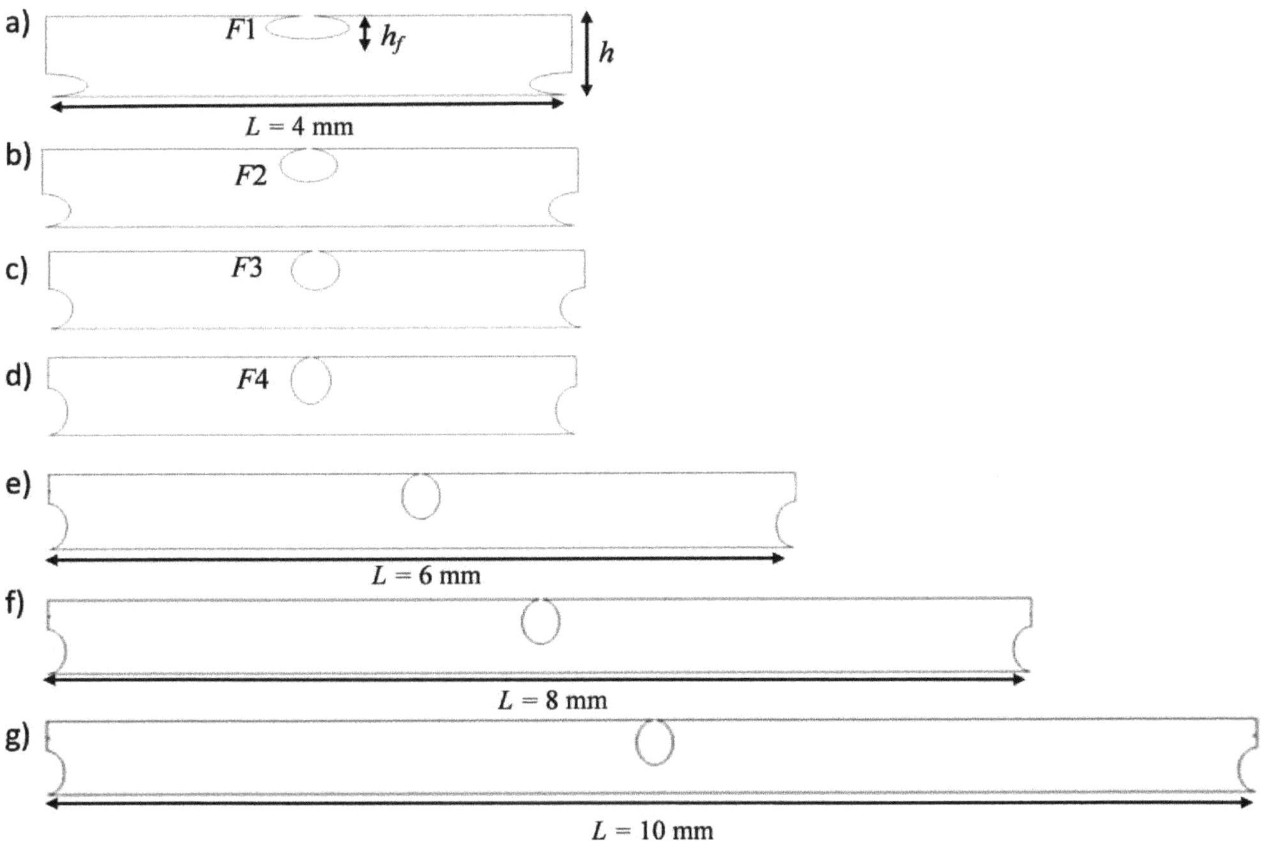

Figure 3. Cell filled with zig-zag spacers for (a) $L = 4$ mm and *F*1; (b) $L = 4$ mm and F2; (c) $L = 4$ mm and F3; (d) $L = 4$ mm and *F*4; (e) $L = 6$ mm and F4; (f) $L = 8$ mm and F4; (g) $L = 10$ mm and *F*4.

3.1. Inlet velocity profiles

Since for normal operation of the SWM, the velocity profiles inside the cell cannot be fully developed due to the filaments periodicity, so that the inlet profiles used in this study were determinate by the aforementioned procedure. As example, **Figure 4** shows the profiles determined for the four elliptical filaments with $L = 4$ and $Re = 100$, for the membrane with impermeable walls. Since permeable velocity is 0.1 mm/s, the inlet velocity profiles are practically unaltered by the permeability of the membrane when compared with the impermeable case, therefore they are not shown in **Figure 4**.

The width profiles decrease from *F*1 to *F*4 corresponding to the increase of flow obstruction by the spacers (**Figure 3**). Note that for the impermeable walls, the inlet and outlet profiles, for each configuration, are the same.

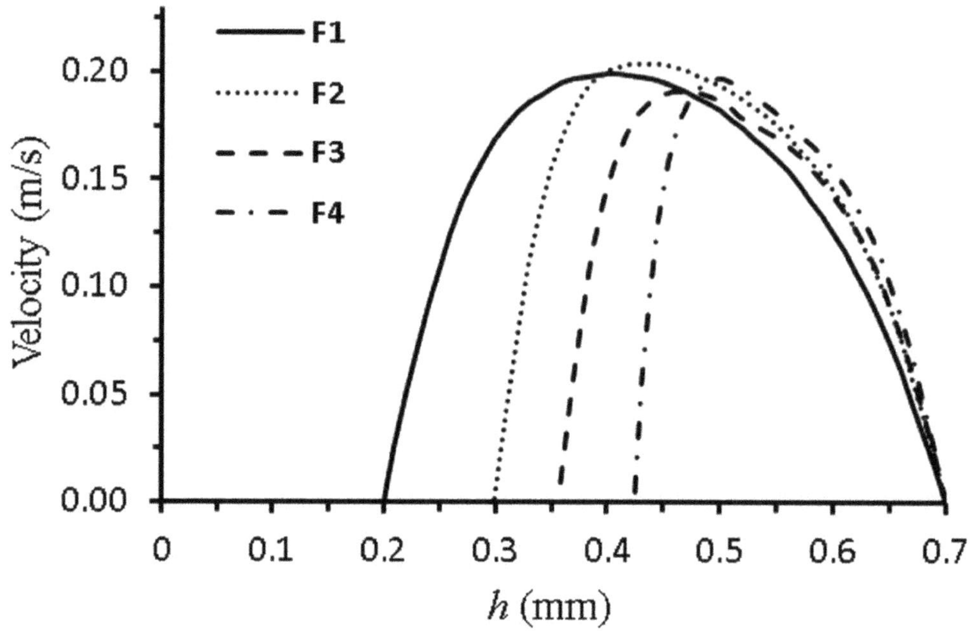

Figure 4. Inlet velocity profiles for *F*1, *F*2, *F*3, and *F*4, with $L = 4$ mm and $Re = 100$.

3.2. Velocity field

Figure 5 illustrates the velocity field for $Re = 100$ and $L = 4$ mm when *F*1 and *F*4 are used for both impermeable and semipermeable cases. A comparison between impermeable and semipermeable cases shows that the maximum velocity decreases by 0.2 and 3.1% for *F*1 and *F*4, respectively. This difference can be explained by the smaller useful area of permeable wall for *F*1 spacer, see Ref. [21]. Note that the average velocity in all cell is higher for *F*1 than for *F*4 because the inlet area is bigger for *F*1 spacer, however the average inlet velocity is the same for the two spacers.

From **Figure 5**, it can be seen that the well-known velocity behavior, that is, for each instant, the maximum velocity inside the membrane is located in the zones between the filament and the adjacent wall, for example Refs. [17, 23, 24]. This behavior is explained by the channel narrowing due to the spacers, with the consequent velocity increase to ensure the mass conservation. From general analysis of **Figure 5**, we can see that the maximum of velocity occurs at T/8. From this instant, velocity decreases until reaches approximately half of the period (T/2) and after that remains almost constant until the start of a new cycle. A more detailed analysis reveals that the permeability has a very little impact in the velocity. For instance, at 5T/8, velocity achieves the minimum value of 0.213 m/s for *F*1 to $v_p = 0$ mm/s while for $v_p = 0.1$ mm/s is 0.212 m/s. For *F*4, this value is 0.214 m/s for $v_p = 0$ mm/s and 0.208 m/s for $v_p = 0.1$ mm/s. **Figure 6** shows the velocity field for *F*1, $L = 10$ mm and impermeable membrane.

With increasing distance between the spacers, it is verified that for each instant, the average velocity within the cell decreases with increases of *L*, this observation are in line with previous studies (e.g. Ref. [17]). On the other hand, as in previous cases, the maximum velocity occurs at

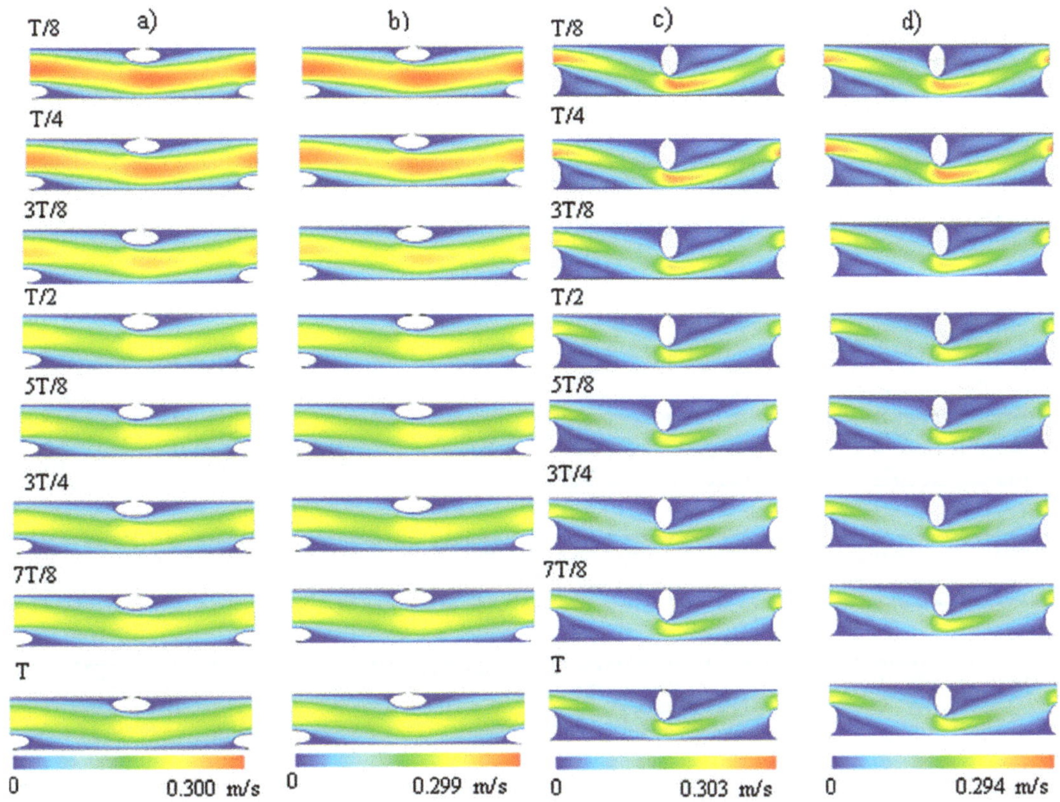

Figure 5. Velocity field during a period at t = T/8, T/4, 3T/8, T/2, 5T/8, 3T/4, 7T/8, and T, for L= 4 mm: (a) $F1$ and v_p = 0 mm/s; (b) $F1$ and v_p = 0.1 mm/s; (c) $F4$ and v_p = 0 mm/s; (d) $F4$ and v_p = 0.1 mm/s.

the t = T/8 and its value is 0.297 m/s. For both L = 10 mm and L = 4 mm, from T/2, looking only at the **Figures 5** and **6**, it appears that from that instant, the velocity remains constant. A more detailed analysis indicates that there are fluctuations, although small, and for t = 5T/8 is when the velocity takes the lowest value, equal to 0.213 m/s.

3.3. Average velocity inside the cell

Figure 7 shows the average velocity inside the cell for both cases, impermeable and semipermeable membrane, for a cell with length L = 4 mm, for all instants and spacers analyzed. The results presented are in concordance with **Figure 5**. The average velocity decreases from $F1$ to $F4$. It is also noteworthy that the highest average velocity occurs at instant T/8 undergoing a decrement up to the minimum value of average velocity at 5T/8. After 5T/8, the velocity remains almost constant but with small fluctuations in concordance with inlet velocity (**Figures 2**, **5**, and **6**).

A comparison between average velocities for impermeable and semipermeable cases also shows that the biggest difference is of 4.49% and occurs at 3T/8, for $F4$. The smallest difference between the two cases is verified for $F2$ at t = T/2, with a velocity decrease of 0.40%. On the other hand, for L = 10 mm, the corresponding biggest difference is of 4.13% for $F4$ at T/4 and smallest difference is of 1.71% for $F1$ at 7T/4, see **Figure 8**.

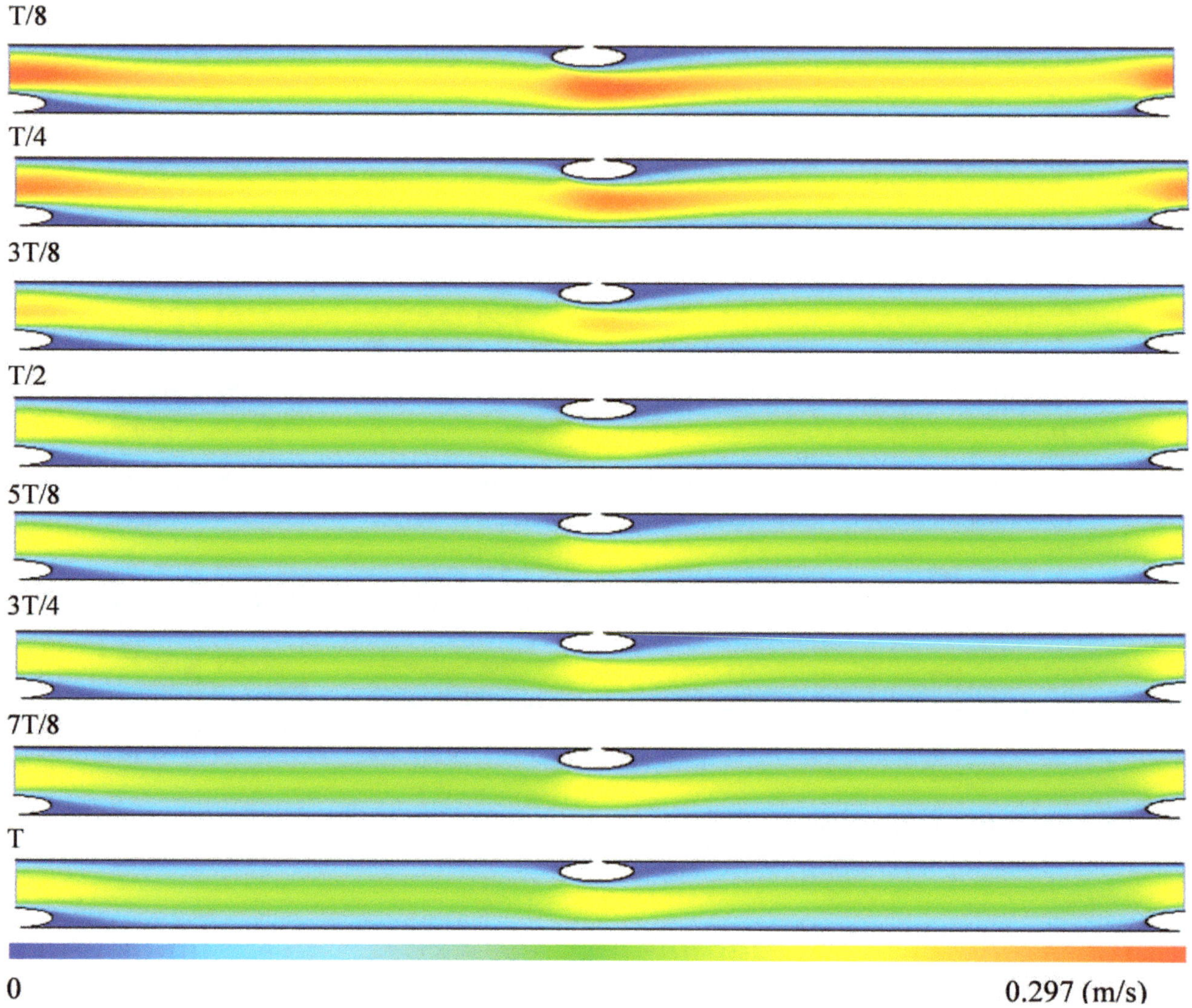

Figure 6. Velocity field during a period at t = T/8, T/4, 3T/8, T/2, 5T/8, 3T/4, 7T/8, and T, for L = 10 mm, $F1$ and v_p = 0 mm/s.

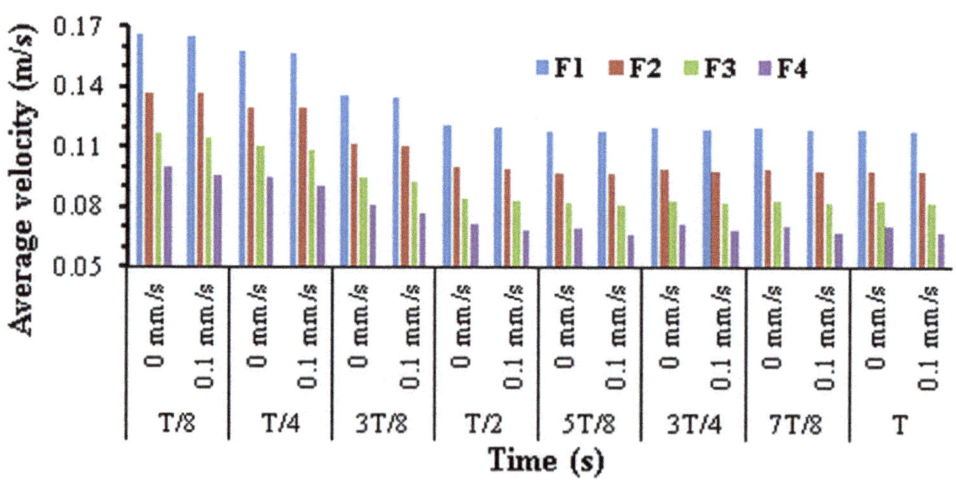

Figure 7. Average velocity during a period for L = 4 mm with v_p = 0 and 0.1 mm/s, for all spacers.

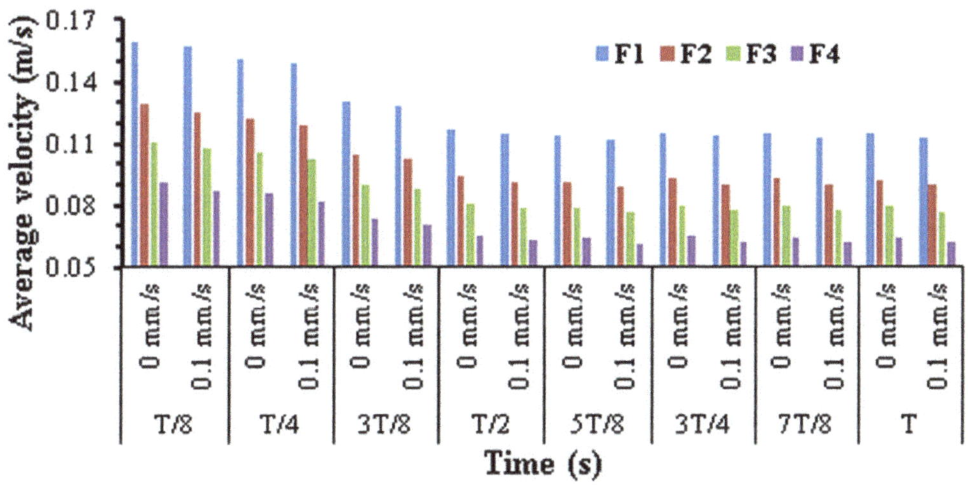

Figure 8. Average velocity during a period for $L = 10$ mm with $v_p = 0$ and 0.1 mm/s, for all spacers.

3.4. Streamlines

Figures 9 and **10** show the streamlines for $L = 4$ mm and the four types of spacers. Since the streamlines' differences for impermeable and semipermeable cases are insignificant, only the results for impermeable case are shown.

The recirculation zones are visible in all instants analyzed for all studied cell geometries. These zones are generally located downstream of the filaments, however, under specific conditions, small upstream recirculation zones can occur as shown in **Figure 9 (a)** at $t = T/8$ and **Figure 9 (d)** at $t = T/8$, T/4, and 5T/8. At $t = T/8$, the eddies occupy the largest area, since this instant is when the average velocity have the maximum value (**Figures 7** and **9**). From T/8 until T/2, the flow is in deceleration; consequently, the velocity decreases and the recirculation zones are reduced. After $t = T/2$, the velocity suffers small oscillations. These oscillations are responsible for the oscillations of the eddies length. This behavior is most easily seen in **Figure 9 (a)**. When $t = T$, it starts a new cycle and the fluid enters into a new stage of acceleration, leading to an increase in velocity which in turn causes a new increase in the recirculation zones. That behavior was observed in all geometrical configurations represented in **Figures 9** and **10**. Comparing the behavior of the different spacers, it should be noted that there is an increase in the area of eddies from *F*1 to *F*4. This increase can be explained by the shape of the spacer filaments. Note that *F*4 is the filament with biggest perpendicular area to the flow. In the present study, the distance between filaments is not sufficiently small to inhibit the growth of eddies. **Figure 10** shows the effects of the cell length, *L*, on the eddy length for the *F*4 configuration and maximum average velocity.

We can infer that the length between filaments has no influence on the eddies formation. Therefore, for all conditions studied, *L* does not affect the length of the recirculation zones.

The variation of the length eddies during a pulsatile flow modeled by a heartbeat shape combined with spacer arrangement may lead to better performance of the desalination membrane. Thus, the formation of desirable flow patterns within the membrane can lead to its better performance, which are in line with literature, for example, Saeed et al. [16].

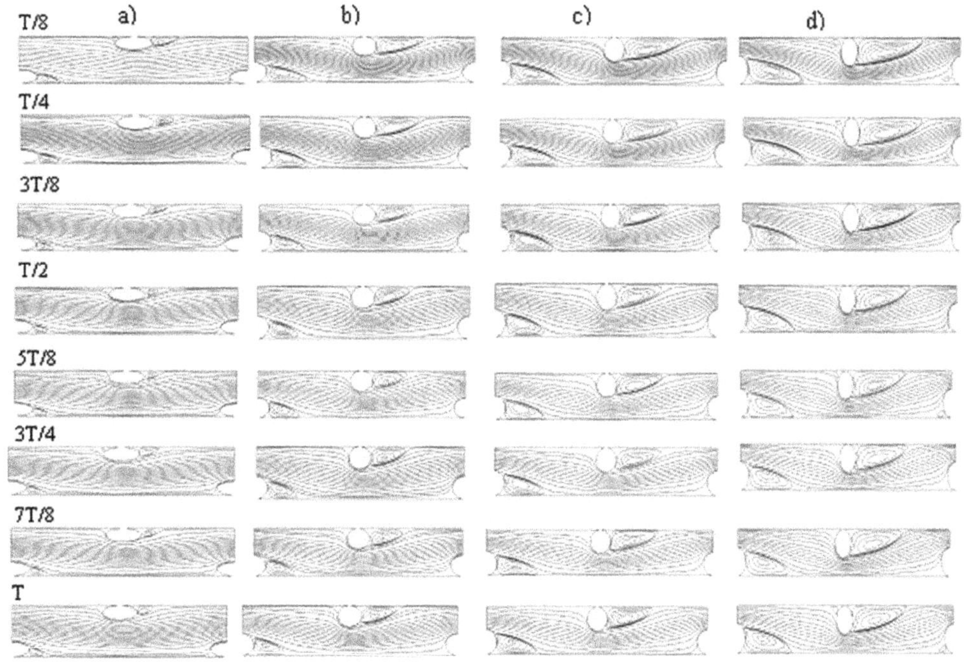

Figure 9. Streamlines during a period T at t = T/8, T/4, 3T/8, T/2, 5T/8, 3T/4, 7T/8 and T for v_p = 0 mm/s and L = 4 mm: (a) F1; (b) F2; (c) F3 and (d) F4.

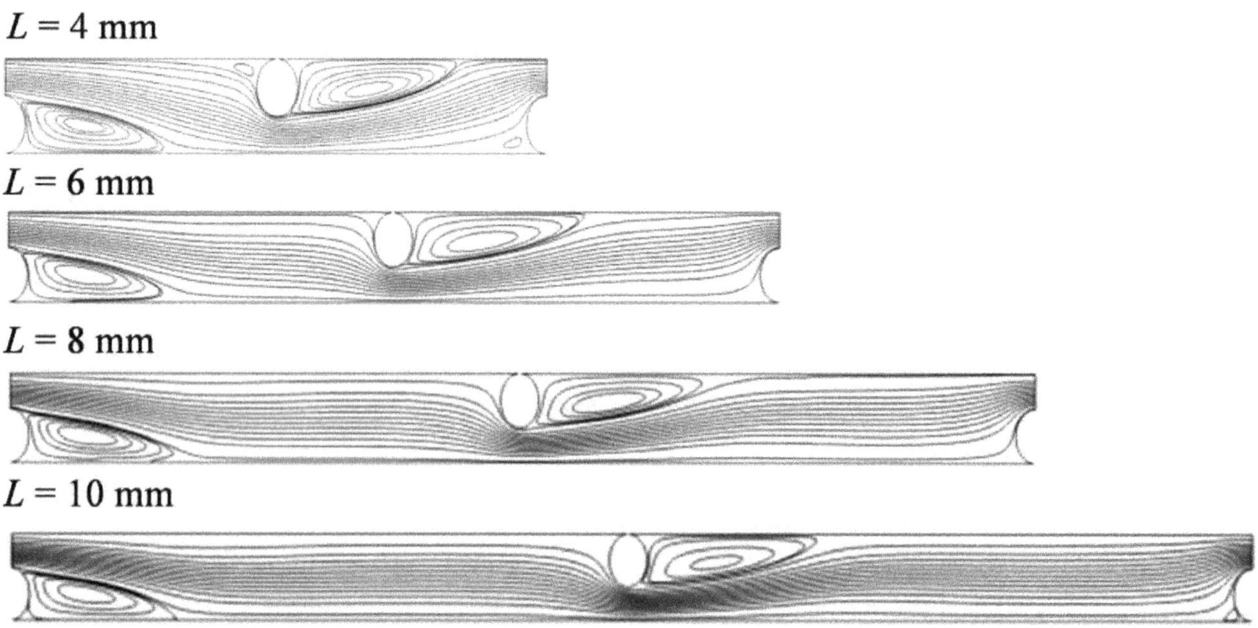

Figure 10. Streamlines at t = T/8 for F4, v_p = 0 mm/s and all values of L.

3.5. Pressure drop per length unit

Figure 11 shows the pressure drop per unit of length ($\Delta p/L$) in a cell membrane of desalination to the both cases, impermeable and semipermeable (v_p = 0.1 mm/s) for all spacers and L = 4 mm.

The maximum value of $\Delta p/L$ occurs at T/8 when the average velocity has its higher value. From there, the flow goes to a deceleration phase, and thus, there is a decrease in $\Delta p/L$ until t = 5T/8. As already mentioned, from that moment, there are small variations in the velocity, which lead to the fluctuation in $\Delta p/L$ until t = T, when starts a new cycle which will lead to a further increase in $\Delta p/L$. However, $\Delta p/L$ has a reverse behavior to that observed for velocity, i.e., $\Delta p/L$ increases from *F*1 to *F*4. This increase can be explained by the geometric shape of the spacer filaments, that is, *F*4 is the spacer with bigger transverse area to the flow and consequently have the geometry with higher resistance to the flow, for that reason is needed a bigger spent of energy to overcome the resistance forces, due to the higher value of $\Delta p/L$.

On the other hand, v_p appears to have a small influence in $\Delta p/L$. The higher effect is verified for *F*4, when t = T/4, with a difference between the impermeable and semipermeable cases of 7.24%. The smallest influence is verified for *F*2, when t = 5T/8, with a difference of 2.45% between the two cases under study.

Figure 12 shows the pressure drop per unit of length ($\Delta p/L$) in a cell of the desalination membrane under impermeable and semipermeable (v_p = 0.1 mm/s) conditions for all spacers. From **Figure 12**, it is possible to infer that with the increase of the distance between the

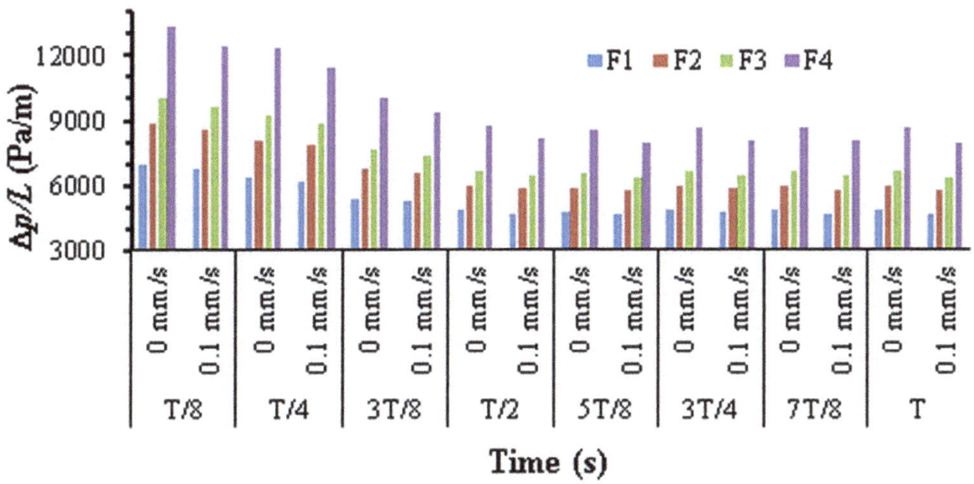

Figure 11. Pressure drop per unit length for L = 4 mm with v_p = 0 and 0.1 mm/s, for all spacers.

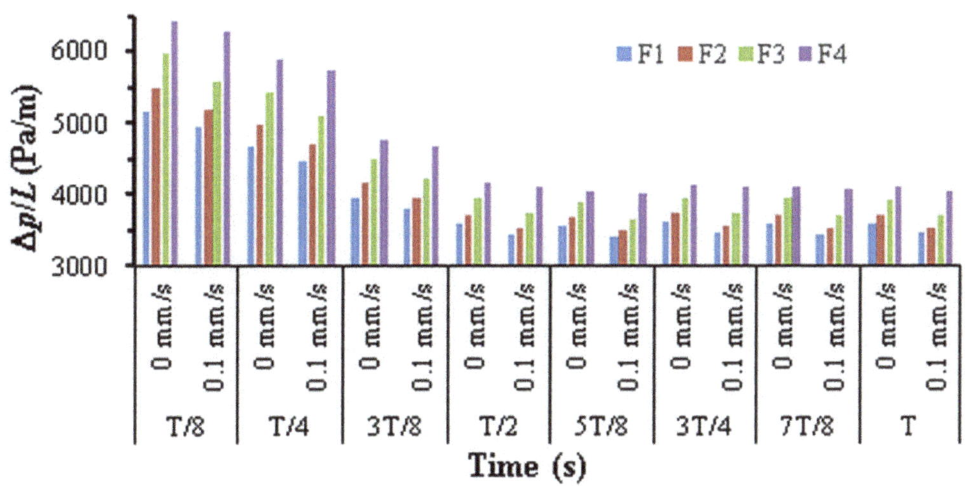

Figure 12. Pressure drop per unit length for L = 10 mm with v_p = 0 and 0.1 mm/s, for all spacers.

filaments there is a decrease in $\Delta p/L$. This phenomenon can be explained by the increased distance between filaments, that is, as the spacers are further apart, these offer a smaller resistance to the flow which is reflected in a smaller value of $\Delta p/L$ (see e.g. Ref. [17]).

The behaviors and conclusions seen in **Figure 11**, on the pressure drop per unit length, are also observed for this case, where $L = 10$ mm and for the other values of L not shown here. The main distinction between **Figures 11** and **12**, which is worth noting, is the difference in the values of $\Delta p/L$ for the different filaments. For instance for T/8 and L = 4 mm, the difference of $\Delta p/L$ between $F1$ and $F4$ is about five times higher than that for L = 10 mm, under the same conditions.

The results also show that v_p have a small influence on $\Delta p/L$. The differences observed are due to the shape of the filaments. The higher relative difference is verified for $F2$, when t = T/2, with a difference between the impermeable and semipermeable cases of 4.95%. The lowest relative difference is verified for $F4$, at t = 5T/8, with a difference of 1.01% between the two cases under study.

In general, an energy gain is expected when an inlet heartbeat function is used, compared to sinusoidal functions because in heartbeat function the flow keeps small fluctuations, at least, during half cycle, so in this case the pressure gradients developed during a cycle are smaller than that for a sinusoidal function.

3.6. Wall shear stresses per unit length

Figure 13 presents, for L = 4 mm, the average wall shear stress per unit length (WSS/L) in the upper wall (impermeable wall) of a desalination membrane for impermeable (v_p = 0 mm/s) and semipermeable cases (v_p = 0.1 mm/s). From **Figure 13**, one can see that for v_p = 0.1 mm/s, the values of WSS/L decrease from $F1$ to $F4$, at all instants analyzed. For the impermeable case,

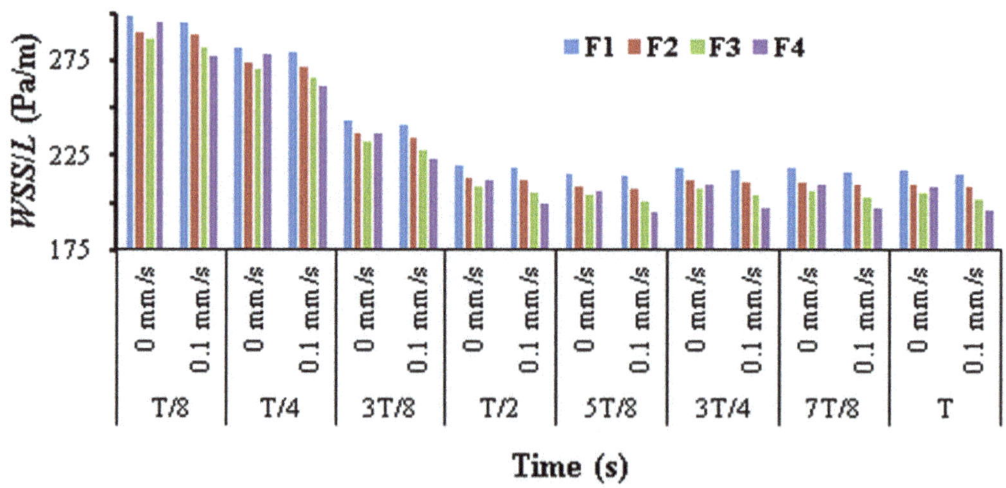

Figure 13. Average wall shear stress per unit length for L = 4 mm with v_p = 0 and 0.1 mm/s, for all spacers.

WSS/L only decreases from *F*1 to *F*3. The maximum values are verified at t = T/8, when the average velocity reaches its maximum values (**Figure 5**). From that instant *WSS/L* decreases in concordance with the average velocity. From T/2, the wall shear stress also presents small fluctuations.

For studied conditions, v_p appears to have small influence in *WSS/L*. The higher effect is verified for *F*4, at t = T/2, with a difference between the impermeable and semipermeable cases of 5.59%. The smaller influence occurs for *F*2, at t = T/2, with a difference between the two cases of 0.57%.

Figure 14 presents, for L = 10 mm, the average wall shear stress per unit length in the upper wall (impermeable wall) of a desalination membrane to the impermeable and semipermeable cases for conditions studied. In general, comparing **Figures 13** and **14**, the behavior of *WSS/L* is similar, with the exception of that for cell length (L = 10 mm), both impermeable and semipermeable cases show the same behavior of *WSS/L* with the type of spacers filaments, that is, for all instant, analyzed *WSS/L* decreases from *F*1 to *F*4.

The *WSS/L* for L = 10 mm also shows that permeability causes a decrease in *WSS/L*. The higher effect is verified for *F*4, at t = T/8, with a difference between the impermeable and semipermeable cases of 4.80%. The lesser influence is verified for *F*1, at t = 5T/8, with a difference between the two cases of 2.30%.

Comparing the results obtained by Cao et. al. [14] and Sousa et al. [17] with the present results, one can infer that the use of inlet flux with heartbeat profiles can benefit the membrane performance because the concentration polarization may be reduced, improving the mass transfer, once it dynamically changes the boundary layer concentrations and prevents clogging of the membrane surface.

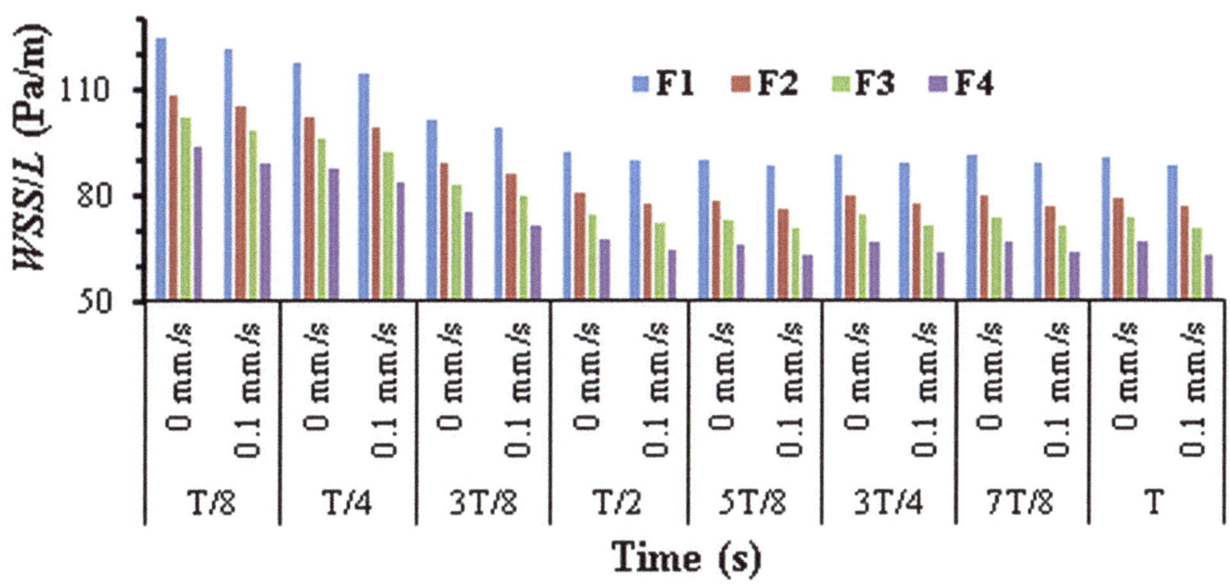

Figure 14. Average wall shear stress per unit length for L = 10 mm with v_p = 0 and 0.1 mm/s, for all spacers.

4. Conclusions

In this chapter, CFD techniques were used to study the hydrodynamics inside the feeding channel of a semipermeable membrane in spiral wound configuration. The main goal of this research was to investigate the effects of a pulsating flow with a profile of a heartbeat on the hydrodynamics of feed channels of a desalination membrane filled with spacers in zig-zag arrangements and transverse to the flow. Both membrane impermeable and semipermeable cases were analyzed. For the semipermeable case, we considered the membrane bottom wall with a typical permeation rate of 0.1 mm/s. The comparison between the two cases shows that the permeability has no significant influence on hydrodynamics. On the other hand, the permeability implies that there is a slight decrease in average velocity, Δp and *WSS*.

Due to the existence of the filaments in the cell, the velocity profiles are not fully developed. Thus, depending on spacer type (*F*1, *F*2, *F*3, or *F*4) and cell length, there is a different characteristic velocity profile. Analyzing the obtained velocity fields is possible to confirm that the maximum velocity occurs in the membrane region between the filament and the opposite wall. For *F*3 and *F*4, the shortest distance between consecutive filaments (L = 4 mm) provides eddies that sweep practically the entire membrane wall downstream of the filament, at t = T/8. The imposed eddies fluctuation by the pulsatile inflow can promote a reduction in the concentration polarization effect and reduce the concentration of solute at the membrane surface, which consequently increases the mass transfer. The reduction in the distance between the filaments increases *WSS/L*. The fluctuation in *WSS/L* can dynamically change the boundary layer concentrations and prevent clogging of the membrane surface. On the other hand, decreasing the distance between the filaments increases $\Delta p/L$ and this should increase the energy consumption. Thus, a combination of the distance between filaments, pulsatile inflow and $\Delta p/L$ is important to optimize separation process.

In summary, this study suggests that the inter-filaments length combined with the flow variation characteristic of a heartbeat can control the development of concentration polarization, and thus reduce the probability of fouling and energy consumption in the process, through an optimization of the recirculation zones.

Nomenclature

F	Filament
h	Cell height (m)
h_f	Filament height (m)
g	Gravitational acceleration (m s^{-2})
L	Cell length (m)
p	Static pressure (Pa)
Δp	Pressure drop (Pa)
Re	Reynolds number
$\overline{v}$	Average inlet velocity (m s^{-1})
v_p	Permeate velocity (m s^{-1})
$v_h(t)$	Normalized inlet velocity
v_x	Velocity component in x-direction (m s^{-1})
v_y	Velocity component in y-direction (m s^{-1})

x	Longitudinal rectangular coordinate (m)
y	Transverse rectangular coordinate (m)
Greek letters	
ρ	Fluid density (kg m^{-3})
μ	Dynamic viscosity (Pa s)

Acknowledgements

This work was supported by the Portuguese Foundation for Science and Technology under UID/SEM/04252/2013

Author details

Armando A. Soares[1,2], João Silva[1,2], Eliseu Monteiro[2,3] and Abel Rouboa[1,2,4]*

*Address all correspondence to: rouboa@utad.pt

1 UTAD, University of Trás-os-Montes and Alto Douro, Vila Real, Portugal

2 CIENER-INEGI/University of Porto, Porto, Portugal

3 C3i - Interdisciplinary Center for Research and Innovation, Polytechnic Institute of Portalegre, Portalegre, Portugal

4 Mechanical Engineering and Mechanics Department, University of Pennsylvania, Philadelphia, Pennsylvania, USA

References

[1] Tsiourtis NX. Desalination and the environment. Desalination. 2001;**141**(3):223–236. DOI: 10.1016/s0011-9164(01)85001-3

[2] Baalousha H. Desalination status in the Gaza Strip and its environmental impact. Desalination. 2006;**196**(1-3):1–12. DOI: 10.1016/j.desal.2005.12.009

[3] Marcovecchio MG, Mussati SF, Aguirre PA, Nicolás JS. Optimization of hybrid desalination processes including multi stage flash and reverse osmosis systems. Desalination. 2005;**182**(1):111–122. DOI: 10.1016/j.desal.2005.03.011

[4] Fritzmann C, Löwenberg J, Wintgens T, Melin T. State-of-the-art of reverse osmosis desalination. Desalination. 2007;**216**(1-3):1–76. DOI: 10.1016/j.desal.2006.12.009

[5] Bódalo-Santoyo A, Gomez-Carrasco JL, Gomez-Gomez E, Máximo-Martín MF, Hidalgo-Montesinos AM. Spiral-wound membrane reverse osmosis and the treatment of industrial effluents. Desalination. 2004;**160**(2):151–158. DOI: 10.1016/S0011-9164(04)90005-7

[6] Wangnick K. IDA Worldwide Desalting Plants Inventory. Report No. 17 ed. Gnarrenburg, Germany: Wangnick Consulting GMBH; 2002

[7] Shakaib M. Pressure and concentration gradients in membrane feed channels: Numerical and experimental investigations [thesis]. Karachi, Pakistan: University of Engineering and Technology; 2008. p. 271. Available from: http://prr.hec.gov.pk/Thesis/353S.pdf

[8] Geraldes V, Semião V, Pinho MN. Flow management in nanofiltration spiral wound modules with ladder-type spacers. Journal of Membrane Science. 2002;**203**(1.2):87–102. DOI: 10.1016/s0376-7388(01)00753-0

[9] Schwinge J, Neal PR, Wiley DE, Fletcher DF, Fane AG. Spiral wound modules and spacers: Review and analysis. Journal of Membrane Science. 2004;**242**(1-2):129–153. DOI: 10.1016/j.memsci.2003.09.031

[10] Baker RW. Membrane Technology and Applications. 2nd ed. Menlo Park, CA: John Wiley & Sons, Ltd; 2004. p. 538. DOI: 10.1002/0470020393

[11] Kim S, Hoek EMV. Modeling concentration polarization in reverse osmosis processes. Desalination. 2005;**186**(1-3):111–128. DOI: 10.1016/j.desal.2005.05.017

[12] Zimmerer CC, Kottke V. Effects of spacer geometry on pressure drop,mass transfer, mixing behavior, and residence time distribution. Desalination. 1996;**104**(1-2):129–134. DOI: 10.1016/0011-9164(96)00035-5

[13] Ahmad AL, Lau KK. Impact of different spacer filaments geometries on 2D unsteady hydrodynamics and concentration polarization in spiral wound membrane. Journal of Membrane Science. 2006;**286**(1-2):77–92. DOI: 10.1016/j.memsci.2005.06.056

[14] Cao Z, Wiley DE, Fane AG. CFD simulations of net-type turbulence promoters in a narrow channel. Journal of Membrane Science. 2001;**185**(2):157–176. DOI: 10.1016/S0376-7388(00)00643-8

[15] Schwinge J, Wiley DE, Fletcher DF. Simulation of the flow around spacer filaments between channel walls. 1. Hydrodynamics. Industrial & Engineering Chemistry Research. 2002;**41**(12):2977–2987. DOI: 10.1021/ie010588y

[16] Saeed A, Vuthaluru R, Yang Y, Vuthaluru HB. Effect of feed spacer arrangement on flow dynamics through spacer filled membranes. Desalination. 2012;**285**:163–169. DOI: 10.1016/j.desal.2011.09.050

[17] Sousa P, Soares A, Monteiro E, Rouboa A. A CFD study of the hydrodynamics in a desalination membrane filled with spacers. Desalination. 2014;**349**:22–30. DOI: 10.1016/j.desal.2014.06.019

[18] Amokrane M, Sadaouia D, Dudeckb M, Koutsouc CP. New spacer designs for the performance improvement of the zigzag spacer configuration in spiral-wound membrane modules. Desalination and Water Treatment. 2015;**57**(12):5266–5274. DOI: 10.1080/19443994.2015.1022003

[19] Al-Bastaki N, Abbas A. Use of fluid instabilities to enhance membrane performance: a review. Desalination. 2001;**136**(1-3):255–262. DOI: 10.1016/S0011-9164(01)00188-6

[20] Lau KK, Bakar AMZ, Ahmad AL, Murugesan T. Feed spacer mesh angle: 3D modeling, simulation and optimization based on unsteady hydrodynamic in spiral wound membrane channel. Journal of Membrane Science. 2009;**343**(1):16–33. DOI: 10.1016/j.memsci.2009.07.001

[21] Silva JC, Soares AA, Rouboa A. A numerical study of the hydrodynamics in feed channels of spiral-wound membrane modules. MEFTE. 2014; 227–232. In: APMTAC, editor. MEFTE 2014—V Conferência Nacional de Mecânica dos Fluidos, Termodinâmica e Energia; 11-12 Set 2014; Porto, Portugal. Porto: FEUP; 2014. p. 227–232

[22] Subramani A, Kim S, Hoek E. Pressure, flow, and concentration profiles in open and spacer-filled membrane channels. Journal of Membrane Science. 2006;**277**(1):7–17. DOI: 10.1016/j.memsci.2005.10.021

[23] Fimbres-Weihs GA, Wiley DE. Review of 3D CFD modeling of flow and mass transfer in narrow spacer-filled channels in membrane modules. Chemical Engineering and Processing: Process Intensification. 2010;**49**(7):759–781. DOI: 10.1016/j.cep.2010.01.007

[24] Radu AI, Bergwerff L, Loosdrecht MCM, Picioreanu C. A two-dimensional mechanistic model for scaling in spiral wound membrane systems. Chemical Engineering Journal. 2014;**241**(1):77–91. DOI: 10.1016/j.cej.2013.12.021

10

Marmara Seawater Desalination by Membrane Distillation: Direct Consumption Assessment of Produced Drinking Water

Coskun Aydiner, Derya Y. Koseoglu Imer,
Salim Oncel, Esra Can Dogan, Ali Oguzhan Narci,
Serif Cakmak, Tugba Nur Yilmaz,
Emin Ender Celebi and Yasemin Melek Tilki

Abstract

Drinking water was produced from Marmara seawater by membrane distillation (MD). The best operating conditions were determined by batch experiments as: 0.45 μm PTFE, 30°C distillate temperature and temperature difference, and 270–360 L/h cross-flow rates in feed-distillate. Seawater desalination was carried out with 99.93% solute rejection and 17.2 L/m^{2}h permeate flux in 66% concentration ratio by lab-scale pilot system. Since the desalinated water contained no organic carbon, turbidity, and nitrate, it seemed to be very suitable for immediate service with quality of 7.3 pH, clear, odor-free, 76.0 μS/cm, 47.1 mg TDS/L, <0.001 color, and 0.01 mg boron/L. The product water lacked of vital cations, especially Na^+, K^+, Ca^{2+}, Mg^{2+} that are essentials for promoting osmotic balanced body liquid and healthy development. A holistic management approach towards satisfying specific water quality requirements in direct service of MD effluents to human consumption was proposed that jointly included in injecting into urban potable water, adding appropriate chemicals into the effluent, and mixing effluents with raw or concentrated seawater (1:250/1:1000 for Marmara seawater) or brackish natural waters under hygienic precautions.

Keywords: seawater desalination, drinking water production, membrane distillation, product water quality, human consumption suitability

1. Introduction

The data of United Nations indicate that the rate of increase in water usage has been higher than twice the rate of increase in population over the past century. It is estimated that there will be 1.8 billion people living in areas with scarce water resources by 2025, while two-thirds of the global population reside in regions of water stress due to use, growth, and climate change [1]. Potable water production has also become a worldwide concern, for many communities, increasing of industrial processes, population growth, climate change, over exploitation of ground water and nearby river systems as well as demand exceed of conventional available water resources [2–7]. For firmly preservation and sustaining stable development of life on earth at present situation, there needs to use plentiful salty water to produce freshwater supplies capable of meeting the increasing demand [7, 8]. Therefore, countries are progressively turning to desalination technologies as a solution to obtain direct potable water from seawater [4, 6, 9].

Commercial desalination technologies are mainly categorized under two main categories as pressure-driven membrane separation by reverse osmosis (RO) and thermal distillation by multi-stage flash (MSF) and multi-effect distillation (MED) [4, 10]. In spite of their growing popularity and improved technological applications in seawater desalination, available plants have potential negative impacts on coastal environments and marine ecosystems, which can be characterized with concentrate and chemical discharges to the sea, limiting effects on land use, salting of groundwater, noise pollution, air pollutants discharges to the atmosphere and high energy consumptions [11–17]. Total capacity of seawater desalination plants in the world about 50% is operated by RO-based membrane technology [4]. The decline in RO performance resulting from unavoidable membrane fouling, which needs expensive pretreatment, higher operation pressures, and frequent cleaning with chemicals which give harm to membranes, impairs the quality of permeate and accelerate membrane replacement, which pushes up the cost of water treatment and energy consumption [6, 18].

Membrane distillation (MD) could be an alternative to RO desalination process to overcome its negative impacts. MD is not highly affected by salt concentration in saline feed solutions, and hence, this technology can achieve good quality distillate with minimal brine discharge [19, 20]. MD, a thermal integrated membrane process, incorporates transporting vapor through microporous hydrophobic membranes, and its operation is based on the principle of vapor-liquid equilibrium as a basis for molecular separation. This process uses a gradient of temperature between the two sides of a porous membrane in order to establish a difference in vapor pressure that actually drives process [5, 19, 21–25]. Its four different configurations that consist of the type of the condensing design have been proposed, but direct contact membrane distillation (DCMD) is the most widely preferred technology because the step of condensation is performed within the membrane module, which brings about a simple mode of operation with no need of external condensers [19, 22, 24]. Additionally, more convenient membranes in MD process became available. Several materials such as polytetrafluoroethylene (PTFE), polypropylene (PP), polyethylene (PE), and polyvinylidene fluoride (PVDF) are used for producing hydrophobic MD membrane for desalination uses [24, 25]. The different commercial membranes have been used in several recent pilot plant studies, but there are still some problems that have not yet been clearly solved to use MD process under real scale [19].

In practice, the use of MD for desalination is still largely limited to pilot-scale studies. The commercial launch of MD desalination technology will first require certain technical issues to be resolved, such as its high energy consumption and the wetting of membrane pores [26]. Feng et al. [27] used different nanofiber microfiltration membranes produced from PVDF material to obtain drinking water through the air gap MD process. This was the first time that an electrospun nanofiber membrane was used in MD. At temperature differences ranging between 25 and 83°C, they obtained flux values between 5 and 28 kg/m^2 h. They also reported that this new approach could potentially compete with other conventional desalination systems. Fard et al. [28] investigated bench-scale performance of DCMD process using flat sheet PTFE membrane under various conditions of inlet flow rates, temperatures, and salinity composition. A permeate flux of 35.6 L/m^2 h could be generated at a different temperature of 50°C between hot and cold stream sides. The rate of salt rejection during the conducted tests was really high with 99.9% and virtually independent of any operational parameters studied. It has been seen that DCMD is a viable and effective technology, which can produce high quality distillate in a consistent way from a very high salinity feed even with dramatic differences in quality compared to other methods of desalination like RO and MSF. Bouguecha et al. [29] attempted to run a DCMD pilot plant powered by solar energy using collectors plus PV panels under actual weather conditions in Jeddah, KSA, throughout two selected sunny days. It had been aimed to assess how the operating parameters affected the process performance in which the trans-membrane temperature difference (ΔT) and the fluid mass flow rates (hot and cold) constituted the most remarkable operating parameters. A maximum permeate flux of 8.87 L/m^2 h was achieved at a ΔT of 60.5°C. According to their findings, it seems that the DCMD is a promising solution for the desalination of brackish water and also for seawater, particularly in distant places and/or whenever affordable low temperature sources are accessible.

In this paper, six different commercially available membranes were used to investigate the effects of different operating conditions including cross-flow rate, membrane type and pore size, solution temperatures, and membrane trans-temperature differences on dissolved ions rejections and permeate flux of DCMD process. Some characteristics of membranes such as roughness and wettability were additionally tested to more comprehensively understand the performances. At suitable operation conditions determined based on batch experimental runs, the process was operated along a 30-h period of time in which raw seawater was concentrated at approximately 70% to examine the direct usability of the MD output water as drinking water. The effluent quality established for direct supply to human consumptions was evaluated in terms of appropriateness to maximum allowable concentrations in the national standard and international guidelines, and a general framework for pragmatic solutions toward practical water quality management applications was proposed for facilitation to serve the MD desalination effluents to the human drinking directly.

2. Materials and methods

2.1. Seawater characterization

Raw seawater used as feed stream in all membrane distillation experiments was collected from Muallimköy Coast of Marmara Sea beside the coastal city of İzmit/Kocaeli in Turkey. It was taken from under one meter of seawater level and then pre-filtered through roughing filtration to remove large particles. The detailed characteristics of the seawater are given in **Table 1**.

Parameters[a]	Raw seawater		
	I[b]	II[b]	Average
T	24.1	24.2	24.15 ± 0.05
pH	8.69	8.62	8.66 ± 0.04
Ec	40,400	41,500	40,950 ± 550
TDS	25, 250	25, 772	25,511 ± 261
TOC	4.3	5.1	4.7 ± 0.4
DOC	4.1	4.9	4.5 ± 0.4
UVA_{254}	0.050	0.040	0.045 ± 0.005
SUVA	1.220	0.816	1.018 ± 0.202
Na^+	8880	8940	8910 ± 30
K^+	443	422	433 ± 11
NH_4^+	<0.1	<0.1	<0.1
Ba^{2+}	0.030	0.029	0.0295 ± 0.0005
Ca^{2+}	395	404	400 ± 5
Mg^{2+}	696	692	694 ± 2
Mn^{2+}	0.018	0.019	0.0185 ± 0.0005
Sr^{2+}	5.50	5.70	5.60 ± 0.10
B	2.95	3.05	3.00 ± 0.05
Si	0.89	0.91	0.90 ± 0.01
Fe	0.007	0.007	0.007 ± 0.000
Cl^-	13,996	14,481	14,239 ± 243
HCO_3^-	300	303	302 ± 2
NO_3^-	0.5	0.5	0.5 ± 0.0
CO_3^{2-}	0.0	30	15.0 ± 15.0
SO_4^{2-}	2314	2440	2377 ± 63
TKN	0.6	0.7	0.65 ± 0.05
Total nitrogen	1.1	1.2	1.15 ± 0.05
Alkalinity	300	333	317 ± 17
Total hardness	4986	5236	5111 ± 125
Color			
436 nm	0.0300	0.0300	0.0300 ± 0.0000
525 nm	0.0020	0.0100	0.0060 ± 0.0040
620 nm	0.0030	0.0020	0.0025 ± 0.0005

[a]Units of all parameters are mg/L except for temperature (°C), electrical conductivity (μS/cm), UVA_{254} (1/cm), SUVA (L/mg.m) and color (1/cm).

[b]Seawaters I and II are raw seawater samples that were used in batch and continuous MD experiments, respectively.

Table 1. Characteristics of Marmara seawater.

2.2. MD membranes

Six different types of flat sheet hydrophobic microfiltration membrane with pore sizes of 0.2, 0.45, and 1.0 μm that were made of PTFE and PVDF materials (Membrane Solutions Inc., China) were used in the experiments. The characteristics and intrinsic performances of the membranes are presented in **Table 2**. The membrane thicknesses were provided by the manufacturer. The results of contact angle measurement below 90° indicate that the PVDF membranes demonstrated lower hydrophobicity than the PTFE membranes. Liquid entry pressure (LEP) parameter also shows that the PTFE membranes are more suitable for the MD processing of seawater due to its lower wettability and higher entry pressure values.

2.3. Experimental setup

The experimental setup of cross-flow lab-scale pilot DCMD system is schematically shown in **Figure 1**. The setup was composed of two thermostatic cycles, that is, feed and permeate, which were connected to a membrane module made by kestamid having an effective membrane area of 140 cm^2. The compartment cells of the module consisted of two machined parts compressing the rectangular MD membrane. Connected to a heating resistant, the feed flow side is kept under high temperature; on the other hand, the permeate flow side, connected to a cooling system, is maintained at a low temperature.

During the experimental investigations, rough-filtrated raw seawater from the Marmara Sea was preheated to intended temperatures and circulated through one side of the membrane (feed side), while de-ionized water was circulated through the other side of the membrane (distillate or permeate side) in a simultaneous way in a counter-current flow mode. The condensation of permeate vapor that was diffused across the membrane occurred in the cold distillate side. The temperatures of feed and permeate streams were controlled at desirable levels using a heater and chiller, respectively. The spacer placed in the feed flow side only to promote turbulence

Membranes	Material	Membrane thickness (μm)	Nominal pore size (μm)	Contact angle (±10°)	Liquid entry pressure (±2.5 kPa)	Water flux[a] (m^3/m^2h)
PTFE	Polytetrafluoro ethylene	160 ± 40	0.22	121	121.3	0.250
			0.45	126	81.1	0.333
			1.0	123	131.4	0.417
PVDF	Polyvinylidene fluoride	100 ± 10	0.22	68	58	6.0–9.0
			0.45	81	47.5	22.2–36.0
			1.0	84	22.5	–[b]

[a]The values given by manufacturer that measured at trans-membrane pressures of 0.2 and 1.0 bar for PTFE and PVDF membranes, respectively.

[b]Not available.

Table 2. Various properties of MD membranes used.

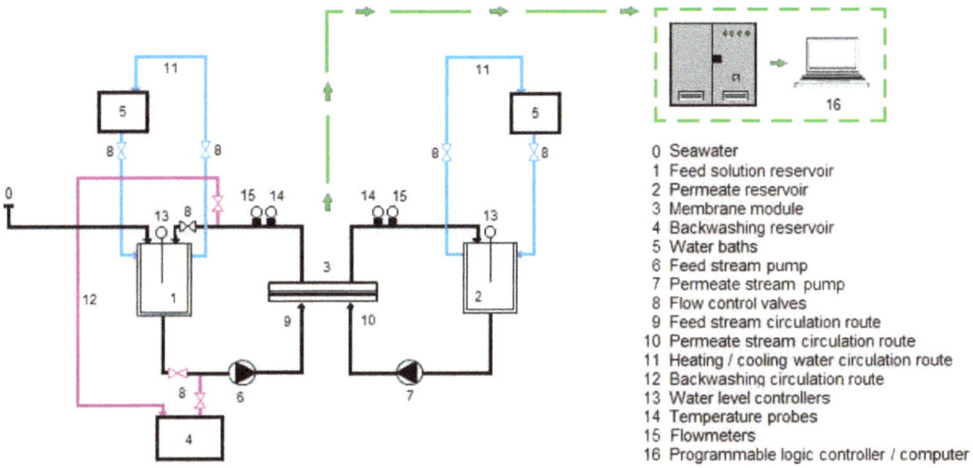

Figure 1. Schematic flow diagram of experimental MD setup.

and to support the membrane was obtained from Sterlitech Inc., USA. In the batch experiments, the storage tanks for the feed and distillate streams in the system were kept at equal volumes of five L. In the continuous experiments operated at concentration mode of the feed seawater of 12 L, the system was worked for 6 hours per day in a period of 5 consecutive days. During these experimental runs, the membrane active layers were cleaned in-place inside the module at the end of each operation day by means of using 1 L solutions in each one of the applications, which were operated by a flushing order of first applying *"1% HCl + distilled water"*, and then followed by *"1% NaOH + distilled water"* except for the first operation day.

2.4. Surface morphology of membranes

The surface morphologies of clean and fouled membranes were marked by means of atomic force microscopy (AFM) to assess the vertically distribution of the fouling on the top layer or membrane surface roughness. After each membrane sample was let dry in air, the visual observations were provided using NanoScope IV AFM system (Digital Instruments, USA) operated in contact mode. The mean roughness (R_a), the root mean square (R_{rms}) of the average height of membrane surface peaks, and the mean difference in height among the five highest peaks and the five lowest valleys (R_z) were identified in order to compare the roughness of clean and fouled membranes. Due to its more representative findings for the surface foulings, the analyses were evaluated only via the variations of the measured R_z values of the specimens. In the AFM analyses, the value of mean roughness (R_a) stands for the mean value of surface in relation to centre plane and was calculated using the following equation [30]:

$$R_a = \frac{1}{P}\sum_{i=0}^{p}\left|z_{cu} - z_{av}\right| \tag{1}$$

where z_{av} is the average of the z values in the particular area, z_{cu} is the present value of z, and p refers to the number of points in a certain area. The root mean square of z values (R_{rms}) was calculated using the equation given below [30]:

$$R_{rms} = \sqrt{\frac{\sum (z_{cu} - z_{av})^2}{p}} \tag{2}$$

The average difference in height among the five highest peaks and the five lowest valleys, R_Z, was calculated in relation to the mean plane, on which the image data have a minimum variance [30, 31]. The cross-sectional morphologies of fouled membranes were monitored by scanning electron microscopy (SEM) (Philips XL30 SFEG) to visualize the fouling formations on the membrane surface. All the SEM observations were performed at 5 kV using Au-coated membrane specimens. Energy dispersive X-ray spectrometry (EDX) analysis of the components found on the membrane surface after the fouling was also done with EDX detector of the same device. The hydrophilicity or wettability of the membrane surfaces was analyzed with contact angle (°) measurements according to sessile-drop technique using a goniometry instrument (Attension Theta Lite Optical Tensiometer) [31]. In the analyses, 2 μL of pure water in tight syringe was manually dropped on the membrane surface. By means of the software processing the measurement data obtained from the camera of the device, the results were determined as the averages of contact angles at both sides of drops fall on five arbitrary places of the membrane surfaces.

2.5. Analytical procedure

Water quality analyses in both MD streams were carried out based upon the parameters that included in temperature, pH, conductivity, turbidity, total dissolved solids TDS, total organic carbon (TOC), dissolved organic carbon (DOC), ultraviolet absorbance at 254 nm (UVA_{254}), specific ultraviolet absorbance (SUVA), total Kjeldahl nitrogen (TKN), NH_4^+, alkalinity, total hardness, Na^+, K^+, NH_4^+, Ba^{2+}, Ca^{2+}, Mg^{2+}, Mn^{2+}, Sr^{2+}, B, Si, Fe, Cl^-, HCO_3^-, NO_3^-, CO_3^{2-}, SO_4^{2-} and color at 436, 525, and 620 nm wavelengths that were measured according to "*Standard Methods for the Examination of Water and Wastewater*" [32].

Temperature, conductivity, TDS and pH were analyzed with the desktop multi-parameter with Hach HQ440 d (Hach-Lange GmBH). TOC was measured at 750°C by carbon analyzer equipped with a high pressure NDIR detector (Hach Lange IL550 TOC-TN) in which *5310 B-a high temperature catalytic oxidation method* was applied. DOC analyses were conducted on the samples filtered by Whatman syringe filter 0.45 μm using the TOC analyzer apparatus. UVA_{254} was measured by *5910 B-UV-absorbing organic constituents method,* and SUVA was calculated with values of UVA_{254} and DOC. TKN was analyzed by *4500-*N_{org}*B-macro-Kjeldahl method,* and ammonia was determined using the measurement probe of Hach HQ440 d by *4500-*NH_3*E-ammonia-selective electrode method.* Alkalinity and total hardness were determined in accordance with *2320B-titration method* and *2340C-EDTA titrimetric method.*

Na^+, K^+, Ba^{2+}, Ca^{2+}, Mg^{2+}, Mn^{2+}, B, and Fe concentrations were measured by Perkin-Elmer ELAN Optima 7000 DV with *3120-B inductively coupled plasma mass spectrometer* (ICP-MS) (Perkin-Elmer SCIEX Instruments, Canada). Si and Sr^{2+} were measured using flame atomic absorption spectrometer (Perkin Elmer 1100). Prior to the analyses, samples were filtered through a

0.45 μm pore-size membrane filter, and pH values of the samples were adjusted to <4.0 using HNO_3. $HCO_3^{2-}$and CO_3^{2-} were analyzed by *2320-B titrimetric method,* and SO_4^{2-} and Cl-were measured by *4500-*SO_4^{2-}*-E turbidity and 4500-*Cl^-*-D potentiometric methods,* respectively. NO_3^{2-} was determined by *4500-*NO_3^{2-}*-C* spectrophotometric method, while color measurements at 436, 525, and 620 nm wavelengths were conducted by *2120C* spectrophotometric method for making possible the time-dependent calculations of MD water fluxes, the measurements of osmolalities of the feed and permeate solutions were executed at definite time intervals in duplicate for each data point using the Advanced Osmometer instrument (Model 3250—Advanced Instruments Inc., USA) by the method of freezing point depression.

2.6. MD performance analysis

The technical performance of MD was analyzed using not only the permeate flux values but the rejection values of the parameters as well, which are given in national and international guidelines for water reuse in drinking water quality. The flux was calculated from the time-dependent variations between the pretest and posttest values of TDS and osmolality as an explicit indicator for solute ions transports from seawater to distillate stream. So, after TDS and osmolality at both sides were individually measured for definite time intervals, V was calculated from the differences of sequential time points according to TDS/osmolarity-based mass balances calculations. From the average values of V volume calculated based on TDS and osmolality data measured at each time point, permeate water fluxes were totally obtained for the whole time scale of relevant experiment using Eq. (3).

$$J = \frac{1}{A}\frac{dV}{dt} \tag{3}$$

where J is the permeate flux; A is the effective membrane filtration area; V is the total permeate volume; and t is the filtration time.

The rejection performances in MD process for each water quality parameter were calculated from the water quality analyses results using Eq. (4):

$$R(\%) = \left(1 - \frac{C_p}{C_f}\right) \times 100 \tag{4}$$

where R, C_p, and C_f are the rejection, concentration in the permeate, and concentration in the feed stream, respectively [33].

3. Results and discussion

3.1. The effects of operating conditions on MD performance

3.1.1. Membrane type and size

The first part of the study was performed under batch experimental runs by using different membranes to examine the effect of different pore sizes and materials on the MD performance.

The experimental performance levels were determined based on the passing water volume, the seawater concentration, and the water flux parameters. Initial experiments were performed with a 30°C difference between the solution temperatures, which are the generally recommended level of difference in the literature [21], by maintaining seawater and permeate water at temperatures of 55 and 25°C, respectively, and by using seawater and permeate water flow rates of 270 and 360 L/h, respectively. In general, available polymeric materials for manufacturing hydrophobic membranes suitable for MD are, typically, PP, PVDF, and PTFE [19, 21].

With PTFE membranes, water flux values for 0.22, 0.45, and 1.0 μm sized pores were determined as 27.7, 40.7, and 28.5 L/m^2 h, respectively. With PVDF membranes, these values were determined as 11.3, 19.4, and 29.6 L/m^{2}h, respectively. TDS values in output water for the PTFE membranes with 0.22, 0.45, and 1.0 μm pore sizes were determined as 4.3, 7.9, and 9.8 mg/L, respectively. **Figure 2** shows permeate flux and solute concentration for the MD membranes with different materials (PTFE and PVDF) and pore sizes (0.22, 0.45, and 1 μm) under batch operating conditions.

An evaluation of the permeate solute ion concentrations with the three different pore sizes revealed that increasing pore size led to higher concentrations in the permeate. Based on the observed water flux performances, it was determined that the PTFE membrane with 0.45 μm pore was the most suitable type of membrane for obtaining potable water from seawater. The PTFE (0.45 μm) membrane was hence chosen for the next stage.

Distillate contamination resulting from wetting of membrane pore is among the chief factors, which impede the broader application of the MD technology. In order to avoid pore wetting, it is necessary that the membrane material be hydrophobic with as high a contact angle as possible. Moreover, it is recommended that the membrane has a comparatively small maximum pore size [34]. The hydrophobic membrane can obstruct the penetration of liquid through surface tension force, but not that of vapor. Consequently, water vapor will be capable of passing from the hot solution side of higher vapor pressure to the cold distillate side of lower vapor pressure [35]. **Figure 3** shows measurement results of contact angle (θ°) and surface roughness (R_z) for surface characterizations of clean (θ_c and $R_{z,c}$) and fouled (θ_f and $R_{z,f}$) membranes.

Results of AFM and SEM evaluations performed before and after filtration on the active layers of the MD membranes are shown in **Figure 4**. The high pore sizes used in the experiments led to greater accumulation of inorganic pollutants inside the pores than on the surface. In addition, the surface fouling layer was observed to be soluble. Consequently, no significant differences were observed on the SEM images. An evaluation of the EDX results revealed that, depending on the structure of polymers used for producing the membranes, and the added chemicals, the membrane surfaces contained the elements boron (B), fluorine (F), carbon (C), nitrogen (N), and oxygen (O). It was observed that the fouling of the surface was largely local and lacked a large variety of ions, and when a prevalent distribution of ions was actually present, this was found to be due to the crystallized precipitates ($CaCO_3$, $CaSO_4$) formed under the effect of temperature [36]. Inorganic foulant materials were not observed on the membrane surfaces. It is hence possible to say that no different elements were observed on

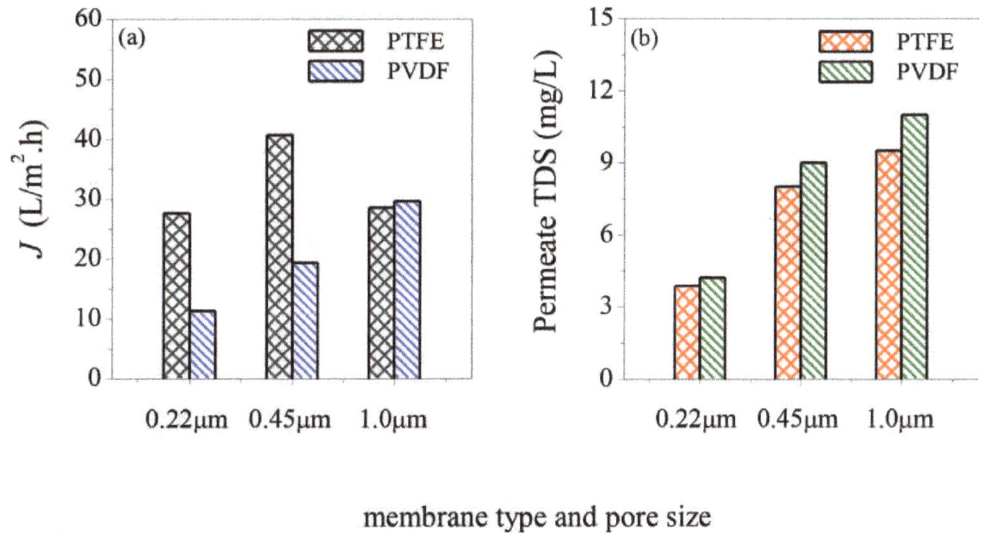

Figure 2. Permeate flux (a) and solute concentration (b) for MD membranes with different materials (PTFE and PVDF) and pore sizes (0.22, 0.45 and 1.0 μm) under batch operating conditions (seawater/permeate: 55/25°C (ΔT: 30°C) and 270/360 L/h cross-flow rates at seawater/permeate streams).

the membranes at the end of the process. The presence of such low and dilute levels of fouling during the experiments clearly indicated that under effective pass vapor pressure and low levels of pollution, the MD membrane is fairly efficient in providing drinking water from seawater.

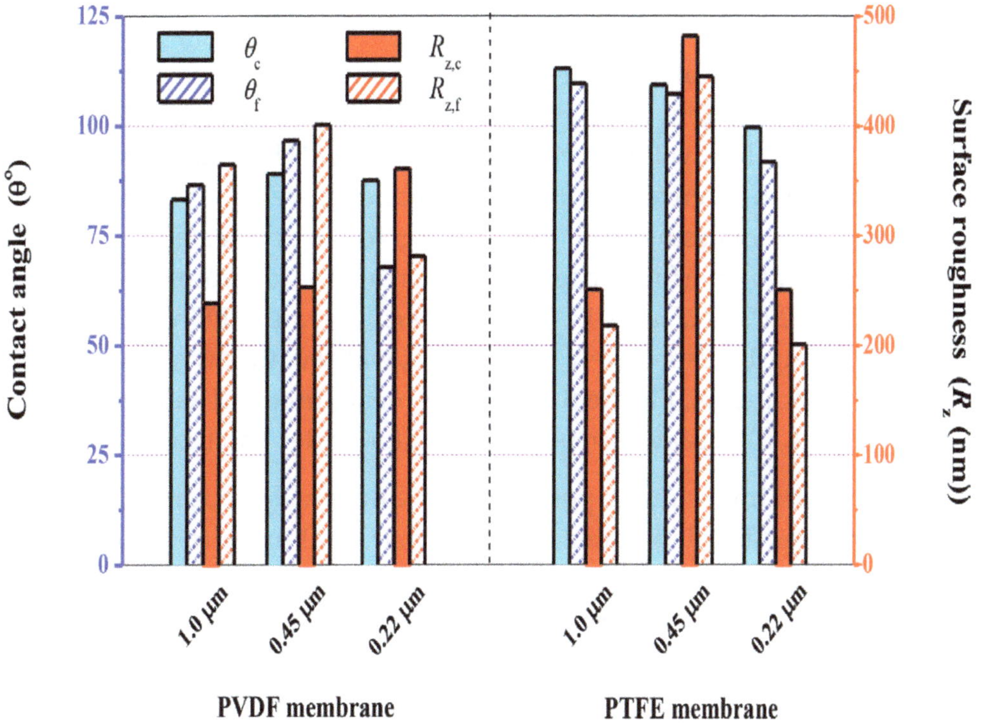

Figure 3. Measurement results of contact angle (θ°) and surface roughness (R_z) for surface characterizations of clean (θ_c and $R_{z,c}$) and fouled (θ_f and $R_{z,f}$) membranes.

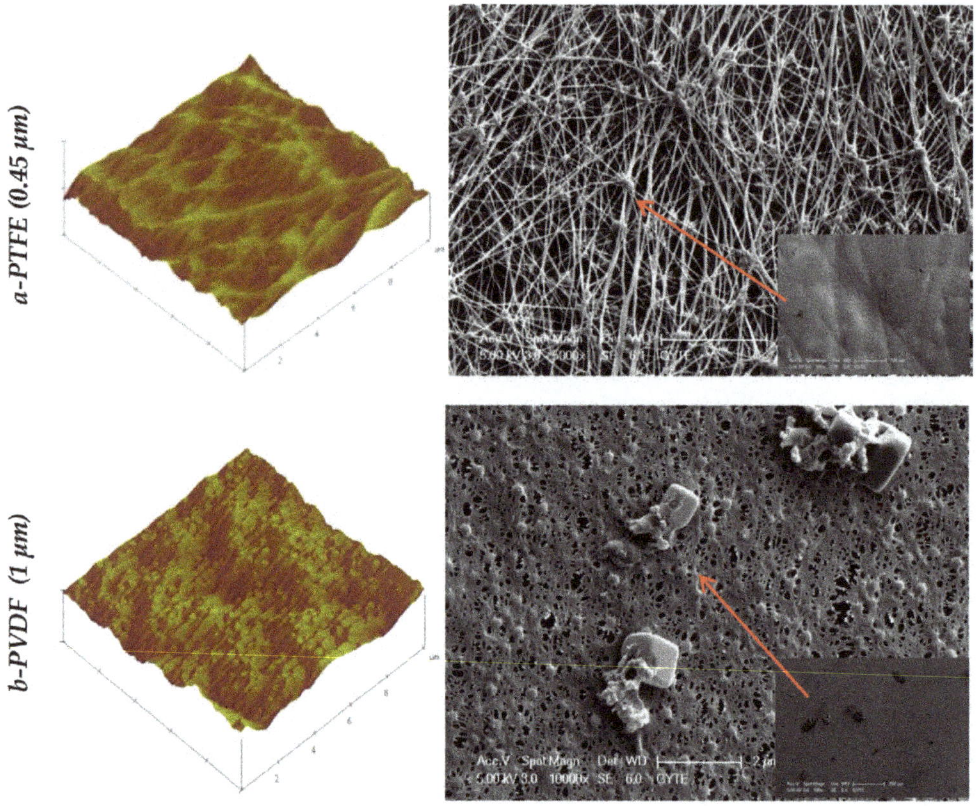

Figure 4. AFM (10 µm × 10 µm) images and SEM (1.4 mm × 0.9 mm) microphotographs of fouled membranes (PTFE-0.45 µm and PVDF-1 µm).

3.1.2. Cross-flow rate

The effect of the flow rates in membrane channel during the MD process was evaluated by applying flow rates with different Reynolds values (90/120, 180/240, 270/360, and 360/480 L/h seawater/permeate water flow velocity values) through a PTFE membrane at a fixed membrane trans-temperature difference of 30±0.5°C. It was observed that increasing flow rate led to a higher volume of water passing from the seawater to the permeate flow, which resulted in an increase in permeate flux performances. With increasing flow rates, water flux levels consecutively reached 23.4, 29.9, 40.7, and 43.4 L/m^2 h. Output water TDS values were determined as 5.3, 6.0, 7.9, and 9.75 mg/L, respectively, and output water with low solute ions was obtained in all flow velocity experiments. **Figure 5** shows permeate flux and salt concentration for different cross-flow rates (L/h) at seawater/permeate water under batch operating conditions.

Under the turbulence regime, similar performance values were observed in the 270/360 L/h and 360/480 L/h seawater/distilled water flow rate pairs. Taking into account that increasing flow rates might increase costs in practice, suitable flow rates were determined as 270 L/h (Re_{av}: 4320) for seawater and 360 L/h (Re_{av}: 4222) for permeate water.

3.1.3. Solutions' temperatures

In the next part of the study, experiments were conducted with 0.45 µm PTFE membranes, a fixed membrane trans-temperature difference value (30°C), a seawater and permeate water

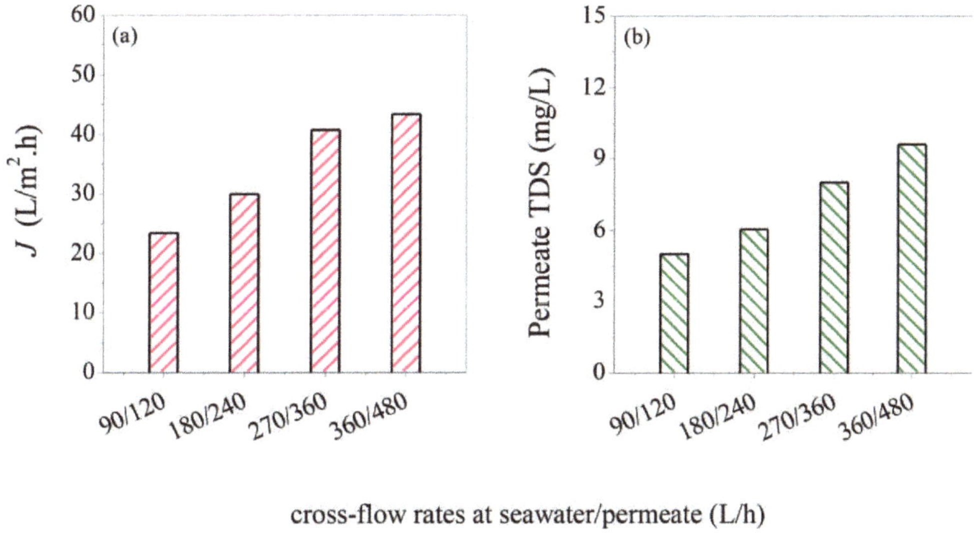

Figure 5. Permeate flux (a) and solute concentration (b) for different cross-flow rates (L/h) at seawater/permeate water under batch operating conditions (PTFE—0.45 μm and seawater/permeate: 55/25°C (ΔT: 30°C)).

flow rates of 270 and 360 L/h, respectively, and distilled water temperatures of 15, 20, 25, and 30°C in order to determine the effect of solution temperature on process performance. The performance results of the experiments were evaluated at seawater-permeate water operation temperatures of 45–15, 50–20, 55–25, and 60–30°C based on the water flux and output water TDS parameters.

It was determined that the water flux values of the process gradually increased in parallel with increasing solution temperature, as 40.3, 32.7, 36.8, 40.7, and 48.0 L/m^2 h, respectively. Total solute concentrations in MD output water were determined as 7.2, 7.5, 8.0, and 8.7 mg/L, respectively, and it was clearly observed that the output water had low dissolved solids. **Figure 6** shows permeate flux and salt concentration for different temperatures at seawater/permeate solutions under batch operating conditions.

The $CaCO_3$ and $CaSO_4$ compounds, which have lower solubility at high temperatures, can cause a serious reduction in water permeation performances of MD membranes by fouling occurring on the membrane surface during the processing. A previous study reported that increasing the input feed flow temperature from 80 to 90°C led to a fourfold decrease in flow level [36]. At membrane, considering the possibility of Ca^{2+} precipitation that might result continuous operation and constant membrane trans-temperature difference, the most suitable was to obtain high flux outputs to continually operate the membranes at fixed permeate water and seawater flow temperatures of 30 and 60°C, respectively.

3.1.4. Membrane trans-temperature difference

It was observed that increasing membrane trans-temperature difference (20, 30, and 40°C) led to an increase in the volume of water passing from the seawater to the permeate water. At a permeate flow temperature of 15°C, the water flux values for membrane trans-temperature differences of 20, 30, and 40°C were determined as 17.0, 32.7, and 53.3 L/m^2 h, respectively. At a

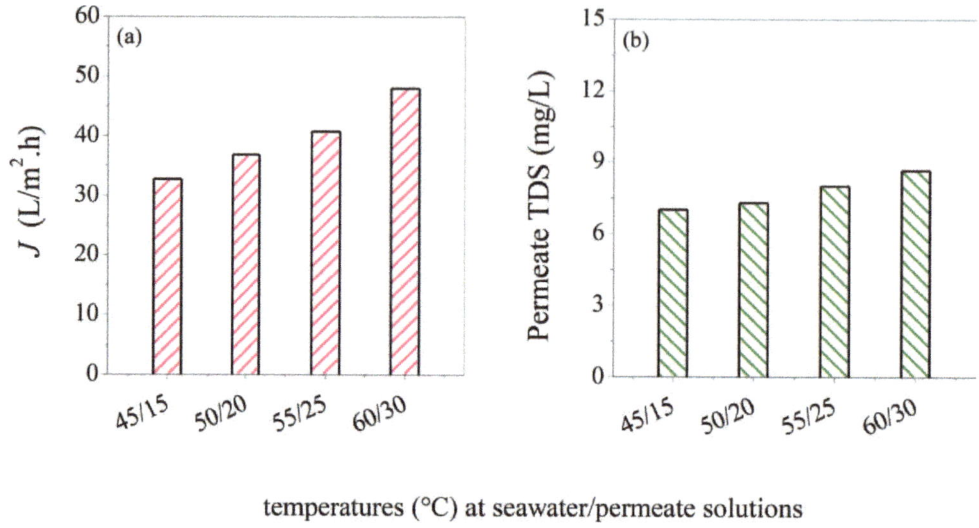

Figure 6. Permeate flux (a) and solute concentration (b) for different temperatures (°C) at seawater/permeate solutions under batch operating conditions (PTFE—0.45 μm, 270/360 L/h cross-flow rates at seawater/permeate, ΔT: 30°C).

permeate flow temperature of 30°C, these values were increased to 31.3, 48.0, and 63.0 L/m^2 h, respectively. At a permeate flow temperature of 15°C, the output water TDS values for membrane trans-temperature differences of 20, 30, and 40°C were determined as 6.5, 6.8, and 7.4 mg/L, respectively. On the other hand, at a permeate flow temperature of 30°C, the output water TDS values for membrane trans-temperature differences of 20, 30, and 40°C were determined as 6.75, 8.3, and 10.3 mg/L, respectively. In all temperature difference experiments, the output water contained low levels of dissolved solids. **Figure 7** shows permeate flux and effective permeate flux per unit of membrane trans-temperature difference and total solutes concentration for varying trans-temperature differences.

As described in the literature, continuous operation at higher seawater temperature may result in $CaCO_3$ and $CaSO_4$ precipitates that foul/clog the pore entrances on the membrane active layer, as well as the pore spaces within the membrane [36]. Based on this consideration, 30°C can be evaluated as the suitable trans-difference among the solutions for the continuous

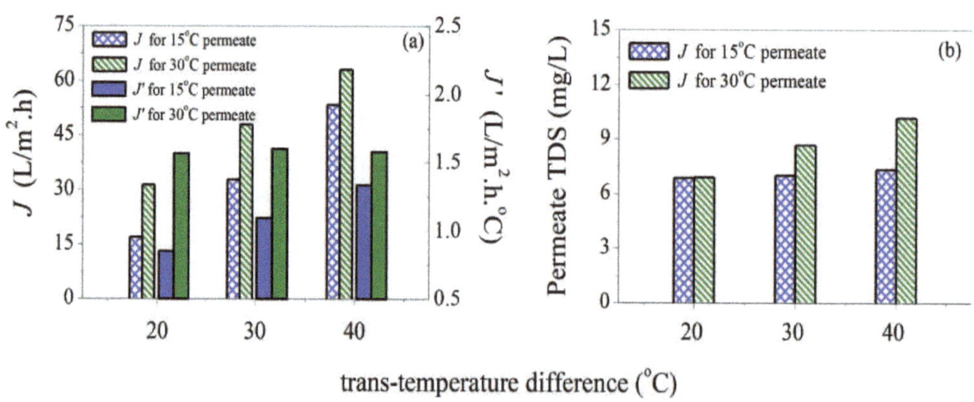

Figure 7. Permeate flux and effective permeate flux per unit of trans-temperature difference (a) and solute concentration (b) for varying trans-temperature differences (PTFE—0.45 μm, 270/360 L/h cross-flow rates at seawater/permeate, 15 and 30°C temperatures at permeate).

operation in which seawater or feed stream would be heated up to max 60°C. Taking into account the fact that energy costs represent the most important obstacle standing in the way of the rising uses of MD process compared to RO seawater desalination plants, it was comprehended that 30°C and some below or above differences would render further possible to arrive the optimal cost solutions in on-land MD plants due to lower heat energy consumption per unit of potable water to be produced.

3.2. MD processing of Marmara seawater

Under controlled membrane cleaning conditions per day cycle where 1% HCl, 1% NaOH and distilled water were consecutively used for flushing the active layer of the membrane while inside the module, and the permeate flow was replaced with distilled water within a suitable time frame, the continuous processing experiments were performed to identify the design flow that would serve as a basis for the field application of the DCMD. In the study, the DCMD was operated in order to concentrate seawater flow to obtain daily concentration performances of 63–66%, and at the end of the 5-day operation period, a 60% volumetric concentration ratio was reached. To ensure that the continuous operation performance for the periods shown in **Figure 8** remained consistent, the initial volume of the feed flow was completed (i.e. brought to the desired level) every day by adding raw seawater.

At the end of days I, II, III, IV, and V, the permeate water fluxes were determined as 38.9, 31.1, 22.6, 16.7, and 15.5 L/m^2 h, respectively. It was found that the flux changes in the first 4-day period were generally characterized by a consistent decrease, while the flux value on day V was very close to the flux value of day IV (despite the fact that the permeate was changed with distilled water). Based on this observation, it is possible to say that the water flux performance almost reached a constant operation level by the end of the day V.

An MD optimization study using PP membrane operated under conditions of $\Delta P = 0.355 \times 10^{-5}$ Pa, $Q_f = 73.6$ L/h and $Q_p = 18.8$ L/h determined an MD process flux of 4.192 L/m^2 h [24]. A previous

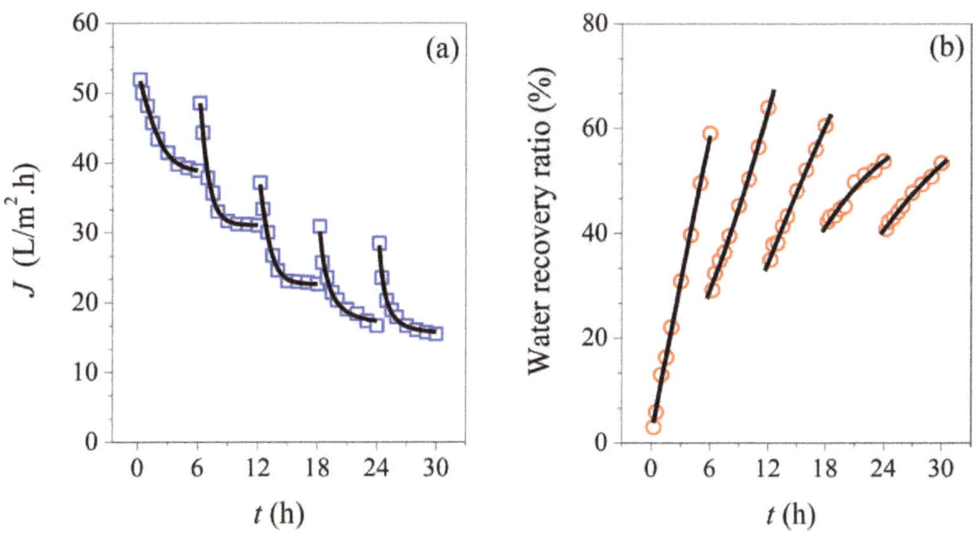

Figure 8. MD process performance under continuous operating condition ((a): permeate flux, (b): water recovery ratio).

study using continuously operated MD process for producing drinking water from seawater made use of a PTFE membrane (0.22μm). In this study, where the feed temperature was 60°C and the permeate temperature was 20°C, the process flux level was reported to decrease from 23.76 to 14.36 L/m^2 h following a 1-month operation. Further, following membrane washing procedures, the same membrane was successfully reused in the DCMD system [22]. In another study using PTFE membranes, a flux level of 25 L/m^2 h was observed following the 70 h operation of the DCMD process with a temperature difference of 60°C [37].

Although the literature describes the use of various different acid and base derivatives for cleaning membranes, HCl use is common in MD [36, 38]. HCl is a quite effective chemical in cleaning of scaling and fouling from basic salts such as $CaCO_3$ [36]. In a study being used synthetic seawater, it was observed that citric acid and NaOH completely remedied for both the flux and the membrane's hydrophobicity [39]. In another study, washing MD membrane with HCl resolved the reduction in flux values caused by $CaCO_3$ fouling, bringing the flux values to their original level [38].

Within the context of the study experiments, the membrane cleaning at the end of day I was performed using only distilled water. On the following days, the cleaning was done using 1% HCl for 20 min, then 1% NaOH for 20 min, and lastly distilled water for 20 min. After flushing, the pH-conductivity values of the acid and base washing solutions changed from 2.05–51.1 and 12.7–52.9 to 2.04–44.9 and 12.71–46.6 mS/cm, respectively. Based on these results, it can be said that no significant changes were observed in the pH values. By the end of the day IV of operation, the changes in the conductivity of the acid and base washing solutions were 11.9 and 12.1%, respectively. For 1 month of a continuous cleaning at the same conditions per the applied operation cycle, it was calculated that the amounts of acid and base lost will reach to about 89.3 and 91.0%, respectively. With regard to the membrane flushing step that will be applied in a real-scale facility, it appears a necessary to renew the 1% acid and base solutions at certain time scales.

By maintaining seawater TDS value within an interval from 26,400 to a max of 78,900 mg/L throughout the processing (66,800 mg/L at the end of the operation), it was ensured that the DCMD system could be permanently operated with steady-state concentrating conditions at a constant temperature difference. In the operations, where distilled water is kept within the system for 24 h, conductivity values of the permeate flow at the end of days I, II, III, and IV were always below the threshold level of 500 μS/cm, being 14, 78, 117, and 375 μS/cm, respectively, while the TDS values for the days I, II, III, and IV permeates were 7, 49, 54, and 179 mg/L, respectively (*data were not shown*). Thanks to the distilled water replace at the beginning of day V, it was noted that these values decreased to 76.0 μS/cm and 47.1 mg TDS/L as adjacent values to the values at the end of the first 2 days. The qualities of concentrated seawater and potable water produced at the end of continuous operation are given in **Table 3**, together with MD rejection performances.

According to **Table 3**, the output water obtained was in very good quality in which the retentions of dissolved organics, boron, silicium, and iron as well as anions and cations were at above 99%. Too low color presence at 525 nm wavelength was observed resulting mostly probably from the trace level amounts of dissolved organics. Based on all results obtained from the continuous operation, it can be suggested for successful field operations of DCMD seawater desalination plants that the distillate stream or product output water should not be remained for a longer operating time than about 10–12 h in the product water storage tank. Certainly, more accurate heuristic knowledge for safe plant operation would be ensured by means of taking into account the performance information to be retrieved from the data of the pilot and/or field scale installations.

Parameters*	MD concentrated seawater	MD permeate (output water)	*R* (%)
T	24.2	24.3	–
pH	8.54	7.30	–
Ec	108, 200	76.0	99.93
TDS	66, 800	47.1	99.93
TOC	20.5	0.08	99.61
DOC	18.4	0.06	99.67
UVA_{254}	0.160	0.001	99.38
SUVA	0.870	1.667	–
Na^+	20, 445	0.114	99.99
K^+	1143	0.005	99.99
NH_4^+	<0.1	0.0	–
Ba^{2+}	0.058	<0.01	–
Ca^{2+}	1097	3.5	99.68
Mg^{2+}	1770	1.4	99.92
Mn^{2+}	0.044	0.0	100.00
Sr^{2+}	15.20	0.02	99.87
B	4.39	0.007	99.84
Si	1.22	0.008	99.34
Fe	0.017	0.0	100.00
Cl^-	46, 236	12.6	99.97
HCO_3^-	244	0.58	99.76
NO_3^-	1.3	0.0	100.00
CO_3^{2-}	112	0.0	100.00
SO_4^{2-}	7050	7.2	99.90
TKN	1.8	0.0	100.00
Total nitrogen	3.1	0.0	100.00
Alkalinity	356	0.58	99.84
Total hardness	16,035	27.1	99.83

Parameters*	MD concentrated seawater	MD permeate (output water)	*R* (%)
Color			
436 nm	0.003	0.000	100.00
525 nm	0.004	0.001	75.00
620 nm	0.001	0.000	100.00

*Units of all parameters are mg/L except for temperature (°C), electrical conductivity (μS/cm), UVA_{254} (1/cm), SUVA (L/mg.m) and color (1/cm).

Table 3. Quality of potable water produced by continuous MD operation together with MD rejection performances.

3.3. Health impacts assessments of produced drinking water

Almost 75 million people across the world acquire drinking water from the sea by treating seawater in desalination plants, and it is expected that this number will go up as a result of the steadily growing demand for water [40].

Seawater abounds with certain ions like calcium, magnesium, sodium, chloride, and iodine but has a lower content of essential ions such as zinc, copper, chromium, and manganese. As a result of seawater desalination applications, the amount of almost all of these ions and minerals, which are essential to the human health and to the agricultural productivity, is dramatically reduced in drinking water [14, 40]. Some essential metals like Cu, Mn, and Zn are required for normal body growth and function [41]. For healthy body, calcium and magnesium are both essential elements, and they play protective roles in the human body development [14, 40, 42]. Calcium, which is a primary constituent of the skeletal structure, is also important for different key physical functions. Due to decreases in bone mass and mineral content, calcium deficiency results in an increased risk of fractures. Deficiency of magnesium, an essential element for physiological processes such as mineralization and skeletal development, cardiac excitability and vascular tone, contractility, reactivity, and growth, can give rise to the pathophysiology of hypertension [40, 43]. Depending on the beneficial health effects rendered by intake of Ca and Mg through drinking water, it has been recommended by World Health Organization (2007) that the minimum and optimum Ca and Mg levels in drinking water should be 20 and 40–80 mg/L and 10 and 20–30 mg/L, respectively [43].

Otherwise, it is also likely that high quantities of these metals will inflict harms for human health. High concentrations of Cd, Cr, Ni, and Pb are regarded to be extremely toxic for humans and aquatic organisms [41]. Furthermore, demineralized water is more aggressive to piping, and therefore, additional risks could be posed through exposure to extracted trace elements such as lead and cadmium. According to certain studies, the use of demineralized water for cooking escalates the loss of essential mineral content of foods, bringing about detrimental effects on health [14, 42]. When the quality parameter values were evaluated in the output water obtained during this study, produced water is low in ions such as Ca, Mg, Na, Cl and essential ions like Zn, Cu, Cr, and Mn. In addition to this, its low solute content of 47.1 mg TDS/L paves the way for corrosive effects to the metal distribution piping and thus the risks by exposure to extracted trace elements like lead and cadmium into the pipe will increase. The allowable limit values for these ions in the relevant national and international drinking water standards were presented in **Table 4** together with the other drinking water quality parameters specific to the produced water.

As described in the literature, to fulfill the desirable values for these parameters under limit concentrations, one possible solution involves making certain additions to the distilled product water by taking into account hygienic requirements. This will make it possible to supply drinking water that does not pose any problems with regard to public health. It is possible to overcome mineral deficiencies directly through addition of minerals at the MD desalination plant or locally by means of chemical injection at specified locations into water distribution systems by also considering hygienic requirements. In an alternative way, quality standards can be reached by mixing high-quality and low-quality water [47], which lead to the sig-

Parameters*	Permeate	TSE 266 [44]	WHO [45]	EPA [46]
T	24.3	–	–	–
pH	7.30	6.5–9.5	6.5–8.5	6.5–8.5
Ec	76.0	2500 μS/cm	250 μS/cm	–
TDS	47.1	–	–	500 mg/l
TOC	0.08	–	–	–
DOC	0.06	–	–	–
UVA_{254}	0.001	–	–	–
SUVA	1.667	–	–	–
Na^{+}	0.114	200 mg/l	200 mg/l	20 mg/l
K^{+}	0.005	–	–	–
NH_4^{+}	0.0	–	–	–
Ba^{2+}	<0.01	–	0.7 mg/l	2 mg/l
Ca^{2+}	3.5	–	–	–
Mg^{2+}	1.4	–	–	–
Mn^{2+}	0.0	50 μg/l	0.4 mg/l	0.05 mg/l
Sr^{2+}	0.02	–	–	–
B	0.007	1 mg/l	0.5 mg/l	–
Si	0.008	–	–	–
Fe	0.0	200 μg/l	0.3 mg/l	0.3 mg/l
Cl^{-}	12.6	250 mg/l	250 mg/l	250 mg/l
HCO_3^{-}	0.58	–	–	–
NO_3^{-}	0.0	50 mg/l	50 mg/l	10 mg/l
CO_3^{2-}	0.0	–	–	–

Parameters*	Permeate	TSE 266 [44]	WHO [45]	EPA [46]
SO_4^{2-}	7.2	250 mg/l	500 mg/l	250 mg/l
TKN	0.0	–	–	–
Total Nitrogen	0.0	–	–	–
Alkalinity	0.58	–	–	–
Total hardness	27.1	–	150–500 mg/L	–
Color				
436 nm	0.000			
525 nm	0.001	–	–	–
620 nm	0.000			

Table 4. Maximum allowable concentrations of specific water quality parameters in use of produced water as drinking water supply.

nificant savings since remineralization facilities and chemicals are no longer needed [40]. The most affordable option to increase the dissolved content of specific ions in desalinated water is blending remineralized desalinated water with treated brackish groundwater or treated seawater [48]. When the MD output water produced in this study is evaluated in terms of physicochemical water quality for directly human consumption, it was found to be much lower than necessary levels especially in terms of dissolved minerals concentrations, which also give corrosive properties to the demineralized water. Based also on the literature knowledge and land experiences, a general management approach for field-scale pragmatic solutions toward to directly use the MD desalination effluents as drinking water was holistically proposed in **Figure 9**; as being the first, adding into existing urban potable water at desirable ratios; the second, injecting appropriate chemicals into effluent by completely mixing; and the third, mixing effluents with raw/concentrated seawaters (1:250/1:1000 for Marmara seawater) or clean brackish natural waters under hygienic precautions.

As the third distinctive solution, the MD product water can be readily mixed with seawater concentrate at a ratio of 1/1000 or with raw seawater at a ratio of 1/250 as shown in **Table 5** after MD desalination of raw seawater. In the calculations, dissolved solutes concentration increased to 180–200 mg/L in the mix ratios of 1/500 and about 1/200 for concentrated and raw seawaters, respectively. But, Ca^{2+} and Mg^{2+} concentration levels increased by 1.7 and 3.5 times although they could not fulfill the desired levels. Low level of total hardness, which is reached to ~60 mg/L, needs to be still increased to the desired standard levels by applying carbonation to the produced water. Hence, the obtained mixtures would also still lack of the desired levels of minerals and ions to be targeted for healthy drinking water production. For this reason, it is deduced that stand-alone use of this solution for an absolute success would not be sufficient. In addition, the practical ways of the mixing applications can be evaluated as an insignificant improving factor for decreasing of the negative effects of the concentrated seawater discharges that importantly affect the sea life near discharge points.

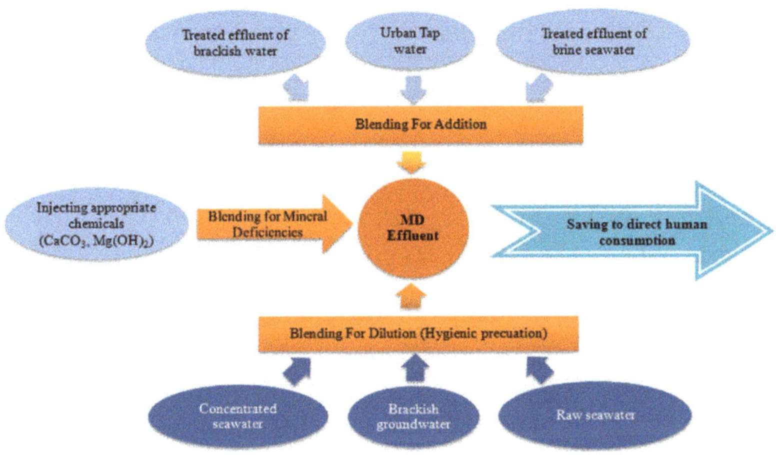

Figure 9. Holistic management approach for satisfying specific water quality requirements in long-term use of MD effluents demineralized from seawater as drinking water.

			Mixing ratio with							
Parameters	**Units**	**Permeate**	**Concentrated seawater**				**Raw seawater**			
			1/100	**1/250**	**1/500**	**1/1000**	**1/20**	**1/50**	**1/100**	**1/250**
T	°C	24.3	24.3	24.3	24.3	24.3	24.3	24.3	24.3	24.3
pH	–	7.30	7.31	7.30	7.30	7.30	7.3	7.3	7.3	7.3
Ec	μS/cm	76.0	1157	508	292	184	2092	882	479	222
TDS	mg/L	47.1	715	314	181	114	1254	530	289	134
TOC	mg/L	0.0	0.2	0.1	0.0	0.0	0.2	0.1	0.0	0.0
DOC	mg/L	0.0	0.2	0.1	0.0	0.0	0.2	0.1	0.0	0.0
UVA_{254}	1/cm	0.004	0.01	0.0	0.0	0.0	0.01	0.0	0.0	0.0
SUVA	L/mg.m	1.67	1.66	1.67	1.67	1.67	1.64	1.66	1.66	1.33
Na^+	mg/L	0.114	205	82	41	21	444	178	89	36
K^+	mg/L	0.01	4	2	1	0.0	22	9	4	2
Ca^{2+}	mg/L	3.5	14	8	6	5	23	11	7	4
Mg^{2+}	mg/L	1.4	19	8	5	3	36	15	8	4
Mn^{2+}	mg/L	0.001	0.001	0.001	0.001	0.001	0.002	0.001	0.001	0.001
Sr^{2+}	mg/L	0.02	0.17	0.08	0.05	0.04	0.29	0.13	0.07	0.04
B	mg/L	0.01	0.05	0.02	0.02	0.01	0.21	0.09	0.05	0.02
Si	mg/L	0.02	0.03	0.02	0.02	0.02	0.07	0.04	0.03	0.02
Fe	mg/L	0.0	0.0	0.0	0.0	0.0	0.0	0.0	0.0	0.0
Cl^-	mg/L	12.6	475	197	105	59	712	292	152	66
HCO_3^-	mg/L	0.58	3	2	1	1	16	7	4	2
NO_3^-	mg/L	0.0	0.01	0.01	0.0	0.0	0.0	0.0	0.0	0.0
CO_3^{2-}	mg/L	0.0	1	0.0	0.0	0.0	0.0	0.0	0.0	0.0

			Mixing ratio with							
Parameters	**Units**	**Permeate**	**Concentrated seawater**				**Raw seawater**			
			1/100	**1/250**	**1/500**	**1/1000**	**1/20**	**1/50**	**1/100**	**1/250**
SO_4^{2-}	mg/L	7.2	78	35	21	14	123	53	30	15
TKN	mg/L	0.0	0.02	0.01	0.0	0.0	0.0	0.0	0.0	0.0
Alkalinity	mg/L	0.58	4	2	1	1	16	7	4	2
Total hardness	mg/L	27.1	187	91	59	43	275	126	77	42
Color										
436 nm	1/cm	0.0	0.0	0.0	0.0	0.0	0.0	0.0	0.0	0.0
525 nm		0.001	0.001	0.001	0.001	0.001	0.0	0.0	0.0	0.001
620 nm		0.000	<0.001	<0.001	<0.001	<0.001	<0.001	<0.001	<0.001	<0.001

Table 5. Changing values of output water quality parameters according to the ratios of mixings with concentrated or raw seawaters.

4. Conclusions

In this chapter, suitable operating conditions for continuous DCMD processing of seawater were first determined based on water flux and solutes rejection performances. Thereafter, by relative long-term MD desalination in seawater concentration mode at lab pilot-scale system, the water qualities of MD product water and seawater concentrate were investigated in light of physicochemical parameters specific to the seawater characteristics. In the final, it was elaborately examined whether the MD produced water is suitable for direct human use as drinking water.

Within the scope of the DCMD experiments under batch conditions, it was determined that the best operating conditions involved the use of hydrophobic PTFE membrane with 0.45 μm pore diameter; a flow velocity for seawater and permeate (distilled water) of 270 and 360 L/h, respectively (corresponding to mean Re numbers of 4320 and 4222); and flow temperatures for seawater and permeate streams of 60 and 30°C, respectively, which are associated with a membrane trans-temperature difference of 30°C.

By means of 30 h MD processing of rough-filtrated raw seawater from the Marmara Sea that was operated with steady-state permeate flux of 17.21 L/m^2 h and solutes retentions of >99% at seawater concentrating level reached to 66% in the constant temperature difference, the production of MD output water below threshold level of 500 μS/cm was continually carried out without replacing intrinsic distilled water of distillate stream. After its replacement was applied, ultimate product water was supplied with 76.0 μS/cm conductivity and 47.1 mg/L dissolved solids. A replacement time of about 10–12 h for initial clean distillate in the output water storage tank would be sufficient for field-scale operations of DCMD seawater desalination plants.

In case the MD product water is to be used as drinking water, it will be necessary to ensure that the dissolved minerals that are essential for a healthy life are found in the water in at least the minimum recommended levels, and to prevent the various trace elements (Cu, Fe, and Mn) that can have toxic effects on living beings above certain levels or cause damage to the processing lines. Possible management options recommended in the literature to remedy these issues include the direct addition of minerals to final waters, the addition of chemicals to specific locations on process lines, and the blending of demineralized water with treated brackish groundwater/seawater. However, stand-alone generalization of each one of these practical solutions would not be made possible for all the challenges to be encountered in all application varieties of seawater desalination plants. In this frame, there need to develop unique approaches oriented on novel pragmatic solutions toward to the direct use of MD demineralized effluents as drinking water. In this respect, under an integrated conservative approach, a general management framework toward satisfying the specific water quality requirements in long-term use of MD effluents was proposed. With an aim of fulfilling the deficiencies of minerals and ions to be targeted for healthy drinking water supply, the developed holistic approach is jointly dependent on injections to urban water distribution systems at desirable ratios, mixings with raw/concentrated seawaters (1:250/1:1000 for Marmara seawater) or brackish natural waters under hygienic precautions, and additions in sufficient amounts of appropriate chemicals by completely mixing.

Acknowledgements

Authors thanks the Scientific and Technological Research Council of Turkey, TUBITAK due to the financial support provided by a national project (*no: 111 Y279*) which was entitled as "Techno-economic analysis of drinking water production from seawater by forward osmosis/direct contact membrane distillation integrated membrane system".

Nomenclature

AFM	atomic force microscopy
C_f	concentration in the feed
C_p	concentration in the permeate
DCMD	direct contact membrane distillation
DOC	dissolved organic carbon
EDX	energy dispersive X-ray spectrometry
LEP	liquid entry pressure
MED	multi-effect distillation
MD	membrane distillation
MSF	multi-stage flash distillation
PE	polyethylene
PP	polypropylene

PTFE	polytetrafluoroethylene
PV	photovoltaic
PVDF	polyvinylidene fluoride
Q_p	flow of permeate side (L/h)
Q_f	flow of feed side (L/h)
R	rejection (%)
Re	Reynolds number
RO	reverse osmosis
$R_{z,c}$	surface roughness of clean membrane
$R_{z,f}$	surface roughness of fouled membrane
SEM	scanning electron microscopy
SUVA	specific ultraviolet absorbance
TDS	total dissolved solids
TKN	total Kjeldahl nitrogen
TOC	total organic carbon
UVA_{254}	ultraviolet absorbance at 254 nm
WHO	world health organization
Greek symbols	
θ_c	contact angle of clean membrane (°)
θ_f	contact angle of fouled membrane (°)

Author details

Coskun Aydiner[1]*, Derya Y. Koseoglu Imer[2,3], Salim Oncel[1], Esra Can Dogan[4], Ali Oguzhan Narci[4], Serif Cakmak[1], Tugba Nur Yilmaz[1], Emin Ender Celebi[1] and Yasemin Melek Tilki[1]

*Address all correspondence to: aydiner@gtu.edu.tr

1 Department of Environmental Engineering, Engineering Faculty, Gebze Technical University, Kocaeli, Turkey

2 Department of Environmental Engineering, Engineering Faculty, Istanbul Technical University, Istanbul, Turkey

3 National Research Center on Membrane Technologies, Istanbul Technical University, Istanbul, Turkey

4 Department of Environmental Engineering, Engineering Faculty, Kocaeli University, Kocaeli, Turkey

References

[1] UN Water Scarcity Report [Internet]. 2016. Available from: http://www.un.org/waterforlifedecade/scarcity.shtml [Accessed: 25 March 2016]

[2] Lorain O, Hersant B, Persin P, Grasmick A, Brunard N, Espenan J.M. Ultrafiltration membrane pre-treatment benefits for reverse osmosis process in seawater desalting. Quantification in terms of capital investment cost and operating cost reduction. Desalination. 2007; **203**:277-285. DOI: 10.1016/j.desal.2006.02.022

[3] Gilau AM, Small MJ. Designing cost-effective seawater reverse osmosis system under optimal energy options. Renewable Energy. 2008;**33**:617-630. DOI: 10.1016/j.renene.2007.03.019

[4] Grenlee LF, Lawler DF, Freeman BD, Marrot B, Moulin P. Reverse osmosis desalination: Water sources, technology, and today's challenges. Water Research. 2009;**43**:2317-2348. DOI: 10.1016/j.watres.2009.03.010

[5] Yarlagadda S, Gude VG, Camacho LM, Pinappu S, Deng S. Potable water recovery from As, U, and F contaminated ground waters by direct contact membrane distillation process. Journal of Hazardous Materials. 2011;**192**:1388-1394. DOI: 10.1016/j.jhazmat.2011.06.056

[6] Melián-Martel N, Sadhwani JJ, Malamis S, Ochsenkühn-Petropoulou M. Structural and chemical characterization of long-term reverse osmosis membrane fouling in a full scale desalination plant. Desalination. 2012;**305**:44-53. DOI: 10.1016/j.desal.2012.08.011

[7] Jansen AE, Assink JW, Hanemaaijer JH, Medevoort JV, Sonsbeek EV. Development and pilot testing of full-scale membrane distillation modules for deployment of waste heat. Desalination. 2013;**323**:55-65. DOI: 10.1016/j.desal.2012.11.030

[8] Frioui S, Oumeddour R. Investment and production costs of desalination plants by semi-empirical method. Desalination. 2008;**223**:457-463. DOI: 10.1016/j.desal.2007.01.180

[9] Matsuura T, Rana D, Qutaishat MR, Singh G. Recent advances in membrane science and technology in seawater desalination—With technology development in the middle east and Singapore. In: UNESCO-EOLLS Joint Committee, editors. Water and Wastewater Treatment Technologies. Encyclopedia of Life Support Systems (EOLLS), developed under the Auspices of the UNESCO [Internet]. Oxford, UK: EOLLS Publishers; 2011. Available from: http://www.eolls.net. [Accessed: 12 April 2016]

[10] Mezher T, Fath H, Abbas Z, Khaled A. Techno-economic assessment and environmental impacts of desalination technologies. Desalination. 2011;**266**:263-273. DOI: 10.1016/j.desal.2010.08.035

[11] Fuentes-Bargues JL. Analysis of the process of environmental impact assessment for seawater desalination plants in Spain. Desalination. 2014;**347**:166-174. DOI: 10.1016/j.desal.2014.05.032

[12] Lattemann S, Höpner T. Environmental impact and impact assessment of seawater desalination. Desalination. 2008;**220**:1-15. DOI: 10.1016/j.desal.2007.03.009

[13] Al-Agha MR, Mortaja RSh. Desalination in the Gaza Strip: Drinking water supply and environmental impact. Desalination. 2015;**173**:157-171. DOI: 10.1016/j.desal.2004.06.212

[14] Cotruvo JA. Desalination guidelines development for drinking water: Background. WHO [Internet]. 2004. Available from: http://www.who.int/water_sanitation_health/dwq/nutdesalination.pdf [Accessed: 29 May 2016]

[15] Sadhwania JJ, Vezaa JM, Santana C. Case studies on environmental impact of seawater desalination. Desalination. 2015;**185**: 1-8. DOI: 10.1016/j.desal.2005.02.072

[16] Liu T, Sheu H, Tseng C. Environmental impact assessment of seawater desalination plant under the framework of integrated coastal management. Desalination. 2013;**326**:10-18. DOI: 10.1016/j.desal.2013.07.003

[17] Manawi YM, Khraisheh M, Fard AK, Benyahia F, Adham S. Effect of operational parameters on distillate flux in direct contact membrane distillation (DCMD): Comparison between experimental and model predicted performance. Desalination. 2014;**336**:110-120. DOI: 10.1016/j.desal.2014.01.003

[18] Choi YH, Kweon JH, Kim DI, Lee S. Evaluation of various pretreatment for particle and inorganic fouling control on performance of SWRO. Desalination. 2009;**247**:137-147. DOI: 10.1016/j.desal.2008.12.019

[19] Francis L, Ghaffour N, Alsaadi AS, Nunes SP, Amy GL. Performance evaluation of the DCMD desalination process under bench scale and large scale module operating conditions. Journal of Membrane Science. 2014;**455**:103-112. DOI: 10.1016/j.memsci.2013.12.033

[20] Naidu G, Jeong S, Choi Y, Jang E, Hwang T, Vigneswaran S. Application of vacuum membrane distillation for small scale drinking water production. Desalination. 2014;**354**:53-61. DOI: 10.1016/j.desal.2014.09.026

[21] Al–Obaidania S, Curcio E, Macedonio F, Profio G.D, Al-Hinai H, Drioli E. Potential of membrane distillation in seawater desalination: Thermal efficiency, sensitivity study and cost estimation. Journal of Membrane Science. 2008;**323**:85-98. DOI: 10.1016/j.memsci.2008.06.006

[22] He K, Hwang HJ, Woo MW, Moon IS. Production of drinking water from saline water by direct contact membrane distillation (DCMD). Journal of Industrial and Engineering Chemistry. 2011;**17**:41-48. DOI: 10.1016/j.jiec.2010.10.007

[23] Winter D, Koschikowski J, Wieghaus M. Desalination using membrane distillation: Experimental studies on full scale spiral wound modules. Journal of Membrane Science. 2011;**375**:104-112. DOI: 10.1016/j.memsci.2011.03.030

[24] Boubakri A, Hafiane A, Bouguecha SAT. Direct contact membrane distillation: Capability to desalt raw water. Arabian Journal of Chemistry. DOI: http://dx.doi.org/10.1016/j.arabjc.2014.02.010 (in press)

[25] Guillen-Burrieza E, Thomas R, Mansoor B, Johnson D, Hilal N, Arafat H. Effect of dry out on the fouling of PVDF and PTFE membranes under conditions simulating intermittent seawater membrane distillation (SWMD). Journal of Membrane Science. 2013;**438**:126-139. DOI: 10.1016/j.memsci.2013.03.014

[26] Duong HC, Cooper P, Nelemans B, Cath TY, Nghiem LD. Optimising thermal efficiency of direct contact membrane distillation by brine recycling for small-scale seawater desalination. Desalination. 2015;**374**:1-9. DOI: 10.1016/j.desal.2015.07.009

[27] Feng C, Khulbe KC, Matsuura T, Gopal R, Kaur S, Ramakrishna S, Khayet M. Production of drinking water from saline water by air-gap membrane distillation using polyvinylidene fluoride nanofiber membrane. Journal of Membrane Science. 2008;**311**:1-6. DOI: 10.1016/j.memsci.2007.12.026

[28] Fard AK, Manawi YM, Rhadfi T, Mahmoud KA, Khraisheh M, Benyahia F. Synoptic analysis of direct contact membrane distillation performance in Qatar: A case study. Desalination. 2015;**360**:97-107. DOI: 10.1016/j.desal.2015.01.016

[29] Bouguecha ST, Aly SE, Al-Beirutty MH, Hamdi MM, Boubakri A. Solar driven DCMD: Performance evaluation and thermal energy efficiency. Chemical Engineering Research and Design. 2015;**100**:331-340. DOI: 10.1016/j.cherd.2015.05.044

[30] Kaya Y, Aydiner C, Barlas H, Keskinler B. Nanofiltration of single and mixture solutions containing anionics and nonionic surfactants below their critical micelle concentrations (CMCs). Journal of Membrane Science. 2006;**282**:401-412. DOI: 10.1016/j.memsci.2006.05.047

[31] Aydiner C. A novel approach based on distinction of actual and pseudo resistances in membrane fouling: "Pseudo resistance" concept and its implementation in nanofiltration of single solutions. Journal of Membrane Science. 2010;**361**:96-112. DOI: 10.1016/j.memsci.2010.06.003

[32] APHA, AWWA, WEF. Standard Methods for the Examination of Water and Wastewater. In: Eaton AD, Clesceri LS, Rice EW, Greenberg AE, editors. 21st ed. Centennial Edition, Washington DC: American Public Health Association Publication; 2005

[33] Dogan EC, Yasar A, Sen U, Aydiner C. Water recovery from treated urban wastewater by ultrafiltration and reverse osmosis for landscape irrigation. Urban Water Journal. 2016;**13**:553-568. DOI: 10.1080/1573062X.2014.992917

[34] Yang X, Wang R, Shi L, Fane A.G, Debowski M. Performance improvement of PVDF hollow fiber-based membrane distillation process. Journal of Membrane Science. 2011;**369**:437-447. DOI: 10.1016/j.memsci.2010.12.020

[35] Hsu ST, Cheng KT, Chiou JS. Seawater desalination by direct contact membrane distillation. Desalination. 2002;**143**:279-287. DOI: 10.1016/S0011-9164(02)00266-7

[36] Warsinger DM, Swaminathan J, Guillen-Burrieza E, Arafat HA, Lienhard VJH. Scaling and fouling in membrane distillation for desalination applications: A review. Desalination. 2015;**356**:294-313. DOI: 10.1016/j.desal.2014.06.031

[37] Shirazi MMA, Kargari A, Tabatabaei M. Evaluation of commercial PTFE membranes in desalination by direct contact membrane distillation. Chemical Engineering and Processing. 2014;**76**:16-25. DOI: 10.1016/j.cep.2013.11.010

[38] Gryta M. Polyphosphates used for membrane scaling inhibition during water desalination by membrane distillation. Desalination. 2012;**285**:170-176. DOI: 10.1016/j.desal.2011.09.051

[39] Curcio E, Ji X, Profio GD, Sulaiman AO, Fontananova E, Drioli E. Membrane distillation operated at high seawater concentration factors: Role of the membrane on $CaCO_3$ scaling in presence of humic acid. Journal of Membrane Science. 2010;**346**:263-269. DOI: 10.1016/j.memsci.2009.09.044

[40] Avni N, Eben-Chaime M, Oron G. Optimizing desalinated sea water blending with other sources to meet magnesium requirements for potable and irrigation waters. Water Research. 2013;**47**:2164-2176. DOI: 10.1016/j.watres.2013.01.018

[41] Khan K, Lu Y, Khan H, Zakir S, Ihsanullah, Khan S, Khan A.A, Wei L, Wang T. Health risks associated with heavy metals in the drinking water of Swat, northern Pakistan. Journal of Environmental Sciences. 2013;**25**(10):2003-2013. DOI: 10.1016/S1001-0742(12)60275-60277

[42] Verma KC, Kushwaha AS. Contemporary issue demineralization of drinking water: Is it prudent?. Medical Journal Armed Forces India. 2014;**70**:377-379. DOI: 10.1016/j.mjafi.2013.11.011

[43] Abtahi M, Yaghmaeian K, Mohebbi MR, Koulivand A, Rafiee M, Jahangiri-rad M, Jorfi S, Saeedi R, Oktaie S. An innovative drinking water nutritional quality index (DWNQI) for assessing drinking water contribution to intakes of dietary elements: A national and sub-national study in Iran. Ecological Indicators. 2016;**60**:367-376. DOI: 10.1016/j.ecolind.2015.07.004

[44] MEU (Republic of Turkey Ministry of Environment and Urbanisation). Legislation on waters to be used for human consumption. Official Journal date and number: 17 Feb 2005—25730. Ankara, Turkey; 2005

[45] WHO. Guidelines for Drinking Water Quality Third Edition Incorporating the First and Second Addenda. Vol. 1—Recommendations WHO Guidelines for the Safe Use of Wastewater. Geneva: World Health Organization; 2008. ISBN 978 92 4 154761 1

[46] USEPA. Guidelines for Water Reuse. EPA 822-S-12-001. Washington, DC: U.S. Edition of the Drinking Water Standards and Health Advisories; 2012

[47] Yang Sh-L, Sun Y-H, Yeh W-G. Optimization of regional water distribution system with blending requirements. Journal of Water Resources Planning and Management. 2000;**126**(4):229-235. DOI: 10.1061/(ASCE)0733-9496(2000)126:4(229)

[48] Withers A. Options for recarbonation, remineralisation and disinfection for desalination plants. Desalination. 2005;**179**:11-24. DOI: 10.1016/j.desal.2004.11.051

11

Low-Cost Multi-Effect Solar Still: Alternative Appropriate Technology for Personal Desalination

Pak Hunkyun

Abstract

Multi effect solar still (MES) has a stack of multiple layers for evaporation and condensation. The latent heat dissipated during condensation at the front layers are repeatedly recycled for evaporation at the back layers to increase overall desalination productivity. Despite of high efficiency and long history, MES has not been widely used yet, because of relative high cost. In this chapter, newly designed MES is introduced. Since it has low cost, light weight material and simple structure, it could be easily mass even at less developed country. The cost of production for a 1 m^2 unit is expected to be less than 300 USD. Structural features are introduced with experimental result which was outdoor tested with homemade lab prototype with 0.219 m^2 effective area. 9kg/m^2 per day of fresh water was obtained at sunny day (19.5MJ/m^2) in Seoul, Korea, which is close to WHO's recommended minimal daily water supply for individuals (7.5~15 liters). For more practical implementation, further development on prototype and production process should be made as well as long term outdoor test under actual climate it would be used. Worldwide collaboration would be necessary for speeding up implementation.

Keywords: solar still, multi-effect solar still, desalination, low cost, appropriate technology, water purification, mass production, small scale, renewable energy, solar heat, personal, individual

1. Introduction

Various desalination methods have been developed since clean water is one of the most vital resources for human life. Generally, the unit cost of fresh water production tends to decrease dramatically, as the total scale of production has increased [1]. Therefore, mega-desalination plants have become a popular trend in modern civilization. However, smaller scale desalination is still important for many applications: (i) It could supply water in remote areas where a water supply line is hardly available. (ii) It would be useful for people who cannot afford large investment for a water plant. (iii) It could also be useful as a temporal water source for many people, such as nomads, campers, sailors, or survivors of various disasters.

Thanks to modern technology development, efficient small-scale desalinators became frequently used. Portable reverse osmosis water purifiers powered by electric or fuel engine are commonly used instruments for such purposes. However, such instruments are still too expensive, hard to supply energy source or difficult to maintain for many people in the world, who cannot afford substantial budget, resource or education. For such cases, the desalination method is appropriate technology, which is affordable to the actual users in do-it-yourself (DIY) style and would be a more realistic option than the expensive high-technology solution.

Solar stills are one of the simplest ways of desalination. In solar still, saline water is evaporated by solar heat and condensed to become fresh water. Though it is old, primitive tool, but is still useful for many people. It can be simply made out of low-cost common material, such as plastic sheet or bottles. Therefore, it could be easily installed and operated with do-it-yourself manner. Various solar stills were made and used for long history [2–6]. However, the extreme low productivity has limited its application only to emergency or temporal water source. Considering that the latent heat of water vapour is about 540 cal/g or 2.26 kJ/g and assuming solar energy per unit area as 1 kJ/s m^2 (equivalent to 1 sun, which is nominal full sun intensity on bright clear day on earth), we may get about 1.6 kg of distilled water with 1 m^2 solar still in an hour. However, even this is only a theoretical maximum value. In reality, only few cups of fresh water could be obtained daily, since there is additional energy loss by vapour leakage or heat dissipation. Though several cups of water should be valuable to save thirsty sufferer, but it might not be enough for sustainable use in everyday life. Of course, we can enlarge the area of solar still to increase water production. However, in many cases, sunny land is also finite resource to be occupied only by solar still.

As shown in **Table 1**, WHO, the world health organization, recommended 2.5–3 L of drinking water per day per person for survival, including the basic hygiene and cooking purpose, at least 7.5–15 L per day per person is required [7].

Type of need	Quantity	Comments
Survival (drinking and food)	2.5–3 lpd	Depends on climate and individual physiology
Basic hygiene practices	2–6 lpd	Depends on social and cultural norms
Basic cooking needs	3–6 lpd	Depends on food type, social and cultural norms
Total	7.5–15 lpd	Lpd: Litres per day

Table 1. Minimal amount of fresh water required for a human living, recommended by the World Health Organization.

Therefore, substantial improvement of productivity should be made on solar still for more practical applications. Multi-effect solar still (MES) was developed to overcome this low productivity problem, by recycling wasted latent heat repeatedly [8–27]. As shown in **Figure 1**, in MES, multiple layers of evaporating wicks and condensing surface are stacked together [14].

Saline water flowing through first evaporation wick is partially evaporated, while the condensed saline water is drained out. The water vapour condensed on the first condensation surface would be collected as fresh water. During the condensation process, dissipated latent heat is reused to evaporate saline water running on the second evaporation wick. By repeating this process on multiple layers, the overall water productivity is very much increased. The Tanaka's group proved both theoretically and experimentally that MES can produce over 10–20 L of fresh water per day with a square metre effective area [8], which is comparable amount to that WHO recommended.

Though MES has more than half century history [2, 3] and performance verification, it is not widely used on practical application yet because of relatively high cost. Metal plates covered with fabric, glass cover, metal or wooden frames and airtight vapour sealants were commonly used materials for traditional MES system. Though they enable rigid and effective MES structure, but still they are relatively expensive and heavy material, which increase overall system cost.

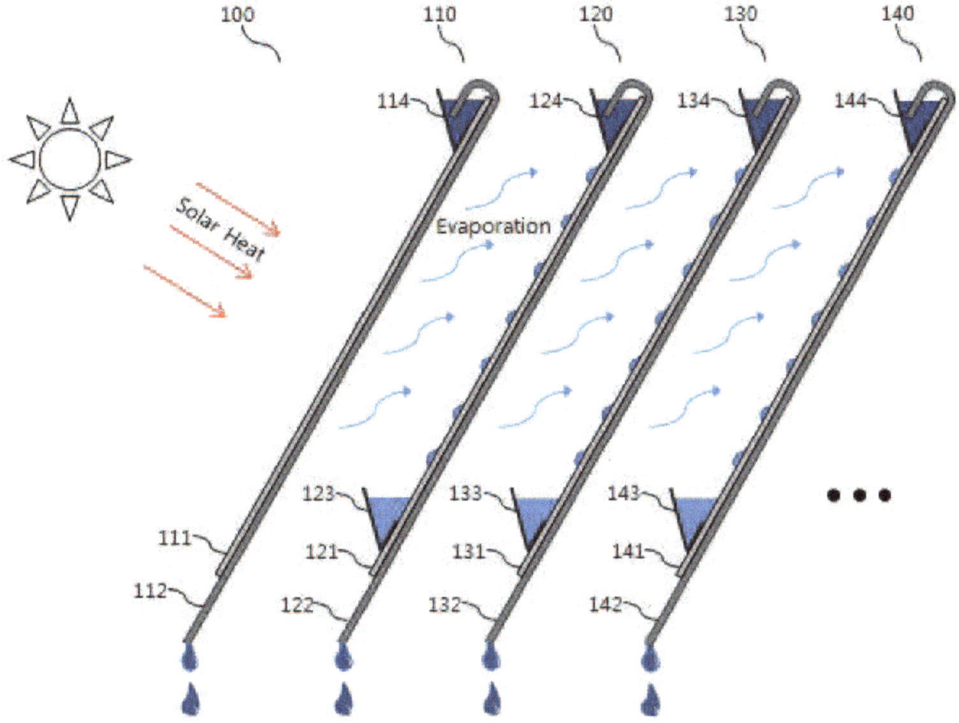

Figure 1. Typical structure of MES. Saline water in upper troughs (114, 124, 134, 144) flows through evaporation wicks (112, 122, 132, 142) and condensed on the condensation surfaces (121, 131, 141) to be collected through the collectors (123, 133, 143).

Figure 1 represents typical inner structure of MES. It may include upper saline water trough, vapour condensing plates, wicks where the saline water evaporates, condensed fresh water collecting guide trough. The structures are packed in sun light passing windows and housing.

For higher productivity in MES, intervals between each evaporation wick and condensing surface should be minimized [9, 11]. However, if the distance is too short, the layers could touch each other, so that water droplet on the condensing surface can move to evaporation wick, which causes loss of distilled water. Similarly, if the saline water overflow into condensing surface, serious contamination of produced water could be occurred. Therefore, the wick and condensing surface should be securely isolated by air gap. Despite air gaps between the layers, there is always a chance of contact within layers because of impact, vibration, and gravitational deformation. To minimize such deformation, various approaches were made. Rigid metallic plates were commonly used for stable condensing layer. However, it resulted in increase of cost and system weight. Spacers were placed between the layers [4]. However, water droplet may flow back to wick along the spacer if too much spacers were applied, which may reduce productivity. The layers were placed vertically to avoid gravitational deformation [8–13]. However, it may increase discrepancy between solar incident angle to the system window, which reduces solar energy usage. Reflecting mirrors [12, 24] or tilted window [8, 9, 13, 27] could be placed in front of the vertical layers to maximize solar intensity. However, it may increase the system cost.

To solve this high-cost problem, new MES was designed and tested. To reduce production cost, alternative light-weight and low-cost material was used instead of such expensive materials. Structure of each component was also designed for easier mass production, which may reduce cost. In the next section, the materials and structure of the low-cost MES would be described.

2. Structure and material of the low-cost MES

2.1. Design of the wick/condenser laminated film layer

In this work, flexible thin plastic film laminated to black fabric wick was used instead of stainless plate covered with fabric wick [14, 15]. The laminated film is obviously low-cost, light-weight alternative than metallic plates. It could be produced by roll-to-roll process, which is mass production favourable. The laminated film was folded like origami as shown in **Figure 2** [14].

Evaporation wick, condensation surface, upper saline water trough, condensed water guiding channel and concentrated saline water draining guide could be included in this single origami structure. Mixing the fresh and saline water at the end of draining guide could be avoided by folding the edge of the film. This simple structure and process may enable mass production and reduce the production cost.

2.2. Design of the spacer layer

Flexible wick/condensing layer is vulnerable for deformation which causes mixing back the fresh and saline water. Therefore, specially designed spacers are required for this system. In

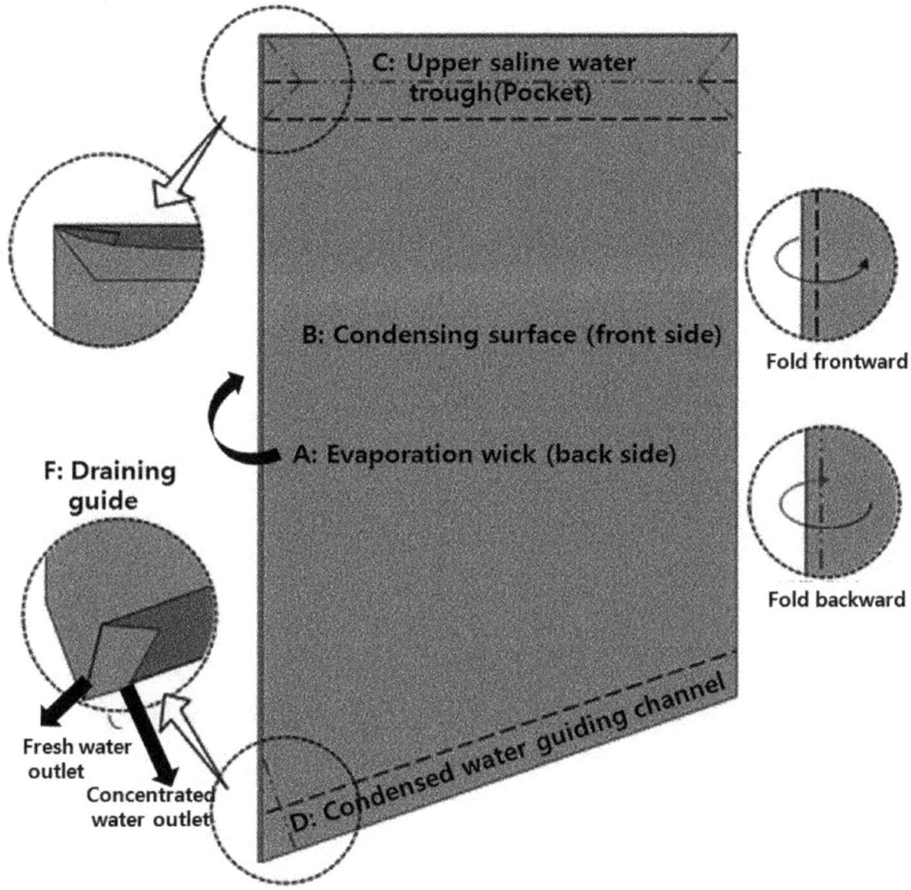

Figure 2. The origami structure of wick/condenser layer. Each part of the structure would have role of the evaporation wick (A), condensing surface (B), saline water feeding trough (or pocket) (C), condensed fresh water guiding channel (D) and draining guide which separate the fresh water and waste water (F).

other hands, tight vapour sealing and thermal insulation are essential for assuring high productivity. In this system, sealing, spacers and inner frames altogether are constructed in one structure, the spacer layer. Its structural features are shown in **Figure 3** [14].

Vertically elongated spacers are connected to upper and lower parts of the frame. The spacer layer has following roles.

Role 1: Internal framework for the MES

It has a role of internal framework structure standing for each layer of evaporation wick/condensation layer film. The spacer layers and wick/condensing films are alternately stacked as shown in **Figure 4**. Though the film is too flexible to be stand alone, the film-spacer combined layer is stiff enough to be stand. The stiffness increases by stacking multiple layers.

Role 2: Fixture for the vertical spacers

The frame holds each vertical spacer in right position and orientation. To avoid a layer's film touching the next layer film, it is important that each spacer on a layer should be exactly superposed on top of the next layer's spacer.

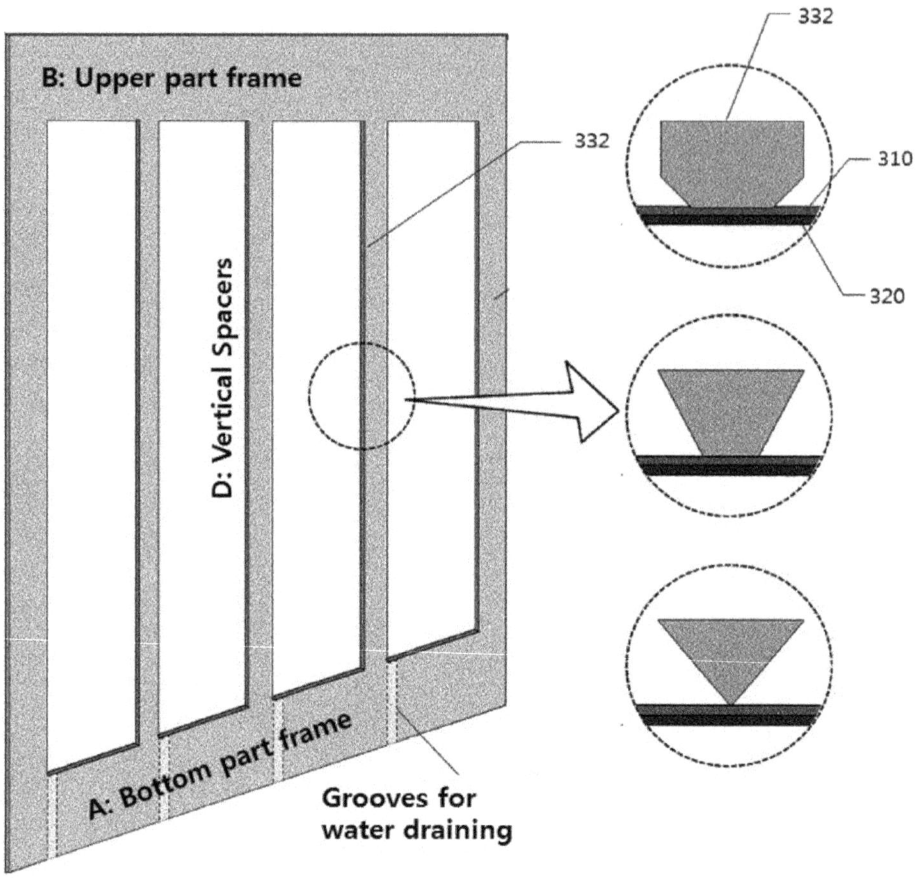

Figure 3. Structure of the spacer layers. The vertical spacers are connected to the perimeter frame. Small pictures in circles represent cross section of the spacers (332) which touches the condensing surface (310) laminated to the wick (320).

Maintaining regular spacer orientation is also necessary. As shown in small circles in **Figure 3**, the cross sections of the spacers were designed to have sharp angles (less than 90°) towards the contacting plane of the vapour condensing surface. Though most of the condensed water droplets drip down straight on the condensing film, some droplet could stagger out while confronting minor defects on the film surface. Staggering motion may increase chance of the droplet hit the spacer. Once touching the spacer, the droplet would be trapped and flow along the sharp-angled crevice by surface tensional force. As long as it was trapped in the crevice, it would not flow back into the saline water running wick. Therefore, it is important that the spacer to be oriented to maintain the crevice located on condensation surface.

For lab prototype, 5-mm thick polystyrene foam board was hand carved with a knife to make the spacing layer, as shown in **Figure 5**. While carving, edges of the spacers were shaped to be 60° angle.

In case of mass production, the foam board could be cut by die press machine (commonly called as Thompson die cutter, a press with pre-shaped knife for cutting soft sheet). While pressing the knife into the foam board, edges of the cutting perimeter are automatically slightly collapsed as round shape, which makes sharp angle at the contact point between the spacer and condensing surface.

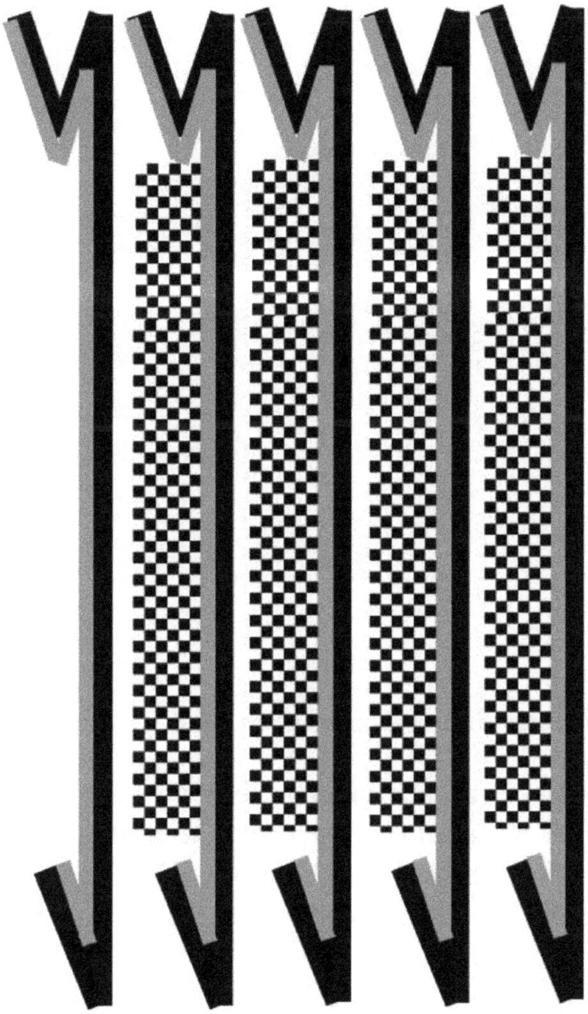

Figure 4. Alternately stacked wick (coloured black)/condenser (coloured grey) film and the spacer layer (coloured as checkerboard).

Role 3: Gasket for vapor and heat

The frame also has a role of gasket and heat insulator which minimize heat and vapour leakage from the system. Since it is made out of soft material, foam board, leakage could be avoided by simply pressing the stacks without using additional sealant. It simplifies assembly process and reduces production cost.

Role 4: Guide of the condensed fresh water

Bottom part of the frame (**Figure 3**) is slightly tilted to the upper part of the frame (**Figure 3**). The upper part of the frame has role of a cross beam, holding saline water containing trough (or pocket) in **Figure 2**. Bottom part of the frame has role of condensed water dripping guide. There are multiple grooves on the bottom frame contacting the condensing film. These grooves are for guiding the collected water to be easily flow into the condensed water-guiding channel. Without grove, condensed water could overflow towards the wick, if too much fresh water produced.

Figure 5. Hand carved spacers.

2.3. Design of the saline water distributer to each wick

Since the evaporation process on each layer repeatedly recycles the latent heat dissipated from the front layer, energy usage of each layer is different. Generally, the layer closely facing sun has more energy for distillation. Therefore, distribution of saline water in proper amount to each layer is critical for high productivity. The Tanaka group did intensive study on how to optimize water distribution, both theoretically and experimentally [8, 11]. They used multiple capillaries with different length to control amount of saline water feed to each layer. Longer capillary reduces flow rate, while short one leads higher flow rate. This is a very useful method, but further simple structure was used in this chapter. A piece of cotton fabric was vertically cut with different width as shown in **Figure 6**.

One end of the piece was immersed in saline water feeding container, and other shredded ends were inserted on the saline water pockets (**Figure 2**) of each layer. Saline water flows along the cotton fabric due to capillary force. Since the flow rate may proportional to the width of the fabric, allocated ratio of saline water to each layer could be simply controlled by controlling shred width of the fabric.

2.4. Durability and maintenance of the system

Heat loss should be minimized for higher productivity in MES. Especially for the current structure, it is important to insulate the front and back sides of the layer stacks, since the sides of stacks are partially insulated already by the frame of spacer layer. Thick (comparable to 10 cm) extended polystyrene foam plate or commercially named as "Isopink" could be placed on the back side of the layer stack. It is commonly used as a construction material for building insulation. For the front window, air-gap window, such as 18 mm thickness, triple layer air-gap polycarbonate could be placed (**Figure 7**). This window material, commercially called

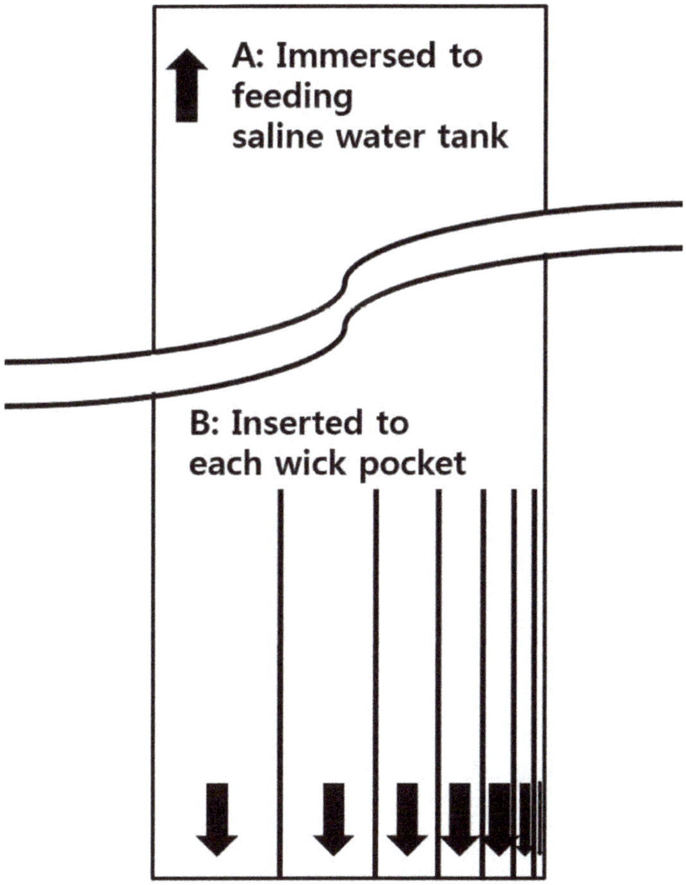

Figure 6. Structure of the low-cost water feeding distributer to each wick.

Figure 7. Cross section of the triple layer air-gap polycarbonate window (bottom) and extended polystyrene foam board (up).

Lexan, is frequently used for greenhouse construction. At the outdoor test, described in the next section, no significant degradation was observed, after leaving the prototype exposed to outdoor climate for more than half year. Though no further long-term stability test was made with this system yet, I expect certain level of environmental durability, since both Lexan and Isopink are already market proven as construction material.

Commercial polycarbonate windows are well known for toughness and UV resistance. Therefore, it might be stable for long period. If polycarbonate window is not available, stacks of glass plated with air gap, which has even higher UV resistance, could be used. However, extended polystyrene foam board is somewhat vulnerable to UV light. Though core parts of the system are protected from direct UV light, external case could be damaged for long period. For such case, the housing could be covered with other UV-blocking materials, such as paint or clay. The wicks are protected by UV-resistant window. However, long-term exposure under strong sunlight with continuous water dripping may lead black colour fading out. For that case, the front-most layer could be replaced with the inner layer periodically, since the inner layers are more protected from the UV light, and all the layers are designed to be exactly of same shape. Assuming that the system has 10 layers, and the front-most layer should be replaced after every 2 years, one system can be used for 20 years [23]. For this purpose, the system should have simple structure for assemble and disassemble. Window, stack of the layers and insulating back plates could be assembled by using common string or tie through holes, premade on each component. Unskilled users would be able to do the maintenance without special tool.

2.5. Installation of the system

Earlier MES models were commonly positioned to receive maximum solar energy during the year. With this structure, the evaporation/condensation layer stack was naturally tilted from the gravitational field. However, gravitational deformation of layers could be caused at the MES. It may cause contact of fresh water droplet to the wick or failure of the system.

As the structure proposed in this chapter, such failure would be minimized because of the spacer layers. Therefore, the system could be placed towards sun side with tilted angle (**Figure 8**).

Following the structure described in this section, dry weight of the 1 m^2 system with 10 layers may not exceed 20 kg, since it has only light-weight material. (Actually, I got 15 kg mock-up system, with the window, the 10-layer stack and insulator layer, as shown in **Figure 9**.) Therefore, the user could easily carry and install in DIY manner.

2.6. Expected cost and further cost down

Production cost for the described MES system is estimated in **Table 2** [15]. Material cost could be much lower, if we have massive economy of scale. Further cost down could be made, if lower cost alternative materials are available nearby. For example, multi-layer (air gap) polycarbonate window is relatively expensive material. It could be replaced with plane window and transparent air-bubble wrap which are inexpensive packing materials for wrapping

Figure 8. MES system installed tilted 45°.

Figure 9. One square metre effective area MES prototype with 10 layers, whose weight is 15 kg.

Item	Expected cost	Comment
Housing	50 USD	Window and insulator housing
Inner parts	180 USD	Stacks with 10 layers of spacers, wick/condensing film
Flow controller	30 USD	Feeding rate controller, hoses and container
Processing cost	40 USD	Labour and instrumental cost
Total	300 USD	Model for 1 m^2 effective area, which supplies about 10 kg/day fresh water

Table 2. Estimated production cost of the MES system.

parcels. The material and design could be modified reflecting the best condition where the system would be produced and used.

3. Lab prototype and its outdoor test result

A MES prototype was handmade using the structure described in previous section for testing its practical feasibility.

Black fabric laminated with plastic film was cut and folded to make the evaporation/condensation layer. Five-millimetre thick polystyrene foam board was used to make the spacer layer. It could be either carved out from one large sheet or assembled from small parts of the frames and thin spacers (**Figure 10**). It may be easier to make the layer by assembling small parts

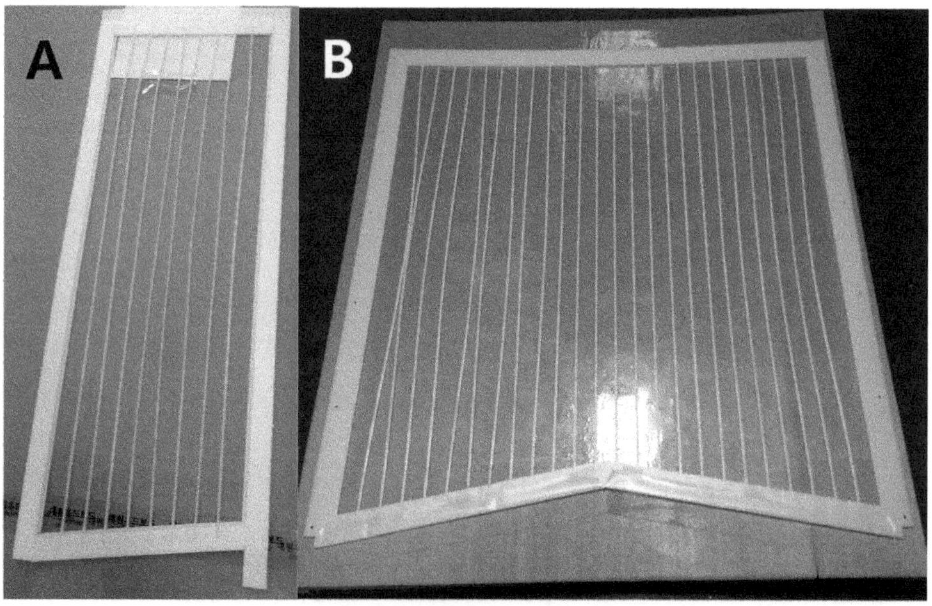

Figure 10. Hand carved and assembled spacer layer (A) and die pressed spacer layer (B).

(and also, save the foam board material), if made with hand one by one. However, for mass production, cutting out from one large sheet by die press would be easier.

The spacer layers were alternately sandwiched with the evaporation/condensation film, while the condensation surface facing the sharp edge of the spacer. Ten identical spacer-film layers were made and stacked. The stack was sandwiched between 1.8-cm triple layer air-gap polycarbonate (Lexan) window and 10 cm polystyrene foam board, by cable tie or box tape. Box tape is useful for short-term usage but may not be a good choice for long-term test, since it would be deteriorated by UV. Common ropes are more recommended. A piece of shred cotton fabric was used for the capillary water distributer. Gutter was placed to collect all the draining fresh water and concentrated waste water in separate beakers (**Figure 11**).

The prototype has 0.219 m^2 active area (27 cm width and 81 cm height) with 10 multi-effect layers. It was placed in relatively non-shaded place in Seoul, Korea (North 37° 34′, East 126° 58′), 45° tilted from the vertical position, facing south (**Figure 12**).

Ambient and inner temperature was measured by thermocouples. For inner temperature measurement, position between first and second layers as well as position backside of the last layer was chosen. Two water collecting beakers were placed on top of load cells to measure weight of fresh and concentrated water drained. A pyranometer was placed as same tilting angle (45°) as the prototype, for measuring solar irradiation. All the data were automatically collected by computer every minute.

Figures 13–15 show the result on a sunny day (On 7 July 2014, the daily accumulated solar intensity was measured as 19.5 MJ/m^2). It shows that it can produce fresh water about 9 kg/L per m^2 day or 0.46 kg/MJ. On a partially cloudy day (30 July 2014, the daily accumulated solar intensity was measured as 13.7 MJ/ m^2), it was reduced to 5.7 kg/L per m^2 day or 0.41 kg/MJ (**Figures 16–18**).

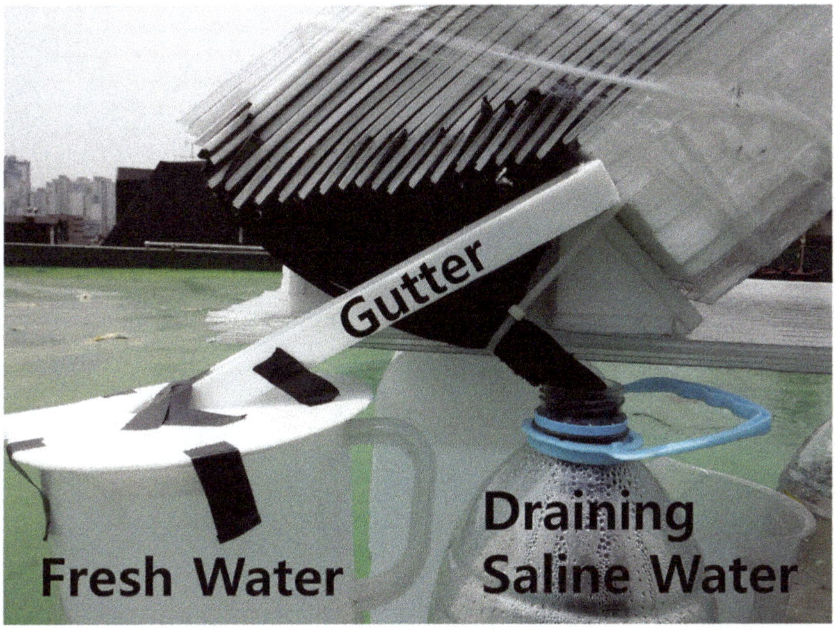

Figure 11. Collecting fresh water from each layer by a gutter.

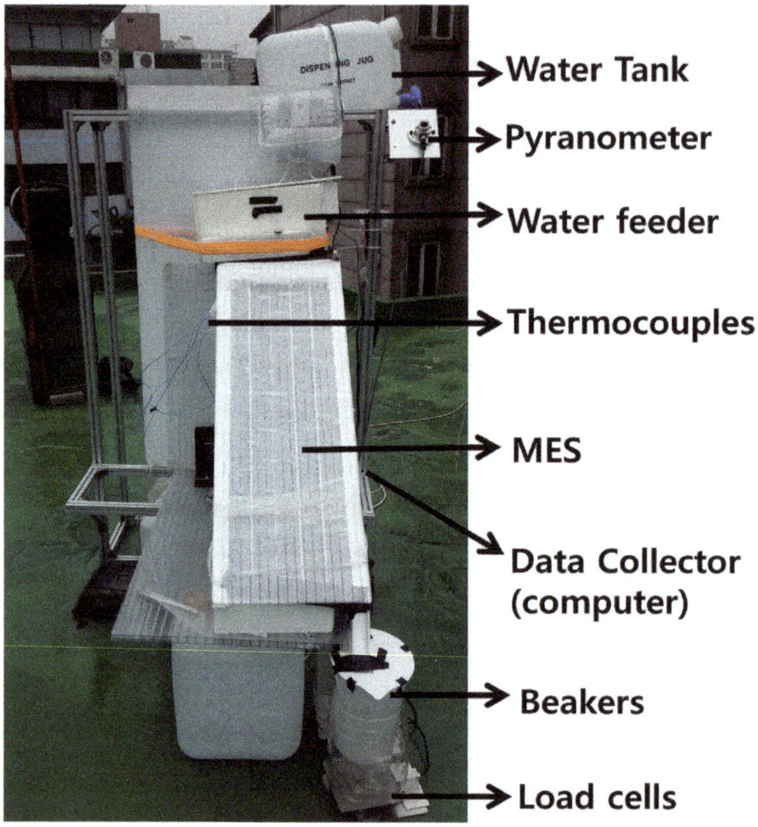

Figure 12. Experimental setup of the prototype MES outdoor test.

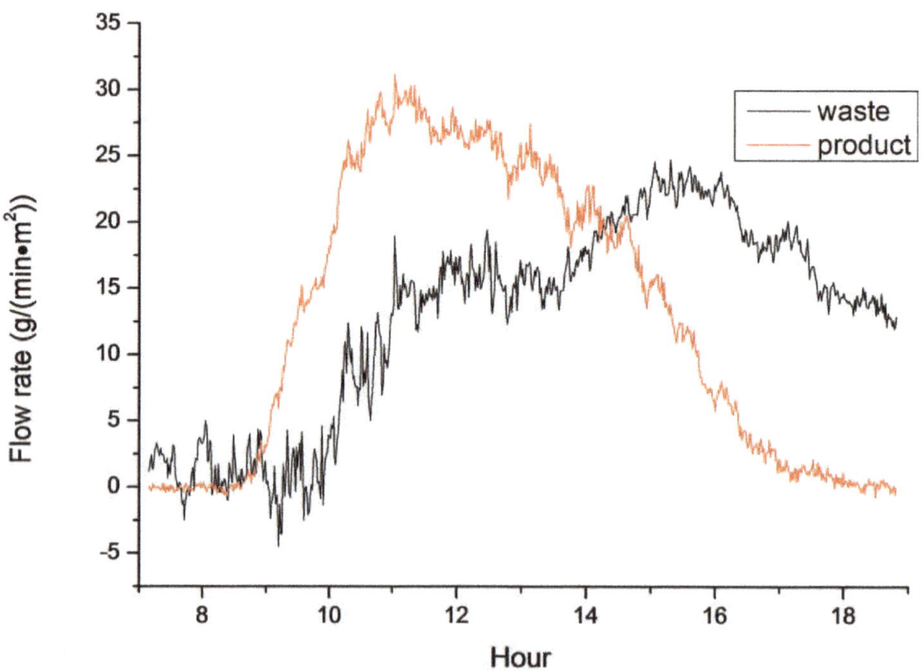

Figure 13. Flow rate (normalized to unit area, 1m2) of the fresh (red, the lower curve at the range after 15:00 PM) and wasted (black, the upper curve at the range after 15:00 PM) water verses time by the MES, measured on a sunny day, July 7th, at Seoul Korea.

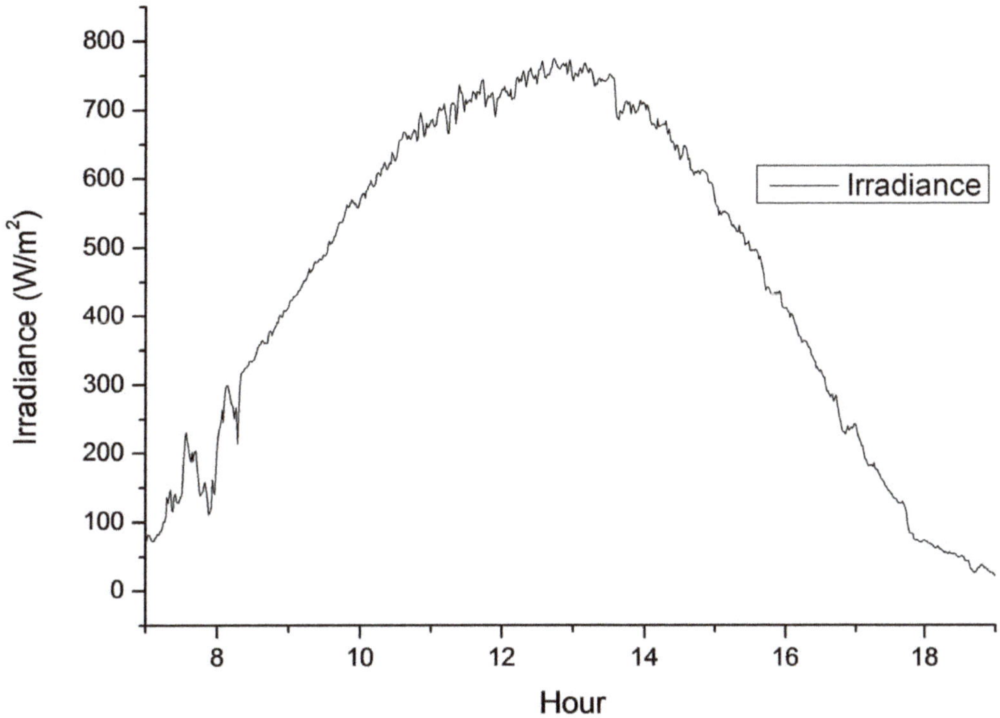

Figure 14. Solar irradiation measured at same day and location in **Figure 13**.

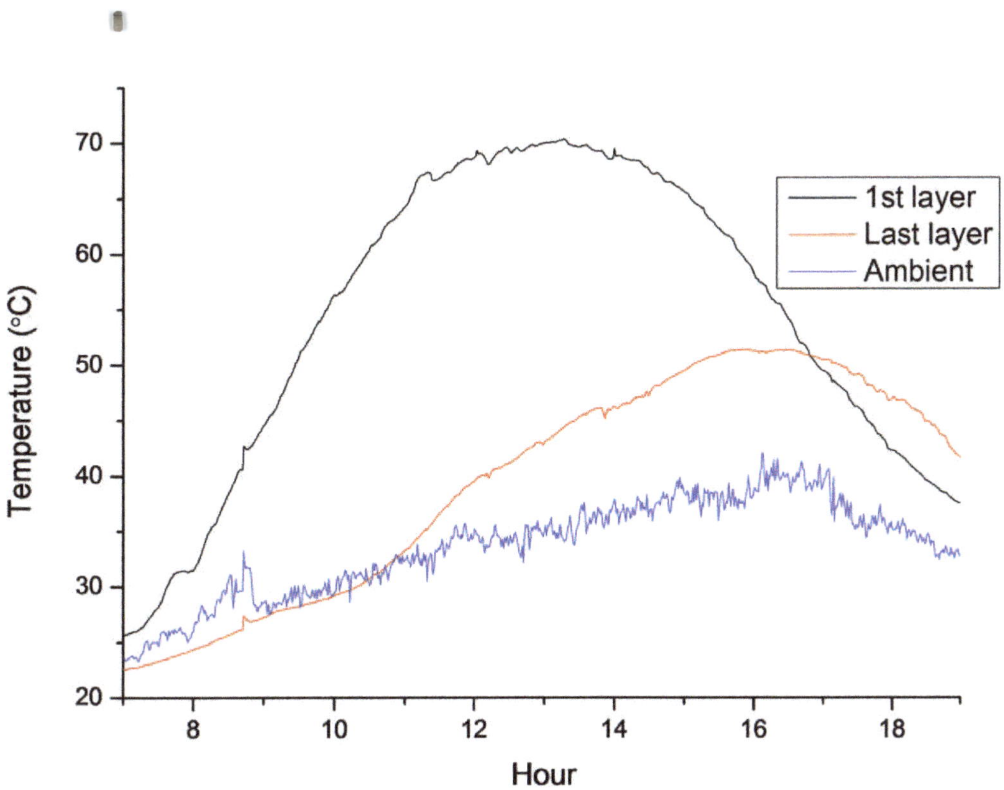

Figure 15. Temperature of ambient (blue, the lower curve at the range between 12:00 to 16:00 PM), backside of the last layer (red, the middle curve at the range between 12:00 to 16:00 PM) and backside of the first front layer (black, the upper curve at the range between 12:00 to 16:00 PM) measured at same day and location in **Figure 13**.

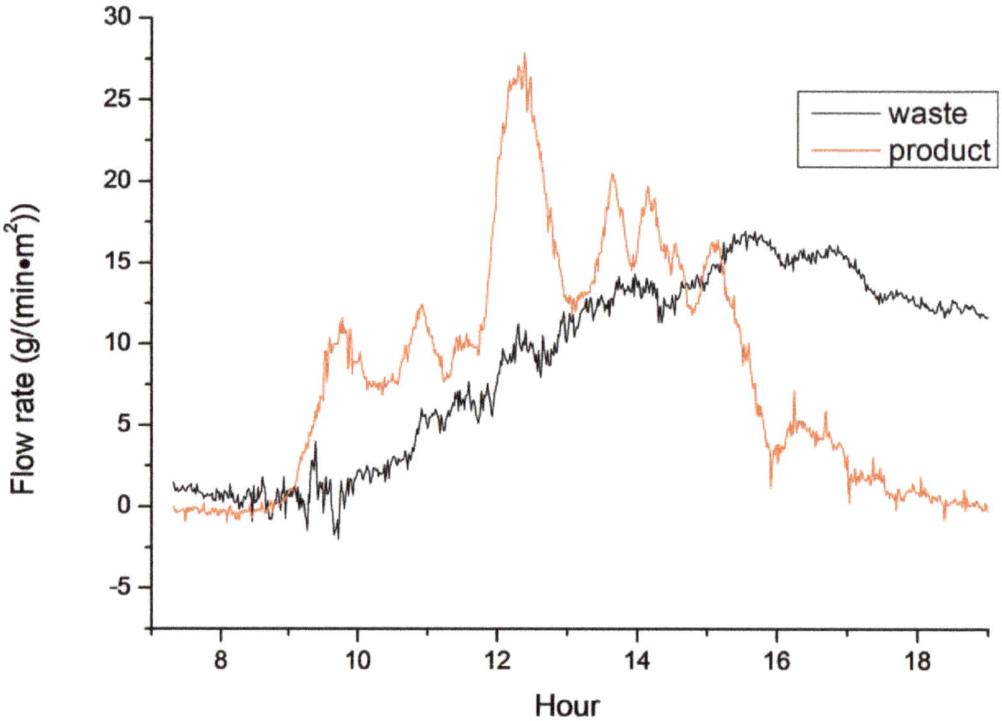

Figure 16. Flow rate (normalized to unit area, 1m2) of the fresh (red, the lower curve at the range after 16:00 PM) and wasted (black, the upper curve at the range after 15:00 PM)water verses time by the MES, measured on a partially cloudy day, July 30th, at Seoul, Korea.

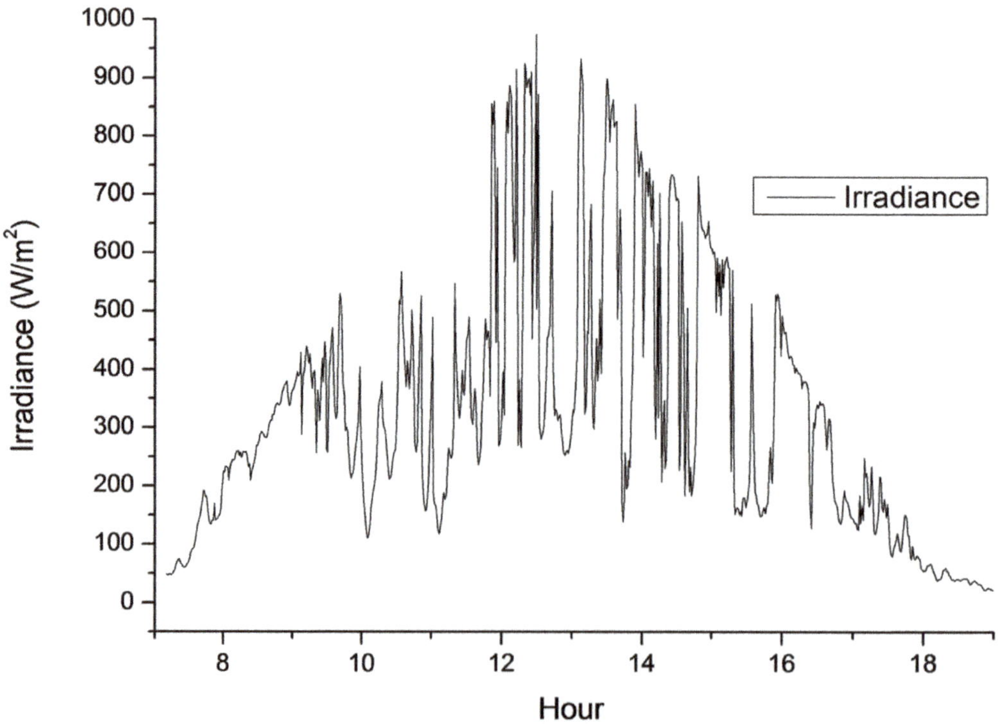

Figure 17. Solar irradiation measured at same day and location in **Figures 16**.

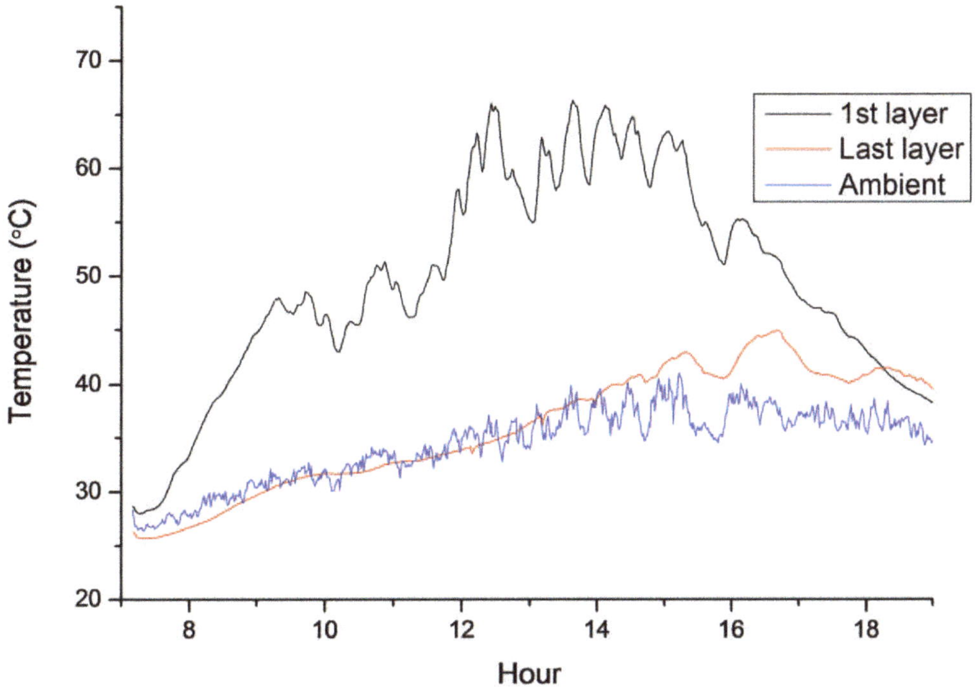

Figure 18. Temperature of ambient (blue, the lower curve at the range between 15:00 to 18:00 PM), backside of the last layer (red, the middle curve at the range between 15:00 to 18:00 PM) and backside of the first front layer (black, the upper curve at the range between 15:00 to 18:00 PM) measured at same day and location in **Figure 16**.

This is somewhat smaller number than previous result reported from other group, who showed that 10–20 kg/m^2 day was obtained [4]. However, considering that the system is not well optimized yet, it would be a good starting point, proving the practical feasibility. There are still many ways that remain to improve productivity. One example is to optimize and control saline water feeding rate. The Tanaka group showed that proper saline water feeding rate is very critical for high productivity [8, 12]. Too much feeding may cause bad productivity, because the saline water would be drained before it gains enough energy for evaporation. Too low feeding rate may cause dry out of the wick, so the system itself could be deteriorated by salt crystal. It was recommended that ratio between fresh water and concentrated water should be around 1:1. As shown in both **Figure 13** and **Figure 16**, this ratio was not controlled well enough yet, especially from around 15:00 O'clock. Better control on water flow may increase productivity. Further optimization on the feeding water distribution towards each layer would also lead higher productivity. Though shred fabric distributer, described in the previous section, was applied on this prototype, there are still room for further optimization of the shred width for each layer.

4. Miscellaneous comments on the experiment

4.1. Checking salt contamination of the condensed water

To avoid salt contamination, saline water in the wick should not flow into condensation layer. Fortunately, in most cases, it is automatically avoided since capillary force in the wick captures the saline water. However, if too much saline water flows into the system abruptly, which could not be kept within the fabric, of course it may overflow into the condensation layer.

To check any trace of salt contamination, simple method could be used [15]. Diluted basic solution, such as aqueous sodium hydroxide solution, could be mixed with the saline water, which would be desalinated by the system. Since even small trace of hydroxide ion increases pH of the solution, it can be easily noticed if the distillates contaminated by salt. Of course, evaporable pH altering material, such as hydrochloric acid, should not be used for such purpose, since it could be evaporated at the wick to change pH of the condensed water.

4.2. Operation of the MES in cold weather

With good insulation and tight seal, inner temperature of the MES was observed to be more than 70–80°C under sun, even below freezing point ambient temperature. This is high enough temperature for water evaporation. However, it is generally not recommended to use it under freezing climate. Though the temperature inside of MES is high under sun, it may freeze during night, which may damage the system if repeated for a longer period. In addition to that, the drained water could be frozen at the outlet of MES at cold weather. At the winter, I found icicles underneath the prototype, which mixed back the purified and concentrated water.

5. Mass production in less-developed country and potential application of the MES

Most of the other competing, small-scale solar desalination instrument requires relatively high-end technology and expensive facility to produce each component, such as nano-porous membrane for reverse osmosis, photovoltaic modules, vacuum tubes etc. It makes less-developed societies difficult to self-supply the tool by themselves. However, MES prototype introduced in this chapter was handmade with elementary tools and materials, such as hand knife, polystyrene foam board and fabrics laminated on plastic film. In other word, it can be relatively easily prepared and studied with minimum budget. In addition to that, it is also easy to be mass produced, not only in advanced countries but also in less-developed countries. Therefore, MES supplying chain could be easily made at the actual countries, where the low-cost small-scale desalination is necessary, regardless of its industrial level.

Wick/condensation layer could be mass produced by roll-to-roll lamination of fabric on plastic film. Lamination process itself is not a high-tech engineering. It would be affordable to most of the underdeveloped countries. However, if it is still too difficult to be manufactured by local industry, laminated material could be produced in more-developed region and then transported to the local producers who do not have expensive machineries. Since the film is light weight and small volume, it is easy to be transported. By the local maker, the sheet material could be cut, folded and assembled with relatively low-labour cost.

Similar approach could be made to the spacer layers. It could be mass produced with foam board by die-cutting press. Any other type of cutters could be used, if the press is not available. Other raw materials, such as foam insulators and windows, are common material for construction, which would be available on most part of the world. All the materials, processes and structures could be modified by the developers reflecting the industrial condition of each one. Because of this industrial flexibility, MES could be useful self-producible desalination tool for many countries.

5.1. Potential applications of the low-cost MES

Low-cost multi-effect solar still could be mostly useful for supplying fresh water for individuals or family who cannot access to public water work. Residents near salty or contaminated water source, remote island or seashore could use this tool. However, it could be useful for public or national level also, not only limited to the individual level. It could be a part of social infrastructure: for example, it can replace highway divider or fences in desert or bridge over sea. Fresh water supplied by MES could be used to cultivate plants along the road. It can be used for fence/wall around buildings or districts to provide fresh water for planting or for citizens. MES modules would also be made in form of tile or curtain walls of buildings. Similar to that photovoltaics module could be part of external surface of buildings (building integrated photovoltaics: BIPV), MES could be part of building (building integrated desalination: BIDSAL), which can produce fresh water for the residents [15]. MES could also be useful as public stockpile against natural disaster or terrorism, in case of existing water line malfunctioning.

6. Suggestion for future work

The MES system introduced in this chapter has both aspects of individuality and publicity. The system could be used as personal water supplier. It could be made, operated and maintained personally without assistance of high-tech industry. However, on the other hands, the system could be more easily produced with lower cost and widely implemented, if mass production industry supports. Once mass produced in low cost, it could be a practical solution to the potable water deficiency problem for large part of the world.

Not only the production and application, research and development of the MES has both individual and public aspects. Initially, I started to develop the low-cost MES as a private project. Because of relatively low material cost and simple measuring instrument, it was an executable project for even an individual. Therefore, I expect that developing MES is relatively easily accessible subject to many other researchers in the world, regardless of their financial status or infrastructure. Every future users, producers or developers of MES in the world may be in different condition: climate; social necessity; industrial level and raw material cost. Therefore, they may need to develop their own optimized system, reflecting their own specific status. Meanwhile, they can share the new findings with others, for worldwide collaboration.

The structure and material introduced in this chapter are just an initial example of low-cost MES. There are many steps that remain before implementation. Further optimization and structural improvement should be made. Practical size prototype should be developed. Long-term stability should be proven by outdoor test under actual climate where the product to be used. Mass production process should be established, reflecting the industrial condition of the producers. Supply chain of raw materials should be established. Local or worldwide distribution of the product should be done with proper instruction of operation and maintenance to the final users. These tasks could be done with international collaboration, especially including the groups in the country where the system has to be implemented. Main purpose of writing this chapter is to suggest active collaboration all over the world. Collaboration between private, public or international supported groups would be helpful for MES implementation. I will also be very happy to be part of the collaboration. Furthermore, it would be helpful to mitigate potable water deficiency problem in the world. Beauty of solar still would be that it may quench thirst of anybody, whether he or she is a drinker or a developer.

Acknowledgements

I appreciate Dr. Park, Chang-Dae in Korea Institute of Machinery and Materials (KIMM), who first introduced MES to me. The subject introduced in this chapter was strongly motivated by his work. I also appreciate Mr. Jin, Byungjin in On-test who helped to equip the measurement system. I appreciate to Korea Water Forum Secretariat (KWFS), since the prize money from 2015 World Water Challenge at 7th World Water Forum was financially helpful on doing this project.

Since it was a privately supported project, I owe many things to my family. My parents-in-law, Mr. Ahn, Jae Chul and Ms. Lee, Young Sook, kindly encouraged me to install the prototype on their rooftop. I was lucky enough to have sunshine test field at the middle of Seoul, where the non-shaded space is rare because of many tall skyscrapers. I developed the origami structure of wick/condensation film while playing origami with my 2-year-old son, Pak, Jongmin I sincerely hope he lives in better future where less people suffer thirst in the world.

Author details

Pak Hunkyun

Address all correspondence to: hunkyunpak@gmail.com

Electronics and Telecommunications Research Institute, Daejeon, Republic of Korea

References

[1] Ghaffour N, Missimer TM, Amy GL. Technical review and evaluation of economics of water desalination: Current and future challenges for better sustainability. Desalination. 2013;**309**:197-207

[2] Talbert SG, et al., editors. Manual on Solar Distillation of Saline Water. Office of Saline Water, United States Department of the Interior; 1970. p. 263 (PDF file available from: http://www.appropedia.org/Solar_distillation#cite_note-20 [Accessed: 2 January 2017])

[3] Defoe C Ginnings. Multiple effect solar still. US Patent 2445350A; 1948

[4] Biswas A, Ruby. Distillation of water by solar energy. VSRD International Journal of Mechanical, Automobile and Production Engineering. 2012;**2**:166-173

[5] Qiblawey HM, Banat F. Solar thermal desalination technologies. Desalination. 2008;**220**: 633-644

[6] Sivakumaran R, Judhesh P. Improvement techniques in performance and productivity of solar stills. International Journal of Engineering Science and Computing. 2016;**6**:2441-2450

[7] World Health Organization (WHO). How Much Water Is Needed in Emergencies [Internet]. [Updated: July 2013]. Available from: http://www.who.int/water_sanitation_health/publications/2011/WHO_TN_09_How_much_water_is_needed.pdf?ua=1 http://www.who.int/water_sanitation_health/emergencies/qa/emergencies_qa5/en/ [Accessed: 2 January 2017]

[8] Tanaka H, Nosoko T, Nagata T. Experimental study of basin-type, multiple-effect, diffusion-coupled solar still. Desalination. 2002;**150**:131-144

[9] Tanaka H, Nakatake Y, Watanabe K. Parametric study on a vertical multiple-effect diffusion-type solar still coupled with a heat-pipe solar collector. Desalination. 2004;**171**:243-255

[10] Tanaka H, Nakatake Y, Tanaka M. Indoor experiments of the vertical multiple-effect diffusion-type solar still coupled with a heat-pipe solar collector. Desalination. 2005;**177**:291-302

[11] Tanaka H, Nakatake Y. Factors influencing the productivity of a multiple-effect diffusion-type solar still coupled with a flat plate reflector. Desalination. 2005;**186**:299-310

[12] Tanaka H, Nakatake Y. Improvement of the tilted wick solar still by using a flat plate reflector. Desalination. 2007;**216**:139-146

[13] Park C-D, Lim B-J, Tanaka H. Development of seawater distiller utilizing waste heat of portable electric generators. Transactions of the Korean Society of Mechanical Engineers B. 2010;**6**:607-613

[14] Pak H. Evaporative saline-water desalination apparatus using solar energy. Patent application number (S. Korea) KR 10-2015-0045121, (PCT) WO2016159625A1

[15] Pak H. Building integrated solar desalination (BIDSAL): Preliminary works with multiple effect solar still. Energy Proceedia. 2014;**62**:639-646

[16] Toyama S, Aragaki T, Salah HM, Murase K, Sando M. Simulation of a multieffect solar still and the static characteristics. Journal of Chemical Engineering of Japan. 1987;**20**:473-478

[17] Adhikari RS, Kumar A, Sootha GD. Simulation studies on a multi-stage stacked try solar still. Solar Energy. 1995;**54**:317-325

[18] Adhikari RS, Kumar A, Sodha GD. Thermal performance of a multi-effect diffusion solar still. International Journal of Energy Research. 1991;**15**:769-779

[19] Rajaseenivasan T, Kalidasa Murugavel K, Elango T, Samuel Hansen R. A review of different methods to enhance the productivity of the multi-effect solar still. Renewable and Sustainable Energy Reviews. 2013;**17**:248-259

[20] Tanaka H, Nosoko T, Nagata T. A highly productive basin-type-multiple-effect coupled solar still. Desalination. 2000;**130**:279-293

[21] Estahbanati MRK, Feilizadeh M, Jafarpur K, Feilizadeh M, Rahimpour MR. Experimental investigation of a multi-effect active solar still: The effect of the number of stages. Applied Energy. 2015;**137**:46-55

[22] Huang B-J, Chong T-L, Chang H-S, Wu P-H, Kao Y-C. Solar distillation system based on multiple-effect diffusion type still. Journal of Sustainable Development of Energy, Water and Environment System. 2014;**2**:41-50

[23] Tiwari GN, Lawrence SA, Gupta SP. Analytical study of multi-effect solar still. Energy Conversion and Management. 1989;**29**:259-263

[24] Goff PL, Goff JL, Jeday MR. Development of a rugged design of a high efficiency multi-stage solar still. Desalination. 1991;**82**:153-163

[25] Xiong J, Xie G, Zheng H. Experimental and numerical study on a new multi-effect solar still with enhanced condensation surface. Energy Conversion and Management. 2013;**73**:176-185

[26] Gräter F, Dürrbeck M, Rheinländer J. Multi-effect still for hybrid solar/fossil desalination of sea and brackish water. Desalination. 2001;**138**:111-119

[27] Park C-D, Lim B-J, Chung K-Y, Lee S-S, Kim Y-M. Experimental evaluation of hybrid solar still using waste heat. Desalination. 2016;**379**:1-9

12

Importance and Significance of UF/MF Membrane Systems in Desalination Water Treatment

Iqbal Ahmed, Khaled S. Balkhair,
Muhammad H. Albeiruttye and
Amer Ahmed Jamil Shaiban

Abstract

The proposed chapter addresses a comprehensive overview of the history and future outlook of ultrafiltration/microfiltration (UF/MF) membrane for desalination water pretreatment. Known theories on UF/MF membrane formation from phase inversion (Dr/wet) systems can be prolonged to define the consequences of high or low molecular weight additives. Also, direct material reengineering and surface modification for high-performance anti-fouling of UF/MF membranes are also highlighted. Before the modern final polymeric film, the characterization techniques, particularly molecular weight cut-off, pore size, pore size distribution, and microbiological activity classification, on to the UF/MF membrane surface were presented, respectively. Lab scale to commercial scale UF/MF membrane configuration and market size of UF/MF membranes for pretreatment desalination are described. The significance of UF/MF provided here as an unconventional approach for desalination water pretreatment is in contrast with the current conventionally used technologies. The recent development made in the integration of established desalination processes, such as spiral wound reverse osmosis (SWRO), multi-stage flash (MSF), multi-effect distillation (MED), electrodialysis (ED) desalination, and UF pretreatment, is addressed. Finally, the influence of UF/MF on desalination water pretreatment step on to the energy cost of desalination process system is discussed.

Keywords: desalination, microfiltration, post-treatment, ultrafiltration

1. Introduction

1.1. An early history of UF membrane

Almost every chemical process involves at least one separation or purification step, and the chemical industry has developed a range of separation techniques to facilitate recovery of the required products. In recent years, membranes and membrane separation techniques have grown from laboratory tool to an industrial process with considerable technical and commercial impact. Instantly, the membrane processes are faster, more efficient, and economical than conventional separation techniques, particularly for desalination water treatment. Among all the membranes process, UF/MF membranes have the largest variety of applications in various industries, because it is a separation technology of high efficiency and low energy consumption [1].

The permeation of water by a thin sheet of animal bladders (diaphragm) was introduced by Abey's Nollet in 1755 (France), and the phenomenon of water permeation was named as osmosis [2]. Later on, Dr. Adolf Eugen Fick from Germany has introduced diffusion law and was developed the first high-pressure synthetic membrane made from nitrocellulose in 1855 [3]. After 50 years, Dr. Bechhold from Germany developed first low-pressure cellulosic membranes (collodion), which is prepared by impregnating filter paper with glacial acetic acid.

The first such low-pressure membranes were produced in 1907, and Dr. Bechhold revealed the term "Ultrafilter" collodion membranes. Since for Bechhold's original membranes were introduced and applied protein solutions by forcing at several pressures against to the atmospheric pressure through his collodion membranes [4]. After Dr. Bechhold breakthrough in UF/MF membrane, there has been continuous effort to develop improved UF membranes, which have resulted in many diverse types of such membranes [5]. Further early developments, principally Zsigmondy and Bachmann [6] and Ferry [7] improved on Bechhold's membrane fabrication method, Elford developed graded porosities UF membranes having the properties "Gradocol" [8]. By the early 1930s, microporous cellulosic membranes such as cellulose nitrate, cellulose di/triacetate were commercially available [9, 10]. With further growth of UF/MF synthetic cellulosic membrane dur-ing 1950s, the synthetic membrane technology was commonly available for removal of bacteria, virus, dextrin, protein from water, in addition to salinated water cleansing, respectively. But, the little flux was the main drawback of such Bechhold's type membranes. These deficiencies put together are too costly and practically inappropriate. The period of cost-effective feasible mem-brane advancement, which was started in the late 1950s and prolongs to this date, may be divided into two time periods. The first generation was from 1959 to 1970 of cellulose acetate integral asymmetric membranes, and the second generation started from 1971 to 1984 of noncellulosic asymmetric membranes [10, 11].

The significant development in artificial membrane technology began in the 1960s and to be headed toward the growth of new UF/MF membranes materials. These events allowed for wider industrial and commercial application of UF/MF [10, 11]. In 1962 Loeb and Surirajan developed a new method of polymeric membrane fabrication, called dry/wet phase inversion process. After the breakthrough of phase inversion the history of the synthetic membrane were entirely changed which was active properties regarding mechanical strength, membrane morphologies, and ten times higher performance than the earlier membrane. **Table 1** shows the first contribution to UF membrane development [10, 11]. The beginning of

Inventor	Development	Year
Bechold	Prepares collodion membranes of graded pore size measure bubble point and use the term ultrafilter	1906
Zsibmondy and Bachmann	Patent collodion filter (German Patent 329-060)	1918
Filter GmbH	Commercializes ultrafiltration membranes	1926
Reid and Breton	Selection of cellulosic material for membrane making	1959
Loeb and Surirajan	Cellulosic acetate integral-asymmetric membranes	1960
Amicon by Koch	Market laboratory-scale UF membranes develop polysulfone, PVDF membranes	1966
Amicon by Koch	Make first UF hollow fiber membrane	1967
Abcor by Koch	Installs commercial tubular UF plant (electro coat paint)	1969
Romicon by Koch	Introduces hollow fiber capillary UF plants	1973
Abcor by Koch	Commercializes spiral wound UF modules	1980
Abcor by Koch	First commercially significant ceramic membrane	1988

Table 1. Key historical development of UF membrane from Market laboratory - scale to commercial scale [10, 11].

thin-film-composite (TFC) polymeric membranes started during 1963, initiated by a research institute and one of its first employees, Peter S. Francis [11, 12]. A significant discovery was made in the art of thin-film-composite membranes by Cadotte in 1970 with the beginning of large-scale commercial polymeric membrane. Cadotte invented two innovative techniques of TFC membrane based on interfacial polymerization and solution coating methods [12]. During 1970–1990, the researchers developed important methods of membrane materials synthesis, membrane fabrication process, membrane geometry, separation, and purification processing techniques. Also during 1985, the synthetic commercial membrane entered into a new era, and by the end of 1990, the MF, UF/dialysis, nanofiltration (NF), RO, ED, and gas separation membranes technology have grown steadily at the industrial level. Moreover, by the end of 19th century, the advancement of membrane growth has enhanced performance, steadiness, and provided lower operating costs, making membranes the preferred technology in the water treatment industry as well as in the food and pharmaceutical industries [1, 10, 13].

1.2. Theoretical background of UF/MF

UF/MF membranes [14, 15] have a porous barrier structure that retains components by a sieving mechanism and used for separation of solutes from the feed solution. The water flux in MF and UF is proportional to the applied pressure (Darcy's law). A pressure gradient (0.5–5 bar) across the membrane transports the solvent through, while the substances larger than the pores are rejected. It is agreed that the important process involved in UF/MF is one of sieve action, tricky by adsorption and other consequences are emerging from the unusually large ratio of pore length to pore diameter in all pressure-driven membranes. The principle of the pure water flux into the UF/MF membranes is due to the capillary pore diffusion model and the mean pore radius can be calculated by Hagan-Poiseuille equation [15, 16],

$$\bar{r} = \left(\frac{8\,J_{\mu}\,\Delta x\,H_{p}}{\varepsilon}\right)^{\frac{1}{2}} \tag{1}$$

whereas $\bar{r}$ is a mean pore radius, $J\mu$ shows pure water viscosity, Δx is membrane thickness, Hp is hydraulic pressure, and ε shows membrane porosity.

However,

$$H_{p} = J_{p}/TMP \tag{2}$$

whereas, Jp shows water permeation, and TMP presents transmembrane pressure (ΔP).

Combine Eq. (2) into Eq. (1),

$$\bar{r} = \left(\frac{8\,J_{\mu}\,\Delta x\,TMP\,\tau}{\varepsilon\,J_{p}}\right)^{\frac{1}{2}} \tag{3}$$

The leading theory of fluid flow through UF/MF membrane in an ideal condition such as consistently sized pores in the membrane, negligible concentration polarization, no fouling, respectively. **Figure 1** summarized the theory of transport phenomenon and fluid dynamics of an UF membrane. Also the pore size, some other factors such as interactions between UF feed components and membrane matrix play a significant role in the transport through the membrane [10, 17–19].

Sakai [19] reported that the Eq. (3) is directly relating to the membrane structure. Sakai and co-worker also indicated that Verniory et al. improved the UF/MF membrane transport parameters. Among membranologist, it has been widely proven that UF/MF membranes flux and rejection are depended upon their structure. Nakao [19–22] has reported that in the case of known relation between flux and rejection, the membrane structure can be characterized

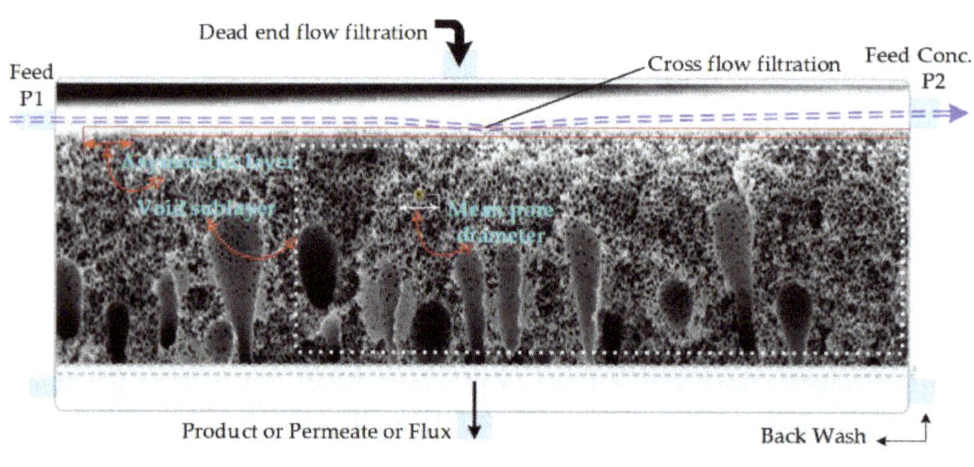

Figure 1. Predicting flux from pore statistics using Hagen-Poiseuille Equation [17].

such as thickness, pore size, pore radius, pore volume, pore density, and tortuosity, respectively. However, Sakai successfully investigated the qualitative attempts to account for tortuosity (τ) in the pore model. Tortuosity τ perhaps described the ratio of pore as,

$$\tau = \frac{\text{Mean pathh length}}{\text{Membrane thickness}} = {}^{L}/_{\Delta x} = {}^{W_c}/_{\Delta a} > 1 \tag{4}$$

where W_c is present water content, which approximates the pore volume in the membrane.

$$J_p = \left(\frac{\varepsilon \, . \, d_p^2 \, . \Delta P_T}{32 \, \Delta x \, . \, \mu} \right) \tag{5}$$

When Eq. (3) is used to calculate the mean pore radius, it becomes complicated because of the introduction of tortuosity, and the result is larger than that obtained from Eq. (1). Unfortunately, it is impossible to determine the pore sizes of asymmetric membranes with the aid of Hagen Poiseuille equation (3), which makes it necessary to use, for this purpose, the data of more complicated methods [23]. Among all above methods, the most informative are the means of electron microscopy, gas pycnometer, which give the possibility to determine pore sizes and pore size distributions (PSD) of asymmetric membranes [18, 24–27].

2. Significant development of low-fouling UF/MF membranes lab scale to commercial scale

Synthetic polymeric membranes can be divided into hydrophobic and hydrophilic classifications, and structure can also be classified. Structural classification is critical because it is the structure which determines the separation mechanisms and the membrane application [10, 28, 29]. Membranes can be further classified as symmetric or asymmetric [10, 29]. The symmetric membranes can be porous, cylindrical porous, and homogeneous (nonporous). The asymmetric membranes can be porous, microporous with top layer, and composite that is consisting of a porous substrate with a dense top layer. The thickness of the top layer in asymmetric membranes is in the range of 0.1–0.5 μm and is supported on a porous sub-layer with a thickness of about 50–150 μm [10, 11, 29]. The development of pressure-driven membrane technology was began after Loeb-Surirajan and Riley et al. [30, 31] and Cadott discovered about the three following significant developments [12],

i. Ability to fabricate particular selective membrane that has high permeation, essentially ultrathin, dense layer, surface barrier layer, integrally supported by a thick, porous, spongy structure and was able to produce at commercial scale.

ii. Ability to form polymeric membranes into compact, high-surface-area, economical membrane configuration.

iii. TFC or ultra thin film (UTF) able to be laminated on microporous or porous support layer.

Modern membrane technology began in late 1990s, the development of polymeric membrane chemistry and processing techniques are used in membrane fabrication. With the developments in polymeric membrane materials, manufacturing technologies, and water treatment processing systems have made this technology an efficient, economical for water treatments, and competitive with traditional water treatment methods [10, 15, 29]. Recently, UF/MF membranes have been succeeded for a range of industrial applications [32]. Each application enforces precise specifications on the membrane material and membrane structure. The revolution in understanding the origin of these structural elements of Loeb and Surirajan phase inversion process was obtained by Wienk et al. and Wu et al. [33, 34],

i. Thermodynamic calculation and kinetics of phase separation of polymer/solvent/nonsolvent.

ii. The role of additives, both high- and low-molecular weight, on membrane formation.

iii. Theory behind macrovoids porous and nodular structures formation.

The latest development revealed "next generation" of membrane materials for UF/MF focus onto:

i. Well-distinct configuration as 'tailored' membrane materials.

ii. Innovative processing of polymers for membranes, particular neatness of membrane industrialized.

iii. Superior functional polymer membranes, qualifying the integration of active barrier structure with 'customized' approach of interactions.

iv. Groundwork of nanoparticles mixed matrix membranes for the synergistic allying of different functions by different polymeric materials.

Also, the development is involved in both organic and interpenetrating multiphase structures with excellent transport properties, agreement to allow membranes with superior chemical/thermal stability, fouling resistance, organic solvent resistance, and unusually high permselectivities and permeabilities [35]. Such kind of polymeric membranes may well circumvent many of these limitations. Similarly, recent developments in UF/MF membrane module design, including rotational membrane devices and cycled flow fluid management for fouling control, use of low-cost refractory monoliths as membrane supports, and use of electric potentials to minimize the fouling rate on to the membrane surface [24]. Today, almost 98% of cross-flow membrane systems installations use polymeric UF/MF membranes [13, 36].

Principally, almost all commercial membrane lifespan faces two serious issues due to natural phenomena during separation (solutes) and purification (fluxes/permeate) such as concentration polarization (solutes) and fouling. A typically asymmetric polymeric membrane as shown in **Figure 2** has random pore sizes. Therefore, concentration polarization plugging up on membrane pores and fouling is happened due to the living microorganism adhesion, gel layer formation, and solute adhesion at the membrane surface (see **Figure 1**) [37, 38]. Several researchers and manufacturers have revealed that the natural phenomenon is responsible for restricting the permeate flux during cross-flow (i.e., permeation followed by cleaning). Throughout the early stage of filtration process within a cross-flow rotation, concentration polarization is one of the prime causes for flux reductions [39, 40].

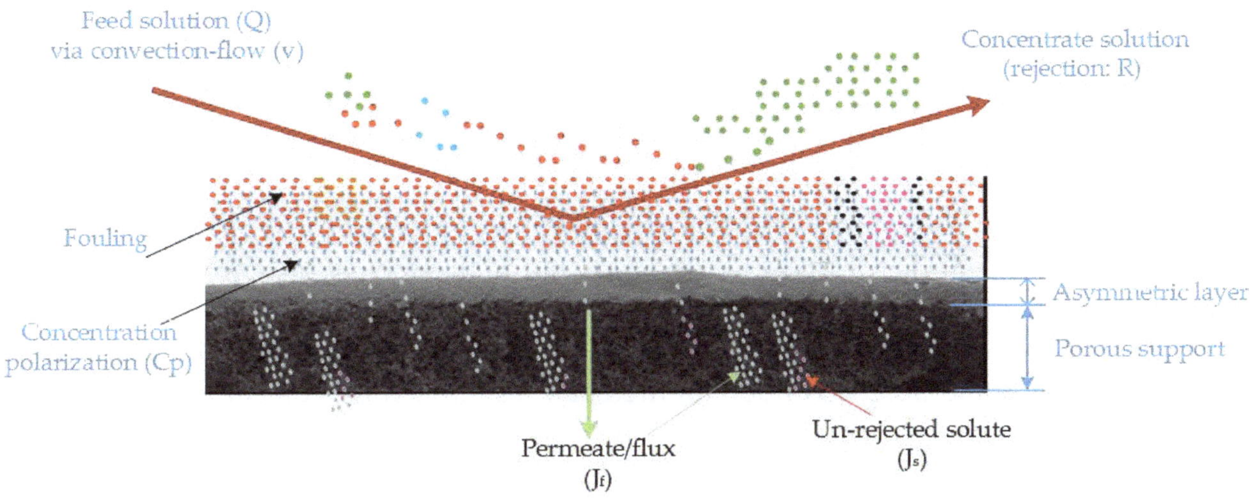

Figure 2. Typical cross-flow process with asymmetric membrane [17].

Commercial scale polymeric UF/MF membrane systems run in a continuous cross-flow mode, where backwash and cleaning operation interchanges with the normal operation. The decrease in the flux for pure water from cycle to cycle, because of fouling, the flux decline within a period due to concentration polarization, and the average flux under steady state level. The latter is also decreasing from cycle to cycle, suggests irreversible solute adsorption or fouling [41–43]. And of the solute retained on a membrane surface leads to increasing permeate flow resistance at the membrane wall region. Strategies to minimize the effect of fouling can be divided into two groups: avoidance and remediation. The remediation is to clean up by the cleanup process, is usually done by chemical cleaning at regular times [44], and this is necessary for all membrane processes in nearly all applications. However, large differences in the cleaning frequency can be found, ranging from daily to yearly, depending on the concentration of foulant and the pretreatment. A large number of cleaning agents are commercially available. The choice of optimal product depends on feed characteristics. Acid cleaning is suitable for the removal of precipitated salts, such as $CaCO_3$, whereas alkaline cleaning is used to remove adsorbed organics. Nearly, all cleaning products contain detergents. Another remediation technique often used in UF/MF is backwashing or back pulsing. A short pulse of water or air from the permeate side to the feed side efficiently removes all fouls blocking the membrane pores [45]. This principle is often applied in a dead-end or semi-dead-end filtration. It is possible to avoid fouling by using adequate pretreatments, such as coagulation precipitation, or slow sand filtration [10, 15, 29, 46].

2.1. Material selection for UF/MF fabrication

Recently, UF/MF membranes become an innovative and dominant technology and have been extensively used in many areas, including wastewater treatment, protein separation, dialysis, and dairy industry [47]. However, the most common applications of UF/MF in downstream processing are protein concentration (i.e., solvent removal), buffer exchange and desalting, virus removal and clarification [41]. Since, the improvements of UF/MF technology to make membrane water treatments economically competitive with traditional water treatment methods

[10, 29, 32], the use of these membranes has increased exponentially for the downstream process of sea water desalination (SWD). Also, the UF/MF membranes employed in SWD have gained significant attention as these methods are efficient in removing the turbidity, the particles, and the microorganisms present in wastewater [10, 29]. With the improvements in this technology to make membrane for separation and purifications economically competitive with traditional separation methods [10, 17, 45], the use of these membranes has increased exponentially.

It has been established that the porous structure and hydrophilicity of UF/MF play crucial roles in membrane manufacturing processes [29, 34, 48]. A suitable porous membrane should be excellent in permeability, hydrophilicity, and chemical resistance to the feed streams. An asymmetric membrane is a good option for high permeability. Thus, currently, much effort is being devoted to improve the performance of the existing membranes regarding anti-fouling properties, high mechanical strength, and excellent chemical resistance. To make a porous or microporous membrane, some mineral or ceramic membranes have been developed. However, polymeric membranes are yet mostly used [49, 50]. Therefore, different polymeric materials have been used for UF/MF membranes and investigated at lab scale to be commercial with changing results, as not all of them produce membranes with suitable performances [50]. Nevertheless, since the first membrane cellulosic and noncellulosic materials were described by Reid and Breton in late 1959 [51], numerous materials have been developed to improve the capacity and performance of membranes filtration [11, 17, 29]. For a given treatment stream, a particular polymeric membrane material can be selected from an assortment of candidates. Till now, there are more than 130 materials (cellulosic, noncellulosic polymers, composite, and inorganic) that have been used to manufacture membranes [11, 17]. The range of materials from which it is possible to create some form of artificial membrane structure is extensive. Each year, number of research papers in polymer and membrane science present many new examples of materials that demonstrate semi-permeable qualities at some scale. However, only a very limited number of these potential candidates make it to the commercial environment [1, 52–54]. Very few materials possess the structural and chemical properties necessary to render them suitable for application in industrial scale membrane processes. **Table 2** shows the various hydrophilic and hydrophobic polymers used for membrane production at lab scale to commercial scale [11, 17, 29, 53]. Furthermore, typical commercial hydrophilic co-polymers are made of polyethylene oxide (PEO), crystallizable polyamide (PA), nylon or aliphatic polyamide (PA6 and PA66), polyurethanes (PU), and polyester (PET). These materials can be used to make a hydrophobic polymer more hydrophilic. Hydrophobic polymeric materials such as PC, PSF, PES, PVDF, PI, PEI, Ar.PA, polyether ether ketone (PEEK), and PAN, are also conventional polymeric materials for the preparation of UF/MF membranes [10, 17, 29, 55].

Among all those materials, CA, PSF, PES, SPES, PAN, and PVDF are the most commonly used polymers for UF/MF membranes at lab scale to commercial scale [1, 17, 29, 52]. Generally, PSF, PES, and polycarbonate (PC), respectively are produced by aromatic bisphenol intermediates such as bisphenol-A-PSF, tetramethyl bisphenol-A polysulfone (TM-PSF), bisphenol-B polysulfone (PSF), and their modified form such as sulfonated polysulfone (SPSF), and sulfonated polyethersulfone (SPES) have been used extensively to fabricate UF/MF [11, 17, 54]. **Figure 3** shows several types of aromatic bisphenol intermediates used for PES, PSF, and polycarbonate (PC) preparation [56]. Typically, all these thermoplastic base materials can easily be dissolved

Hydrophilic polymers			
Poly(viny1 alcohol)	PVAL	Cellulose and its derivative	
Poly(viny1 chloride)	PVC	Cellulose acetate	CA
Polyamide	PA	Cellulose triacetate	CTA
Poly(acry1ic acid)	PAA	Cellulose acetate butyrate	CAB
Poly(ethy1ene oxide)	PEOX	Cellulose acetate propionate	CAP
Polyacrylonitrile	PAN	Cellulose nitrate	CN
Poly(viny1 acetate)	PVAC	Cellulose propionate	CP
Poly(viny1 butyral)	PVB	Ethyl cellulose	EC
Poly (p-hydroxystyrene)	PHS	Carboxymethyl cellulose	CMC
Hydrophobic polymers			
Polysulfone	PSF	Polytetrafluoroethylene	PTFE
Polyethersulfone	PES	Polyethylene	PE
Poly (vinylidene fluoride)	PVDF	Silicone	Si
Polycarbonate	PC	Polyphenylene oxide	PPO
Polypropylene	PP	Polyphenylene sulfide	PPS
poly(methyl methacrylate)	PMMA	Polystyrene	PS

Table 2. Commercial available hydrophilic and hydrophobic polymers for membrane production [11, 17, 29, 53].

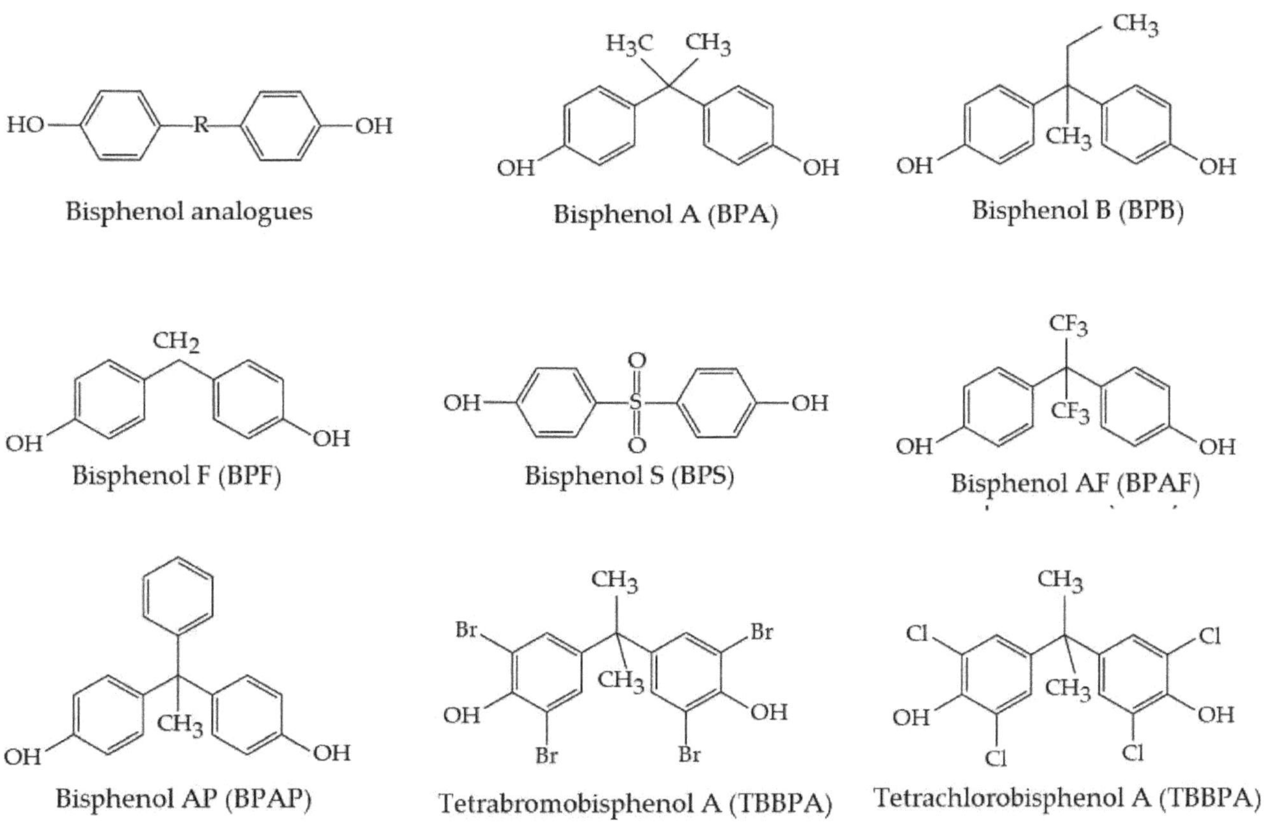

Figure 3. Chemical structure of aromatic bisphenol intermediates of PSf, PES, and PC [55].

in aprotic solvents and produce membranes with excellent thermal, hydrolytic, and mechanical stability properties in both hot and wet environments. **Figures 4** and **5** show chemical structures of CA, CTA [56, 57], and thermoplastic polymers [17, 58–62], and **Table 3** summarized the leading manufacturer of polymer used for membrane fabrication. Almost more than 90% of membrane manufacturer are producing PSF, PES, TM-PSF, SPSF, SPES, PVDF, and PAN membrane in a wide range of UF/MF applications. To prepare membranes for the liquid separation processes using repeated applications with either hot water or sterilization to keep the membrane clean [11, 17, 63]. Sulfonated polyethersulfone membrane has been customized to be drastically more hydrophilic than standard PES, PSF membranes [29, 42, 58–60, 62, 63]. It is biocompatible and has highest opposition to fouling by hydrophobic compounds such as fats, lipids, anti-foams, and other similar highly fouling substances [1, 10, 32, 42, 60, 64].

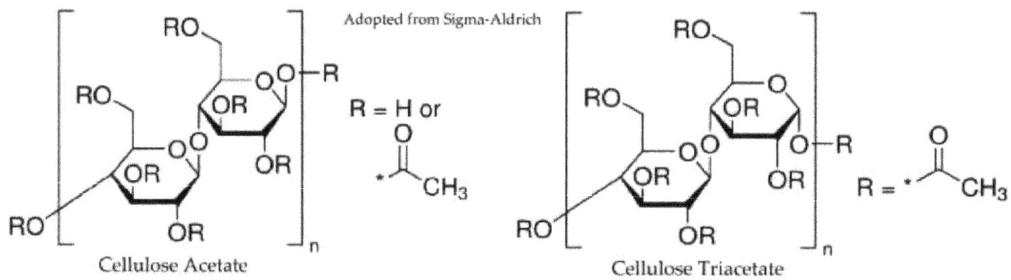

Figure 4. Chemical structure of commercial CA and CTA [56, 57].

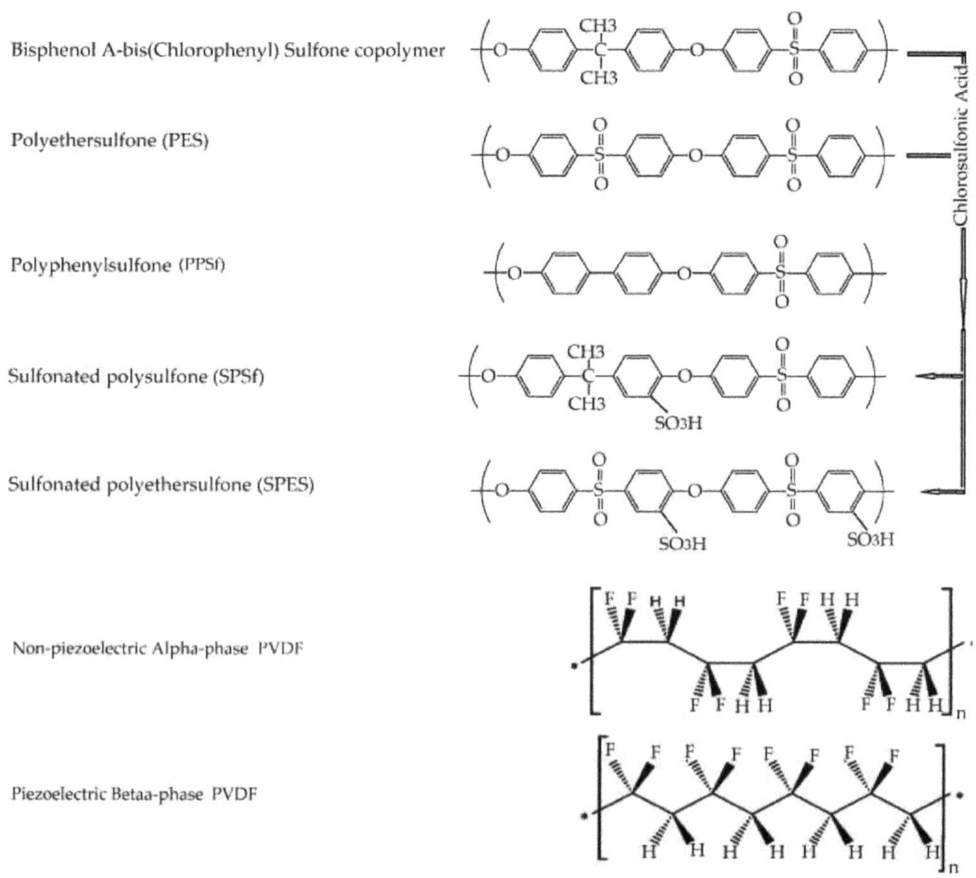

Figure 5. Chemical structure of hydrophobic PSF, PES and modified hydrophilic SPSf, SPES [58, 59].

Manufacturer	Product name	Trade mark	WA	TM	Tg	MW
			%	MPa	oC	g/mol
BASF	Blend GF45	1010 G9	0.65%	15,500	216	np
	PESU	E 1010 NAT	0.80%	2650	222	np
	PESU	E 2010 SW Q31	0.80%	2650	225	np
	PESU-CF30	E 2010 C6	0.60%	22,000	225	np
	PESU-GF20	E 2010 G4	0.60%	7300	225	np
	PESU-GF20	E 2010 G4 MR	0.60%	7301	225	np
	PESU-GF30	E 2010 G6	0.60%	9800	225	np
	PESU	E 2010 HC	0.80%	2650	225	np
	PESU	E 2010 MR	0.80%	2650	225	np
	PESU	E 2020 P	1%	2650	2250	4800
	PESU	E 2020 P SR	1%	np	np	55,000
	PESU	E 2020 P SR	1%	np	np	55,001
	PESU	E 3010	0.80%	2650	228	np
	PESU	E 3010 MR	0.80%	2650	228	np
	PESU	E 6020 P	1%	2650	225	75,000
	PESU	E 7020 P	1%	np	225	92,000
	PESU+PTFE	KR 4113	1.50%	11,000	225	np
	PPSU	P 3010	1.20%	2270	220	np
	PPSU	P 3010 MR	1.20%	2270	221	np
	PSU	S 2010	0.80%	2550	187	np
	PSU	S 2010 G4	np	np	np	np
	PSU-GF30	S 2010 G6	0.60%	8900	187	np
	PSU	S 3010	0.80%	2550	187	52,000
	PSU	S 3010 MR	0.80%	2600	187	np
	PSU	S 6010	0.80%	2550	187	60,000
SOLVEY	Udel PSU	P-1700	0.30%	2480	174	np
	Udel PSU	P-1720	0.30%	2480	174	np
	Udel PSU	P-3500 LCD	0.30%	2480	174	np
	Udel PSU	GF-110	0.29%	3720	179	np
	Udel PSU	GF-120	0.29%	3720	180	np
	Udel PSU	GF-130	0.20%	8690	181	np
	Veradel PES	3000 P	0.50%	np	np	np
	Veradel PES	201 NT	0.50%	2100	np	np
Solvey Solef	PVDF	6020	0.04%	1700	−40	np

Manufacturer	Product name	Trade mark	WA	TM	Tg	MW
			%	MPa	oC	g/mol
Sumitomo chemicals	Sumikaexcel PES	3600P	0.43%	2550		
		4100P	np	np	np	np
		4800P	np	np	np	np
		5200P	np	np	np	np
		5003P	np	np	np	np

np: not provided.

Table 3. Membrane polymer manufacturer.

2.1.1. *Surface chemistry and choice of additives*

The significance of all pressure-driven membranes particularly UF/MF rely on the properties of their surfaces. As reviewed above and has been proven that the thermoplastic polymers specifically PSF, PES, SPES, PVDF, and PAN are dominant materials for membrane making at retail level [10, 17, 63]. However, surface contamination which may lead to deterioration in membrane performance is also known to be governed by the membrane surface properties and obstacle in membrane performance. Therefore, the membronologist has been paid much attention to the membrane surface modification and were identified theoretical and phenomenological reasons behind the hydrophobic reasons of thermoplastic polymers [17, 29, 32, 64, 65]. Also during 50 years, membronologist has developed very innovative methods to modify the hydrophobic membrane surface into hydrophilic. Zeman and Zydney [66] have reported that almost 50% of all MF and UF membranes traded by 1996 were surface-modified. Moreover, many numbers of high cited research manuscripts and books have been published regarding membrane surface modification techniques. Mittal1 has compiled highly cited work in several volumes entitled "Polymer Surface Modification: Relevance to Adhesion." Pinnau and Freeman [65] edited a book of a membrane-based symposium entitled "Advanced Materials for Membrane Separations" [62]. In this book, relevant topics covered in the 2001 ACS Symposium summarize recent advances in various research areas for development of novel materials used in membrane separations. Benham and Kinstle [67] edited ACS symposium entitled "Chemical Reactions on Polymers," the topics included as one of the greatest active fields in polymer science because of its unique ability to produce specialty polymers with desirable chemical and physical properties through modification of readily available polymers [58]. Xu and co-authors [68] published a very comprehensive book on surface engineering of polymer membranes, squeezes those processes which alter membrane surfaces to improve their in-service performance. The book shows the basics of the surface design of UF/MF polymeric membranes, together with membrane surface modification, to minimize fouling, to modulate hydrophilic and hydrophobic. Also, improve biocompatibility, act as a diffusion barrier, provide bio- or chemical functionalities, mimic a biomembrane, fabricate nanostructures, or directly improve the esthetic appearance of the membrane surface. Xu and co-authors [68] also described general techniques of surface modification of membranes via functionalization and macromolecule immobilization methods for membrane surface modification.

Morao et al. [69] also reported a comprehensive literature work on surface modification plans that include ion-beam irradiation [70]. Mulder, Pinnau and Freeman said plasma treatment or grafting [29, 66] and UV-induced grafting [67] on a polymeric membrane surface. Several other authors also reported chemical sulfone enrichment [60, 71], chemical dehydrofluorination by alkaline solution [71–73], coating temperature-sensitive polymeric brushes [74], and grafting with pH- and ionic-strength-sensitive polymeric brushes [10, 29, 72, 74]. Moreover, researchers has been used several other techniques such as, the irradiation-induced grafting [75], physical adsorption of water-soluble polymers [76], a formation of Langmuir-Blodgett films [77], thermal grafting of a hydrophilic polymeric surface coating [78, 79], and photografting with UV irradiation [68], respectively. All these surface modification techniques are usually applied on hydrophobic-casted surface and these are complicated and expensive and require at least one additional step in the membrane preparation process [68]. Besides, the physical techniques, all the methods mentioned above allow the membrane surface to be modified without affecting the bulk properties too much when appropriate conditions (the modification time) are selected. Technically the modification of thermoplastic or thermoset types of UF/MF membrane surface by using physical techniques revealed are easy, economical, green, and improved surface properties. However treating polymeric membrane surface by physical method resulting unstable and mechanical strength drawback [80–82].

Typically, several methods have been widely used to fabricate synthetic membranes using all mentioned polymers (See **Table 2**). The principle of synthesis is to transform the polymeric material using a convenient method to achieve a polymeric membrane structure with a significant morphology for a distinct separation. The techniques that are being employed for the preparation of artificial membrane are phase inversion, stretching of films, irradiation and etching of films, track-etching, sintering of powders, sol–gel process, microfabrication vapor deposition, and coating [11, 29, 33, 49, 81]. The ultimate morphology of the membrane film or fibers will deviate significantly based on the properties of the selected polymeric materials and the operational conditions. The majority of membranes at the lab to commercial scale are prepared by controlled phase inversion process, i.e., dry/dry, dry/wet, and wet/wet; yet, preferably, the dry/wet phase inversion techniques are the most useable process currently [1, 29, 33, 49, 81]. Consequently, polymeric membranes are formed by the so-called phase inversion techniques, which include the main steps of Refs. [33, 49, 81] are:

- **a.** Dissolving the polymer together with pore-forming additives (either organic or inorganic) in an aprotic solvent (preferably NMP, DMAc, DMF, DMSO, acetone, etc.).
- **b.** Flat sheet casting the resulting solution as a thin film on the surface of nonwoven PET fabric.
- **c.** The solution is spinning the self-support as a geometry of capillary or hollow fiber.
- **d.** Dipping the spun hollow fiber or cast film in a polymer nonsolvent bath.

2.2. Comprehensive characterization of UF/MF membrane

The most significant methodological problem in engineering and fabricating of pressure-driven membranes is "tailoring" of membrane surface function into hydrophilic or hydrophobic together with anti-fouling properties, uniform structure, and pore forming properties for the selective sieving process [62, 81, 83]. Therefore, the characterization of surface chemistry is an essential tool in membrane science and technology, because it is well known that membrane performance depends not only on feed hydrodynamics and steric hindrances but also on membrane surface (hydrophobicity, hydrophilicity, and membrane surface charge) and membrane-solute(s)-solvent chemical interactions. Moreover, without characterization, the synthesized pressure-driven membrane cannot predict their properties mainly pore size or molecular weight cut-off (MWCO); thus, it is only a tool for membranologist to emphasize the significance of prepared polymeric membranes. Nevertheless, to tailor UF/MF membranes and eventually use the most important polymeric membrane for a final application, most important membrane has to determine the attribute of MWCO, pore size distribution, using independent characterization methods [10, 11, 29]. Among these are:

i. Surface functionalization analysis

(a) Contact angle

(b) Fourier transform infrared (FTIR)

(c) X-ray diffraction

ii. Structural analysis

(a) AFM and SEM/EDX

(b) Membrane equilibrium water content (EWC) and porosity

iii. Performance evaluation

(a) Membrane affinity test

(b) Molecular weight cut-off (MWCO)

(c) Pore size and pore size distribution

(d) Solute rejection curve

2.2.1. Surface functionalization analysis

2.2.1.1. Contact angle

As reviewed above that the surface hydrophilicity of selected polymer is a major component to predict the membrane affinity with permeate and anti-fouling properties of UF/MF membranes. Hydrophobic polymeric membranes have nonpolar groups and lower surface free energy; this can prevent contact with water and can push out the water molecules adjoining it [36, 75, 84]. The majority of membranologists and manufacturers are using the hydrophobic polymer as a base polymer for UF/MF fabrication. The relative hydrophobicity and hydrophilicity of membrane surface can be qualitatively determined by computing the contact angle of a water drop deposited onto the surface, which has commonly been used to assess the wettability and interfacial energy of the substrate surfaces [76]. **Figure 6** summarized several types and most suitable hydrophobic and hydrophilic thermoplastic polymers for membrane constructions [17, 58–61]. In **Figure 6**, among all polymers, sulfonated polymers are hydrophilic form of PSF, and PES and their contact angle almost 50% lower than virgin polymer [85]. Typically, contact angle measurements should be made inside an enclosed chamber to stop airborne particles and establish an equilibrium vapor pressure of the liquid tested, which is especially preferable when the test liquid is volatile. It has been observed that evaporation can cause the fluid front to retract and that a retreating or an average contact angle is recorded unintentionally. However, the inherent inaccuracy of the direct measurement technique and the use of liquids with high boiling points make the enclosed chamber unnecessary in many cases.

Therefore, the contact angle of the surface of the membrane can be measured at ambient conditions using the contact angle to water the produced membranes was measured by the sessile drop method [85–87]. Sample coupons at horizontal dimension (2.5*6 ± 0.5 cm) should be prepared by cutting the membranes sheet at random locations. Then, the sample should be placed on a glass plate (active surface of the membrane must be upward) and fixed with a double-sided tape. A drop of double-distilled water (5 μl) should be placed on the surface using a microsyringe (Hamilton Company, Reno, NV). The position of the moving bed can be adjusted manually so that the water drop will be fitted to the scale when projected on the screen. After a fixed deposition time (10 sec), the image will be recorded. To measure the contact angle, the height-width method can be used. For the reason that this approach gives reliable data for contact angles less than 90°. The contact angle should be measured at five different spots on each membrane sample so as to ensure reproducibility of data.

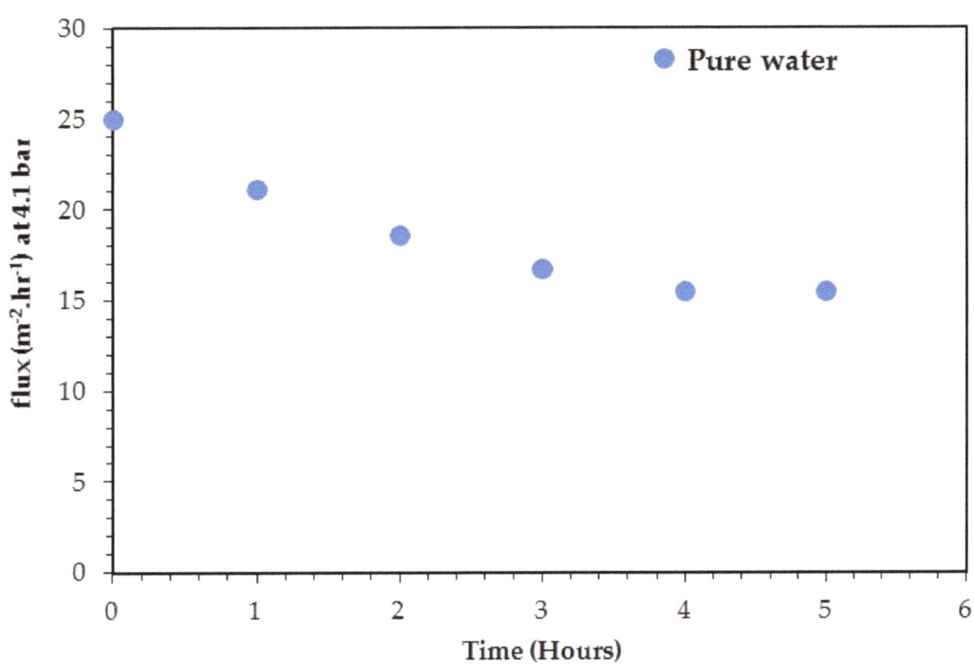

Figure 6. Membrane hydraulic resistance against pressure with respect to time.

2.2.1.2. *Fourier transform infrared (FTIR)*

The analysis of polymeric membrane films in transmission using FTIR is particularly a consistent method for assessing the significant properties of polymers due to additives for achieving quantitative information on modified polymer used for the membrane [53, 88]. There are different methods of polymer sample handling used with FTIR, together with diffuse reflectance (DRIFTS), attenuated total reflection (ATR), and correct specular reflectance/reflection absorption. Established methods of sample holding are more helpful than others for particular specimen types. To analyze and achieve a high-quality spectrum of a sample, it is important to know which handling technique works best for the sample type. Since, FTIR scanning is too responsive, therefore, before analyzing membrane samples make sure the sample should be free of contamination. Therefore, before scanning membrane sample should be clean with 1:1 alcohol/DI-water and then dry in nitrogen chamber at 50–60°C (subject to base polymer properties) for 24 h. Typically, almost all ATR accessory equipped with a Zinc selenide (ZnSe) and diamond crystal at a nominal incident angle of 45°, yielding to 12 internal reflections at the membranes surface with 16 scans at 4 cm^{-1} resolution and ratio to background spectra recorded in the air at abscissa 350–4000 cm^{-1} [17].

2.2.1.3. *X-ray diffraction (XRD) techniques*

X-ray diffraction method is nondestructive analytical techniques, which reveal information about the crystal structure, chemical composition, and physical properties of materials. Also with the help of XRD, the memranologist can be identified unknown material during membrane modification particularly while we use nanoparticles during dope solution preparation and membrane fabrication and can be compared its crystal structure to that of a standard database [83]. XRD can also be used to identify the presence of multiple phases where different crystalline compounds coexist [89].

2.2.2. *Structural analysis*

2.2.2.1. *AFM and electron microscopy*

Atomic force microscopy (AFM) is a comparatively a vital tool for membranologist to understand the surface roughness of membrane. As compared to scanning electron microscopy (SEM), AFM is capable of producing images both in air and liquid without conducting the membrane. Besides, the samples do not require any chemical etching or conductive metal coating. However, due to membrane surface roughness, it is often difficult to obtain a pore size distribution. Nevertheless, since the discovery of AFM, it has been used broadly for investigating the surface roughness of UF/MF membranes [26]. Moreover, a significant attribute of AFM is its tendency to measure force interactions as a function to look into surface separation distance [90]. All AFM measurements can be carried out using multimode AFM with a Nanoscope IIIa controller and contact mode OTR8 silicon nitride probes (Olympus, from Japan, Bruker, USA, Agilent, USA). Before scan polymeric membrane samples, the probes should be wiped in argon plasma [90]. Membrane samples can be fixed to the metal sample discs using epoxy resin. Measurements should be carried out under ambient conditions using the contact mode of imaging. The surface roughness morphology at 2D and 3D topographic images can be extracted. The roughness analysis of mean roughness (Ra), the root mean square of data (Rz), and the average difference in the height between the five highest peaks and the five lowest valleys (Ry), as well as regarding the diameter of the nodules [27] of prepared membrane and additives such as BSA, PEG, and PVP, etc. Z is described as the difference between the highest and lowest points within the given area (nm).

Membrane imaging with scanning electron microscopy (SEM), transmission electron microscope (TEM), and field-emission scanning electron microscopes (FESEM) has become standard method to investigate membrane morphology or structure and the mechanism of membrane formation [91, 92]. However, all these microscopic samples required one additional step of the metal coating before scanning, but a metalization of the membrane sample is necessary to yield high-quality images and for the visualization of the pore structures [93]. Kim et al. [91], have revealed that polymeric membrane coating is essential at low voltage for conductivity and high voltage for contrast. Also, Kim et al. [91], have revealed that FESEM has proven essential tools for membranologist to examining both surface morphology of finely porous membranes and the fouling process in membrane research. Several researchers [26, 27, 94–96] has combine AFM with SEM together with solute transport data and were determined UF/MF membrane pore size and pore size distribution. Earlier researchers used AFM to study the UF/MF membrane surface structures [96–99] and to interpret the mechanisms of fouling in membrane processing [85].

To analyze polymeric membrane with SEM/FESEM, the sample should be washed with ionized water and dried in vacuum oven at 50–60°C. Also, to get an efficient cross-sectional morphology, the membrane samples must be ruptured in liquid nitrogen [46]. The sample needs to be electroconductive for current, and it can be done by coating them with a very lean layer of about 1.5–2.0 nm using gold or gold/palladium, platinum metals. Also, polymeric membranes sample must be capable of upholding at high vacuum and should not change the space. Polymers, metals, and crystals are usually little problematic and keep their structure in the SEM [89].

2.2.2.2. Membrane equilibrium water content (EWC) and porosity

Equilibrium moisture content or water uptake or water swelling is considered to be an essential tool to characterization factor as it indirectly shows the degree of hydrophilicity or hydrophobicity of a membrane [17, 100–102]. EWC is directly related to the porosity, and UF/MF membrane and the porosity play a vital role not only in also characterizing in important in membrane application. Therefore, some researchers have described the porosity and numbers of techniques have been suggested to estimate membrane porosity [35, 103–108]. Some of them give information on the overall membrane porosity. Nevertheless, the porosity of the membrane can be measured by the gravimetric method, and by apparent density method as reported several researchers [11, 100–107]. According to this approach, the average membrane porosity is determined as the overall void fraction, calculated as the volume of the pores divided by the total volume of the membrane. Perfectly dried membrane samples were weighed with a precision balance. Membrane samples, then, immersed in pure water for

24 h and weighed again. The overall porosity (ε) was calculated using the following formula (Eq. (6)):

$$\varepsilon = \frac{(M_{wet} - M_{dry})}{A\,\Delta_x} \times 100 \tag{6}$$

where M_{wet} is the wet membranes, M_{dry} is the weight of dry membranes,

The volumetric porosity,

$$\varepsilon = \frac{V_e}{V_w} = \frac{V_w - V_d}{V_w} = 1 - \frac{V_d}{V_t} \tag{7}$$

$$\varepsilon = 1 - \frac{\rho_m}{\rho_p} \tag{8}$$

where V_e is the empty volume of the membrane, V_d is the dry volume of the membrane, V_w is the wet volume of the membrane, ρ_m shows membrane density, and ρ_p is the density of base polymer used for UF/MF membranes fabrication.

Porosity via pycnometer calculation as follows,

$$\varepsilon = \frac{[M_{dry} + P_l - (P_l + M_{wet})]}{\rho_m V_t} \tag{9}$$

Where M_{dry} is the weight of dry membrane, P_l is the weight of pycnometer totally filled with the pure liquid, such as pure water, kerosene, toluene, etc., and $P_l + M_{wet}$ is the weight of pycnometer filled with liquid and membrane. The density of membrane shows ρ_m [17, 103]

2.2.3. *Performance evaluation*

2.2.3.1. *Membrane affinity test*

Typically, the primary characterization of any membrane is their performance and rest of all above characterization are depended on and to support the performance. Theoretically, the permeate (flux) of any substance per unit driving force which is "transmembrane pressure" is directly proportional to the permeability of the material. Before evaluating UF/MF membrane for protein or waste water treatment, it is critical to examine the membrane affinity test of the pure water. Thus, in membrane affinity, the hydraulic resistance (R_m) is one of the important properties to study the membrane compatibility with clean water against transmembrane pressure (ΔP) with respect to time [11, 95]. To characterize the membrane hydraulic resistance (R_m) and compaction, the pure water (de-ionized water) flux of membranes can calculate at different transmembrane pressures (ΔP) later than compaction. Yasuda and Tsai [100], revealed that after earlier time, the initial flux would be started after the pressurization of a test cell. The de-ionized water permeates usually dropped severely in the starting and flattened steady state after about 6 to 8 h. The resistance of the membrane, R_m, can be examined from the slope of water flux vs. (ΔP) graph, and **Figure 7** shows R_m vs. flux with respect to time.

Following equations can be used to calculate the membrane resistance and membrane compaction against pressure [11, 102]:

$$J_w = \frac{\Delta P}{R_m}, \text{ where } \Delta P = \pi_m - \pi_b \tag{10}$$

$$J_w = \frac{\Delta W_p}{\rho A \Delta t \Delta P} \tag{11}$$

Where R_m shows membrane hydrodynamic resistance and can be determined by the slope of the water vs transmembrane pressure (ΔP) difference graph, π_m is the applied pressure on the membrane (driving force), π_b is the osmotic pressure, and ΔW_p is the permeate weight. The difference, Δt is an interval of time, A is the membrane active surface area, and ϱ is the density of permeate.

2.2.3.2. *Molecular weight cut-off (MWCO)*

Technically, UF/MF is a pressure-driven process designed to remove large macromolecule (>0.001 μm for UF and >0.1 for MF) from selected solution (see **Figure 8a**). MWCO is a pore characteristic of membranes and is related to rejection for a given molecular weight of a solute and the value frequently used by membrane manufacturers to described their porous UF/MF membranes. Dextran, PVP, PEG, and BSA of a range of molecular weights are often used to rate the MWCO of UF/MF membranes [11, 17, 29]. For example, a membrane that can remove dissolved solids with molecular weights of 150+ has a molecular weight cut-off of 150. **Figure 8b** [17] shows a retention vs. molecular mass curve and membranes with this particular MWCO

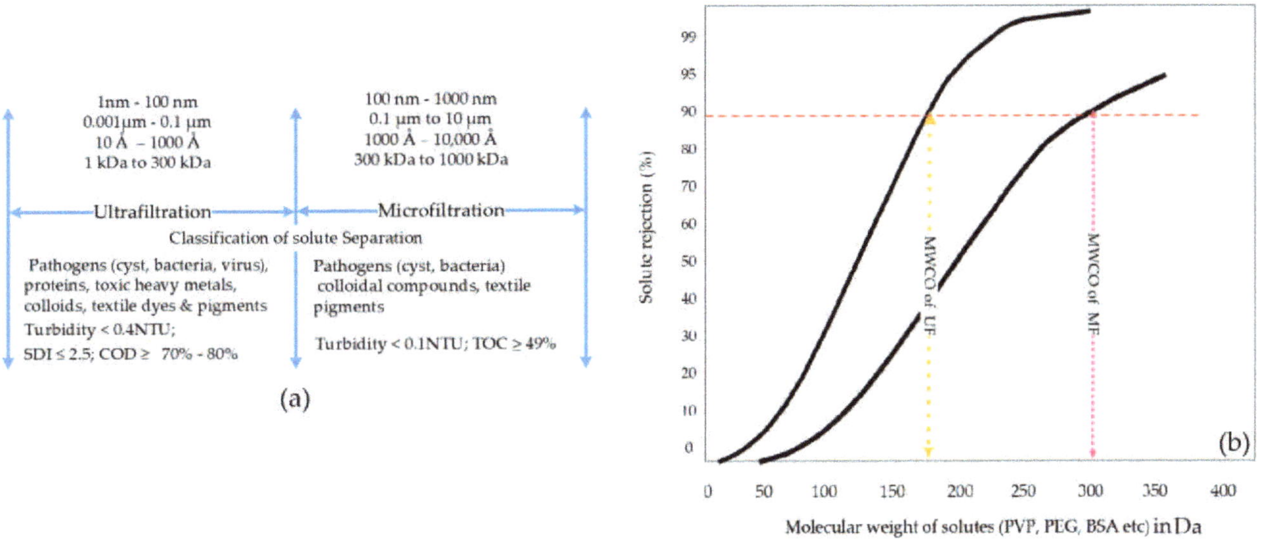

Figure 7. Classification of solute separation and MWCO of UF/MF [17].

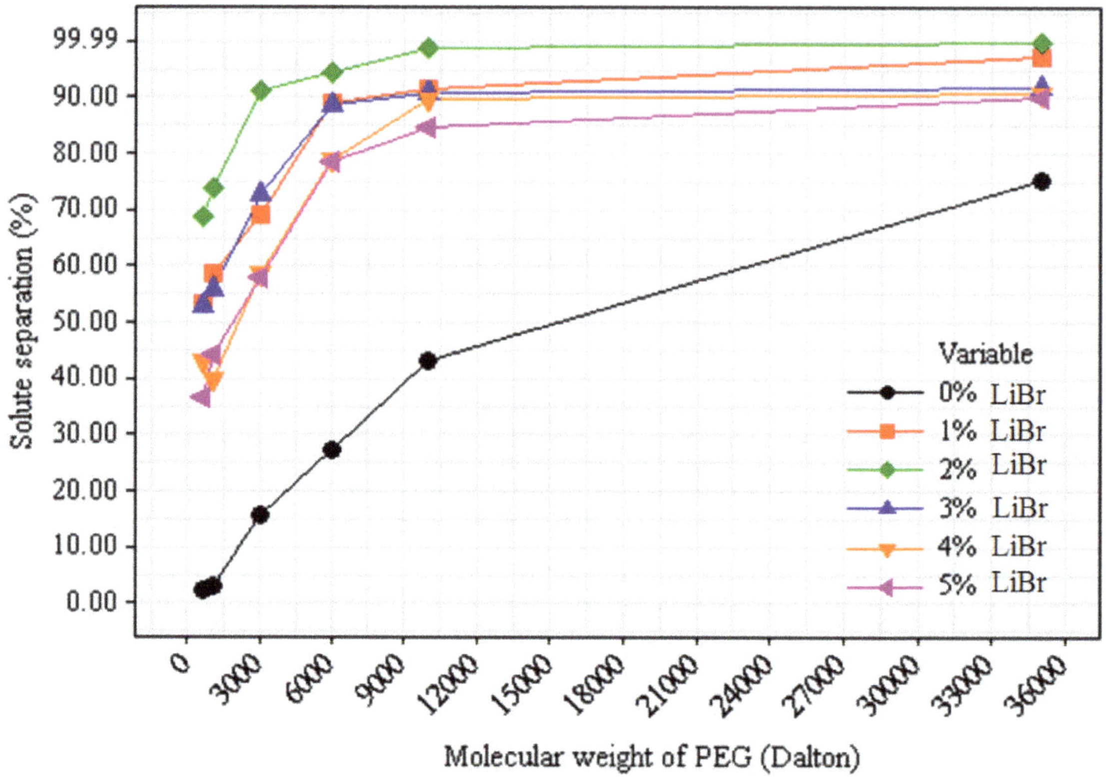

Figure 8. A typical rejection curve, variable shows UF membranes with and without (0%) additives [17].

would be appropriated for applications. Theoretically, UF and MF have the same chemistry and same phenomenon except for pore size difference. **Figure 8a** [17] shows a pore size or MWCO differences between UF/MF membranes. Typically, UF membrane has pore sizes in the range from 10 to 1000 Å and is capable of retaining species in the molecular weight range of 300 to 500,000 Daltons, while MF pore size is usually almost 100 times higher than UF. Typically, the value of MWCO is defined as the molecular weight (MW) which is rejected by 90%. The molecular weight has a linear relationship with the pore radius or pore size of a membrane [10, 11, 25–27, 108]. In general, the MWCO of a membrane is determined by the identification of an inert solute, which has the lowest molecular weight and has a solute rejection of 80–100% in steady state UF experiments. The permeate solutes (bovine serum albumin or polyethylene glycol or PVP) concentration can be determined using UV-vis spectrophotometer at a wavelength of 280 nm. Theoretically, solute rejection performance for UF/MF membranes is usually illustrated as the observed amount of rejection, and it can be defined as,

$$\text{Solute rejection } (R) = \left[1 - \frac{C_p}{C_f}\right] \times 100 \tag{12}$$

Whereas, C_p and C_f are the solute concentration in the flux and feed, respectively. In UF/MF macromolecular solutions, the concentration polarization phenomenon always happened. As a result of this phenomenon, the solute concentration at the membrane surface *Cs* is higher than for UF/MF membranes; therefore, it is expressed as the true rejection, R_t, and it can be evaluated as:

$$R_t = \left[1 - \frac{C_p}{C_m}\right] \tag{13}$$

The value of C_m can be calculated using the boundary layer resistance (BLR) model for the permeate flux. In BLR, the permeate flux is expressed as the following Eq. (14):

$$J_w = \frac{\Delta P}{\mu_{O(R_m+R_{b1})}} \quad (14)$$

Insert Eq. (10) in Eq. (14)

$$J_w = \frac{\pi_m - \pi_b}{\mu_O R_{b1}} = \frac{\pi_b}{\mu_O R_{b1}} \quad (15)$$

where R_{b1} is the hydrodynamic resistance of the boundary layer to the flow of solvent defined as:

$$R_{b1} = \int_0^{\delta} P_s(x)^{-1} d_x \quad (16)$$

If the relationship of transmembrane pressure as a function of solute concentration is known, then the value of C_m can be determined from Eqs. (14) and (15).

2.2.3.3. Pore size and pore size distribution

Typically, the average pore size, pore volume, pore size distribution, pore density, pore geometry, and properties of UF/MF porous membranes are imperative. Also, it is vital to the development of new kind of membrane [19, 21, 23] but also a great help in selecting and using membrane correctly and fast in the application. Kaneko [106] and Nakao [21, 22] has comprehensively reviewed and described various kind of pore size and pore size distribution methods. Sakai was also discussed and provided a general description of the concept of pore size and pore size distribution method for UF/MF. Calvo and co-authors [104] have reported comprehensive reviews on almost all pore size and pore size distribution methods used for UF/MF. However, recently among all the methods, the membranologist is using true solute rejection data and MWCO (solute rejection curve) for pore size and pore size distribution method.

2.2.3.4. Solute rejection curve

The variations of solute rejection with solute molecular diameter yield and shaped curve as illustrated in **Figure 9** (MWCO) and **Figure 10** (log-normal). **Figure 9** suggests that the relation between solute rejection and solute diameter is described by log-normal probability function as reported by several authors [22–27] and yields a straight line on a log-normal probability graph as shown in **Figure 10**. If the solute rejection correlates with solute diameter by the log-normal probability function, this relationship can be expressed as:

$$R_t = \frac{1}{\sqrt{2\pi}} \int_{\infty}^{y} e^{-u^{2/2}} d_u \quad (17)$$

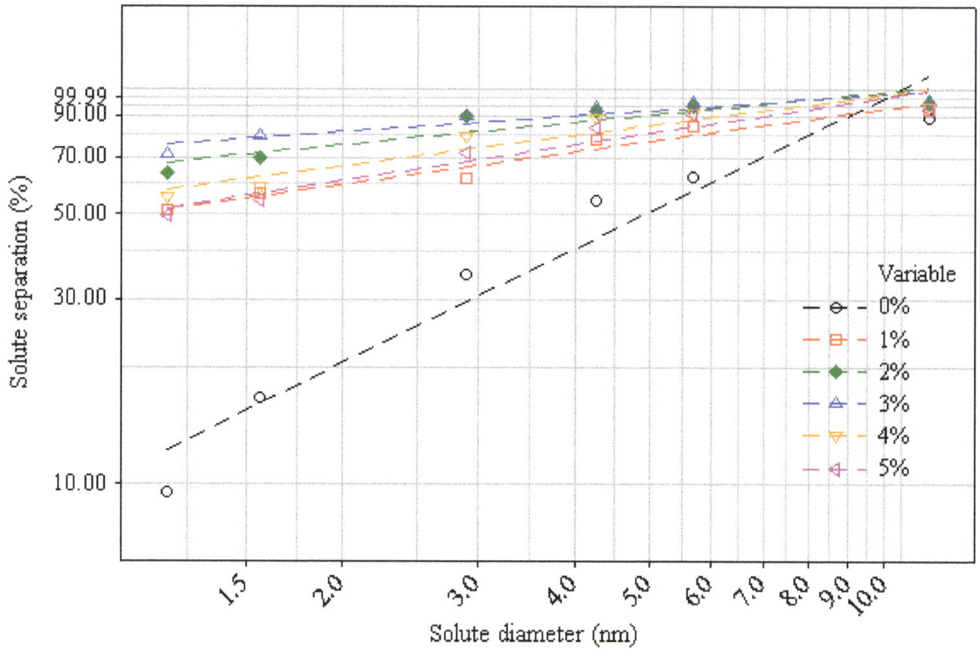

Figure 9. A typical rejection curve on log normal probability graph, variable shows PES UF membranes with LiBr (1%-5%) and without additives (0% LiBr) [17].

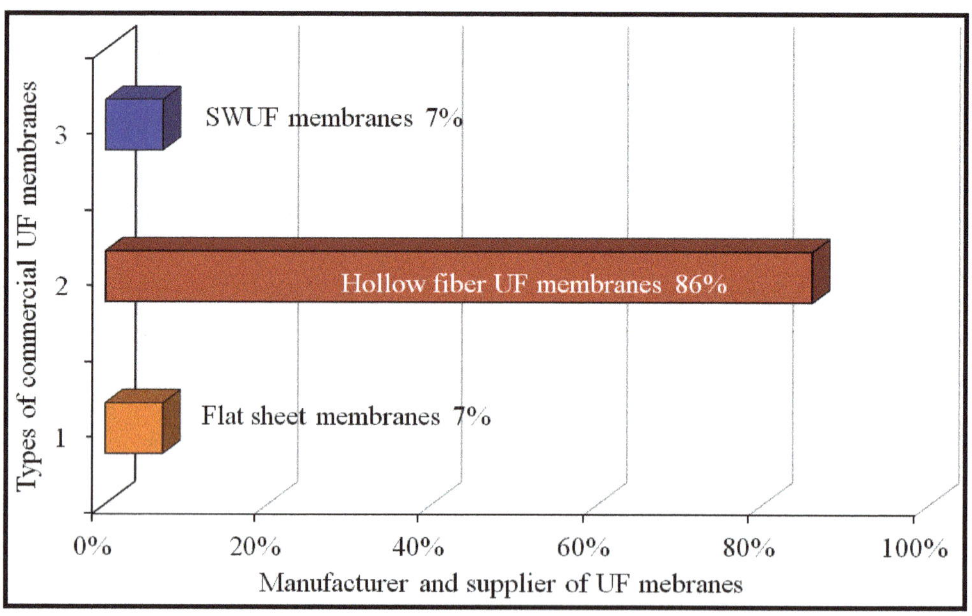

Figure 10. Recent status of commercial HF membranes [63, 108].

$$y = \frac{lnD_s - \ln \bar{D}_s}{\ln \sigma_g} = \frac{\ln D_s - \mu}{\ln \sigma_g} \tag{18}$$

Where D_s is the solute diameter, μ is the geometric mean diameter of solute at $R_t = 50\%$, and σ_g is the geometric standard deviation about the mean diameter. According to the Eqs. (17) and

(18), a straight line in the form of solute yield among R (solute separation in %) and *Ds* (solute diameter) on a log-normal,

$$R(R_t) = A + B(\ln D_s) \tag{19}$$

Where, *A* and *B* are the intercept and the slope, respectively. From this log-normal plot, mean solute size (*Ds*) can be calculated as *Ds* corresponding to *R* =ˆ50% can be determined from the ratio of *Ds* at *R* =ˆ84.13% and 50% [26, 27, 106]. By ignoring the dependence of solute separation on the steric and hydrodynamic interaction between solute and pore sizes [25, 26], the mean pore size (*Ds*) and the geometric standard deviation (σ_g) of the membrane can be considered to be the same as that of solute mean size and solute geometric standard deviation. From Eqs. (18) and (19) and the tables of statistics that were presented by Michaels [25], μ and σ_g are evaluated as:

$$\mu = \ln \bar{D}_s = (\ln D_s)_{R_t=0.5} = -A/B \tag{20}$$

$$\sigma_g = \frac{(lnD_s)_{R_t=0.84} - (\ln\bar{D})_{R_t=0.159}}{2} = \frac{1}{B} \tag{21}$$

2.3. UF/MF membrane configuration and components

To minimize the operational in desalination process, beach well-intake pretreatment is used [10, 14, 28, 32, 36, 46, 109]. However, this is not always precisely promising and not liable to headway due to operating cost. Thus, the pressure-driven membrane processes such as MF, UF, and NF are now the new drift in deceitful n RO or MSF pretreatment systems [46]. Microfiltration (MF) is an open technique for the removal of suspended solids and for lowering the silt density index (SDI). Energy consumption in MF is relatively small so that the total costs for the MF pretreatment are comparable to the beach well intake [83, 84, 109–112]. Whereas, the cost for a corresponding conventional pretreatment is more than double. MF provides an RO feed water of high quality, with (slightly) lower COD/BOD, and smaller SD1 in comparison to the untreated seawater, although there is a significant influence on the feed water quality [111–114]. Good quality seawater may be used for large SWRO plants with a minimal pretreatment and at relatively low cost.

To minimize the energy cost, scaling inhibitor, and fouling lessening, respectively for the useful of the RO, and MSF treatment practice of MF and UF optimizes only the pretreatment given lower capital and operating costs, or on a wider variety of sources [83]. Besides, the implementation of NF as a pretreatment, on the other hand, will lead to a breakthrough in the application of RO or MSF because it has implications for the desalination process itself and not only on the quality of the feed water. Turbidity, microorganisms, and hardness are removed in the NF unit, as well as a fraction of the dissolved salts [14, 109]. The worldwide market for UF membranes has moving parallel with RO, grown to nearly \$3.3 billion in 2016 from \$3.1 billion in 2015. The market is expected to increase at a 5-year compound annual growth rate (CAGR) of 6.9% from 2016 to 2021, increasing to nearly $4.6 billion in 2021 [115, 116]. Thus, the world demand of UF is parallel moving with RO membranes. As described above that majority of membrane manufacturer and supplier currently prefer hollow fiber membrane components for UF and MF application. Recently, MF/UF membrane systems have enjoyed exceptional growth, producing high-quality feed waters to downstream RO systems and as stand-alone filtration methods for a variety of applications. Therefore, the majority of fresh water treatment plants are preferred low-pressure hollow fiber membranes elements configuration [63, 113, 114]. **Figure 10** summarized the recent status of hollow fiber membrane elements as compared to SW setup and flat sheet elements [63, 108]. Several types of HF elements configuration are available commercially depend upon types of water treatment or other applications. **Figure 11** summarized the HF elements for a configuration such bundles, using a simple technique or spiral winding techniques that better shell flow distribution and lower mass transfer resistance [108, 111, 117]. Fiber crimping also has been introduced to reduce mass transfer resistances. The introduction of fluid into the shell region of these devices has received significant attention as it can impact performance significantly [108].

Commercially, available in microfiltration and ultrafiltration, and the tubular membranes operate in tangential, or cross-flow, a configuration where process fluid is pumped along the membrane surface in a full action. Several membrane manufacturer and suppliers provide several offers of strong choice that are effortless to control and clean, serving many industrial and municipal applications [63, 108].The major demand of HF element configuration is usually the strict quality requirement of pharmaceutical, food industry. Besides, HF is also suitable for wastewater treatment, gas separation, etc. **Figure 11** also shows several types of tubular membrane configuration, and **Figure 12** revealed top seller of UF/MF membrane configuration. Tubular membrane elements easy procedure, high suspended solids, and concentrate product competent and often to greater end-point concentration degrees without plugging, making them ideal for recovering wastewaters, and clarifying juices [63, 108, 111].

2.4. Commercial UF/MF materials

Currently, several membrane configurations are available commercially. **Table 4** lists some of the commercially available membranes for UF [1, 17, 63]. The membranes itself must satisfy some mechanical, hydrodynamic, and economic requirements [17, 63]. Automatic membrane condition means the ability to provide the necessary physical support for the membrane including the capacity to undergo the required pressure drop and any back flushing. Hydrodynamic membrane requirement means minimizing pressure drop through the device and thus reduce pumping costs [52, 118]. The solute mass transfer is optimized, and thus, concentration polarization is reduced. Particulate plugging and dead spots for clean design are minimized. Economic membrane requirements reduce membrane packing densities, manufacturing costs, permit easy access for cleaning/replacement and provide sufficient chemical resistance [45, 52, 63, 110, 115]. Recently, Millipore Corporation from the USA has introduced their new generation of UF membranes. These membranes are based on new casting processes that provide void-free membranes and high process flux with enhanced product retention and significantly increased mechanical resistance. These membranes have been commercialized under the trade names of Biomax (void-free PES family) and Ultracel

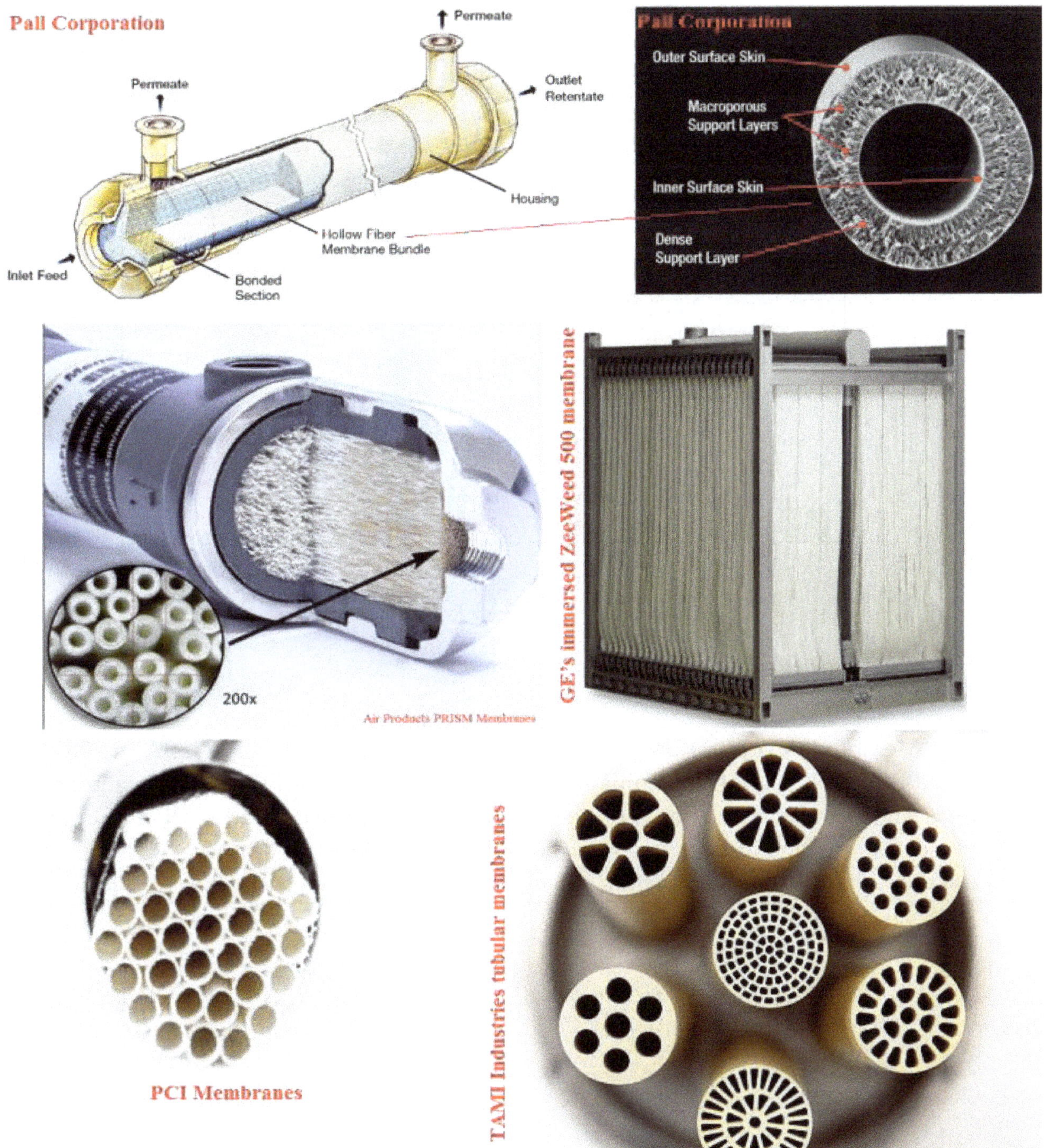

Figure 11. Commercial available hollow fiber membrane elements configuration.

(void-free composite regenerated cellulose family). Applied membrane Inc from the USA is a leading supplier of RO and UF membranes systems and water filtration elements under the AMI label for over 25 years. Previously, FILMTEC produced a broad range of CA thin composite membranes ranges for RO. Dow/FILMTEC is currently commercializing UF membrane series M-U4040 PES and MU2540 PAN for pharmaceutical, food beverage, and wastewater treatment [17, 52, 63].

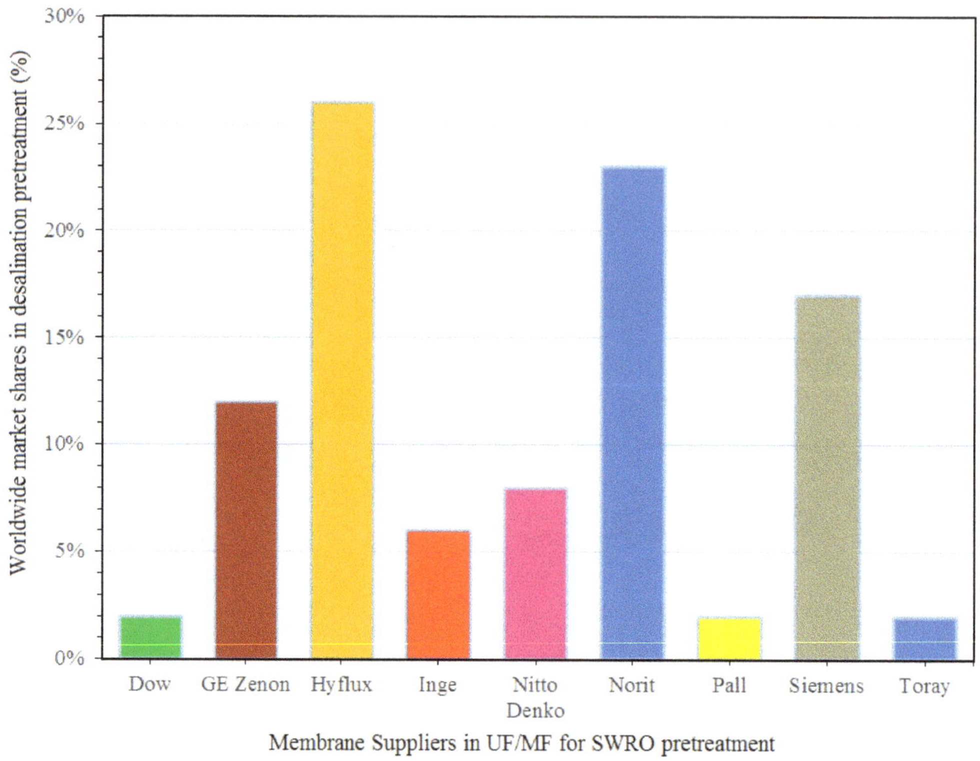

Figure 12. Top seller of UF/MF membrane configuration [17, 63].

American Membrane Corporation (AMC) was incorporated in Michigan USA in 2001 and established its first subsidiary in China in 2002. AMC has launched the Accupor membrane which is a highly microporous membrane composed of modified hydrophilic PES that is a

Manufacturer	Brand	Material & module	MWCO (kDa)	Flux ($L.m^{-1}.h^{-1}$)	Application
Applied Membranes (USA)	M-U4040	PES (SW)	10	4.5–18.2 at 3.2 bar	Pharmaceutical & food
	MU2540	PAN (SW)	20	50 at 3.5 bar	Wastewater
DOW/FLMTEC	SFP & SFD	PVDF, (HF)	0.03 μm	40–90 at 2.5 bar	RO feed water & Wastewater
AMC (USA/ China)	AC 120R01	PES (HF)	15	112 at 4.1 bar	Pharmaceutical & food
AsahiKasei, JP	AP	PAN (HF)	69	16	Pharmaceutical & food
KOCH, USA	HFK	PES,	5–10	24–53 at 3–4 bar	Wastewater
	HF8H7235PMPW	PSF	100	32 at 3 bar	PVC separation
Luxx Ultratech, USA	L series	PES, PSF, PVDF (tubular	5	27–45	Food & Wastewater
PCI, USA	CA, ES, PU, FP, AN, FP	CA, PES, PSF, PVDF, PAN (Tubular)	2–200	NP	MBR, wastewater, Pharmaceutical & food

Manufacturer	Brand	Material & module	MWCO (kDa)	Flux (L.m^{-1}.h^{-1})	Application
Millipore, USA	Biomex & Amicon	CA& PES	5–10	35–97.2 at 1 bar	Protein purification
membrane element	M-series	PAN (HF)	0.03–1 μm	4.5–18.2 at 9.3 bar	Oil water separation
Nitto Denko, JP	Hydracap	PES (HF)	150	51–128 at 3 bar	Wastewater
PALL, USA	BTS	PES (Cassette)	0.5–10	187.2 at 3 bar	Pharmaceutical & food
Polymem	polymem	PSF (HF)	6	313.2 at bar	Protein purification
Synder, Canada	PES 100	PES (SW)	70	51 at 3 bar	Gelatin separation
Trisep, Canada	UE10	PES (SW)	10	2.1 at 2–3 bar	Dairy and Food
Toray, Switzerland	HFU, HFS, HSU	PVDF (HF)	150–200	2.6–8 at 3 bar	RO feed water & Wastewater

Note: Polyethersulfone (PES), polysulfone (PSF), polyvinylidene fluoride (PVDF), polyacrylonitrile (PAN), cellulose acetate (CA), spiral wound (SW), hollow fiber (HF), Japan (JP), note provided (NP).

Table 4. Commercially available membranes for ultrafiltration.

tough, durable, and temperature-resistant aromatic polymer. This membrane is specifically designed for biological, analytical, electronic, pharmaceutical, beverage, and sterilizing filtration applications [17, 63]. Nitto Denko and Asahi Kasei commercialized Hydracap (hydrophilic PES), UF AP series (Hydrophilic PAN), low-pressure UF hollow fiber membranes for wastewater, pharmaceutical, and food industry. Recently, KOCH has supplied a series of UF HFK-131/138 (PES) and HF8H-72-PMPW (PSF) spiral wound and hollow fiber membranes for wastewater treatment [17, 52, 63]. PALL has introduced BTS highly asymmetric membranes; these membranes are cast from PES and PSF by a unique process exclusive to another supplier. The "cut off" layer in the BTS membranes is only about 10 μm thick, vs. traditional membranes with cut-off layers of about 100–125 μm thickness. This difference in thickness gives the BTS highly asymmetric membrane significantly higher flow with much lower pressure drop for pharmaceutical purposes. Besides that Polymer has also launched a very high-performance asymmetric PSF hollow fiber membrane with uppermost permeation rate at very low pressure for Protein Purification.

3. Conclusion

During 50 years, UF/MF membrane technology has experienced a quick growth as it relates to most applications such as pharmaceutical, food, and beverage industries. Also, theoretically, it has been proven that UF/MF is the best post-treatment process (upstream) for RO, eliminating from the feed water nearly all of the possible constituents liable of desalinating membranes fouling. The 30 years of commercialization have viewed new materials of UF/MF membranes employed at ever lower pressures (1–2 bar) and with remarkable performance. However, UF/MF technology still has some limits, and yet membrane fouling is a destructive obstacle for UF/MF membranes. Material customization and membrane configuration (modules) for selective application and molecular transport modeling are efficient tools to minimize the fouling phenomenon, which is used in the advancement of membrane technology. Also, UF/MF system design continues to progress, and current movements in MF and UF in certain are a feedback to industry pressures for a basic MF/UF system design that holds a diversity (not clear) of membrane components and system configurations. Consequently, the remarkable UF/MF market growth leads to continuously increasing demand particularly in feed water treatment for SWD and wastewater treatment process and without doubt, UF is recently becoming a reasonable feed water system for RO in an extensive field of raw water quality.

Acknowledgements

I gratefully acknowledge the Centre of Excellence in Desalination Technology, Department of Mechanical Engineering King Abdulaziz University and for the kind moral support by providing me platform thus enables me to carry out this chapter.

Appendix

See Appendix Table.

Technical words	Abbreviations	Chemical compounds and polymers	Abbreviations
Sea water desalination	SWD	Calcium carbonate	$CaCO_3$
Spiral wound reverse osmosis	SWRO	Zinc selenide	ZnSe
Multi-stage flash	MSF	Polyethylene oxide	PEO
Multi-effect distillation	MED	Polyamide	PA
Electrodialysis	ED	Polyurethanes	PU
Reverse osmosis	RO	Polyethylene terephthalate	PET
Nanofiltration	NF	Polycarbonate	PC
Ultrafiltration/ Microfiltration	UF/MF	Polysulfone	PSU or PSF
Thin film composite layer	TFC	Polyethersulfone	PES
Ultra thin film	UTF	Polyvinylidene fluoride	PVDF
Pore size distributions	PSD	Polyimide	PI

Technical words	Abbreviations	Chemical compounds and polymers	Abbreviations
Molecular weight	MW	Aliphatic polyamide (Nylon)	PA6 and PA66
Molecular weight cut-off	MWCO	Aromatic polyamide	Ar.PA
Fourier transform infrared	FTIR	Polyethylenimine (PEI) or polyaziridine	PEI
Equilibrium water content	EWC	Polyether ether ketone	PEEK
attenuated total reflection	ATR	Polyacrylonitrile	PAN
Atomic force microscope	AFM	Cellulose acetate	CA
Scan electron microscopy	SEM	Cellulose triacetate	CTA
Energy-dispersive X-ray	EDX	Trimethyl polysulfone	TM-PSF
Field emission scanning electron microscopic	FE-SEM	Sulfonated polysulfone	SPSF
transmission electron microscope	TEM	Sulfonated polyethersulfone	SPES
Kilo Dalton	KDA	N-Methyl-2-pyrrolidone	NMP
Ultraviolet	UV	Dimethylacetamide	DMAc
Silt density index	SDI	Dimethylformamide	DMF
Carbon oxygen demand	COD	Dimethyl sulfoxide	DMSO
Biological oxygen demand	BOD	Bovine serum albumin	BSA
Annual growth rate	CAGR	Polyethylene glycol	PEG
Hollow fiber	HF	Polyvinylpyrrolidone	PVP
Equations			
Mean pore radius	$\bar{r}$	pure water viscosity	J_{μ}
Membrane thickness	Δx or Δa	Hydraulic pressure	Hp
Membrane porosity	ε	Water permeation	J_p
Transmembrane pressure	TMP (ΔP)	Tortuosity	τ
Water content	W_c	Pore volume	P_v
Pore size distributions	PSD	Mean roughness	Ra
Root mean square of data	Rz	Highest peaks	Ry
Difference between the highest and lowest points	Z	Wet membranes	M_{wet}
Dry membranes	M_{dry}	Empty volume	V_e
Dry volume	V_d	Wet volume	V_w
Membrane density	ρ_m	Density of base polymer	ρ_p
Pure liquid	P_l	Wet membranes	M_{wet}

Technical words	Abbreviations	Chemical compounds and polymers	Abbreviations
Membrane hydrodynamic resistance	R_m	Applied pressure	π_m
Osmotic pressure	π_b	Permeate weight	ΔW_p
An interval of time	Δt		
Membrane active surface area	A	Density of permeate	ρ
Solute concentration in the flux	C_p	Solute concentration in the feed	C_f
Solute concentration at membrane surface	C_s or C_m	True rejection	R_t
Hydrodynamic resistance	Rb_1	Solute diameter	D_s
Geometric mean diameter	μ	Solute rejection	R_t
Geometric standard deviation	σ_g	Intercept	A
Slope	B	Boundary layer resistance	BLR

Author details

Iqbal Ahmed[1,2]*, Khaled S. Balkhair[1], Muhammad H. Albeiruttye[1] and Amer Ahmed Jamil Shaiban[1]

*Address all correspondence to: iqbalmoujdin@gmail.com

1 Center of Excellence in Desalination Technology, King Abdulaziz University, Jeddah, Saudi Arabia

2 Department of Mechanical Engineering, King Abdulaziz University, Jeddah, Saudi Arabia

References

[1] Nunes SP, Peinemann KV. Membrane Technology in the Chemical Industry. 2nd ed. WILEY-VCH Verlag GmbH & Co. KGaA, Weinheim; 2006, Chapter 1-3.

[2] Nollet JA. Investigations on the causes for the ebullition of liquids. Journal of Membrane Science. 1995;**100**:1-3. DOI: 10.1016/0376-7388(94)00224-M

[3] Fick A. Ueber difusion. Annalen der Physik. 2006:**170**:59-86. DOI: 0.1002/andp.18551700105.

[4] Bechhold H. Kolloidstudien mit der filtration method. Zeitschrift für Elektrochemie und Angewandte Physikalische Chemie. 1907;**13**:509-544. DOI: 10.1002/bbpc.19070133207

[5] Allegrezza AE, Burke ET, Winchester. Composite Ultrafiltration Membranes. 1989; United States Patent 4,824,568.

[6] Zsigmondy R, Bachmann W, Über near filter. Zeitschrift für anorganische und allgemeine Chemie (ZAAC). 1918; **103**:1-256. DOI: 10.1002/zaac.19181030107

[7] Ferry JD. Ultrafilter membranes, and ultrafiltration. Chemical Reviews. 1936;**18**:373-453.

[8] Elford EJ. Principles governing the preparation of membranes having graded porosities. The properties of "Gradocol" membranes as ultrafilters. Trans. Faraday Soc. 1937;**33**:1094

[9] Elford WJ, Grabar P, Ferry JD. Graded Collodion membranes for bacteriological studies. Practical aspects of the mechanism determining the character of the membrane, and the roles of particular solvent constituents. British Journal of Experimental Pathology. 1935;**16**:583-587

[10] Baker RW. Membrane Technology and Applications. John Wiley & Sons Ltd; 2004, England

[11] Lloyd, DR. Materials Science of Synthetic Membrane, Vol. 269. ACS Symposium Series. Washington, DC: American Chemical Society: 1985

[12] Cadotte JE, Petersen BJ. Synthetic membranes. Vol. 153. ACS Symposium Series. Washington, 1981. Chapter 21: pp. 305-326. DOI: 10.1021/bk-1981-0153.ch021

[13] Wilf I. New membrane research and development achievements. Desalination and Water Reuse, 2000;**10**:28-33

[14] Bruggen BV, Vandecasteele C, Gestel TV, Doyenb W, Leysenb RA. Review of pressure-driven membrane processes in wastewater treatment and drinking water production. Environmental Progress. 2003;**22**:46-56

[15] Boddeker, KW. Liquid Separations with Membranes: An Introduction to Barrier Interference. Berlin Heidelberg: Springer-Verlag; 2008

[16] Gooch, JW. Hagen-Poiseuille equation. Encyclopedic Dictionary of Polymers, Springer. 2011;**1**:355-355. DOI: 10.1007/978-1-4419-6247-8_5752

[17] Ahmed I. High-performance ultrafiltration polyethersulfone membrane. Ph.D. Thesis. University Technology Malaysia; Malaysia. 2009

[18] Ahmed A, Idris A, Noordin MY, Rajput R. High-performance ultrafiltration membranes prepared by the application of modified microwave irradiation technique. Industrial & Engineering Chemistry Research. 2011;**50**:2272-2283

[19] Sakai K. Determination of pore size and pore size distribution 2. Dialysis membranes. Journal of Membrane Science. 1994;**96**:91-130

[20] Verniory A, Du Bois R, Decoodt P, Gassee JP, Lambert PP. Measurement of the permeability of biological membranes application to the glomerular wall. Journal of General Physiology. 1973;**62**:489-507.

[21] Nakao S, Kimura S. Models of membrane transport phenomena and their applications for ultrafiltration data. Journal of Chemical Engineering of Japan. 1982;**15**:220

[22] Nakao, S. Determination of pore size and pore size distribution 3. Filtration membranes. Journal of Membrane Science 1994;**96**:131-165.

[23] Cherkasov AN, Polotsky AE. Critical particle-to-pore size ratio in ultrafiltration. Journal of Membrane Science. 1995;**106**:161-166

[24] Michaels AS. Membranes, membrane processes, and their applications: Needs, unsolved problems, and challenges of the 1990's. Desalination. 1990;**77**:5-34

[25] Michaels AS. Analysis, and prediction of sieving curves for ultrafiltration membranes: A universal correlation?. Separation Science and Technology. 1980;**15**:1305-1322.

[26] Singh S, Khulbe KC, Matsuura T, Ramamurthy P. Membrane characterization by solute transport and atomic force microscopy Journal of Membrane Science. 1998;1**42**:111-127.

[27] Idris A, Mat Zain N, Noordin MY. Synthesis, characterization, and performance of asymmetric polyethersulfone (PES) ultrafiltration membranes with polyethylene glycol of different molecular weights as additives. Desalination. 2007;**207**:324-339.

[28] Winston Ho WS, Sirkar KK. Membrane Hand Book. Springer Science+Business Media New York, 1992

[29] Mulder, M. Basic Principles of Membrane Technology. 2nd ed. Dordrecht, The Netherlands: Kluwer Academic;1996.

[30] Riley RL, Lonsdale HK, Lyons CR, Merten U. Preparation of ultrathin reverse osmosis membranes and the attainment of theoretical salt rejection. Journal of Applied Polymer Science. 1967;**11**:2143

[31] Riley RL, Lonsdale HK, Lyons CR, Merten U. Preparation of ultrathin reverse osmosis membranes and the attainment of theoretical salt rejection. Journal of Applied Polymer Science. 1967;**2**:2143-2158.

[32] Basile SA, Nunes UP. Advanced Membrane Science and Technology for Sustainable Energy. Woodhead Publishing Limited, Cambridge, UK

[33] Wienk IM, Boom RM, Beerlage MAM, Bulte AMW, Smolders CA, Strathmann H. Recent advances in the formation of phase inversion membranes made from amorphous or semi-crystalline polymers. Journal of Membrane Science. 1996;**113**:361-371

[34] Wu T, Zhou B, Zhu T, Shi I, Xu Z, Hua C, Wang J. Facile and low-cost approach towards a PVDF ultrafiltration membrane with enhanced hydrophilicity and antifouling performance via graphene oxide/water-bath coagulation. Royal Society of Chemistry Advances. 2015;**5**:7880. DOI: 10.1039/c4ra13476a

[35] Hosseini SS, Bringas E, Tan NR, Ortiz I, Chahramani M, Shahmirzadi MAA. Recent progress in the development of high-performance polymeric membranes and materials for

metal plating wastewater treatment: A review. Journal of Water Processes Engineering. 2016;**9**:78-110.

[36] Koch Membrane Systems, Inc. Application Bulletin, An Overview of Membrane Technology and Theory. December 1, 2011. Available from: http://www.kochmembrane.com/PDFs/Application-Bulletins/KMS-Membrane-Theory.aspx

[37] Zheng QZ, Wang P, Yang YN, Cui DJ. The relationship between porosity and kinetics parameter of membrane formation in PSF ultrafiltration membrane. Journal of Membrane Science. 2006;**286**:7-11

[38] Zhang H, Li H, Yang F, Wang T, Gao J, Jiang T. Preparation of cross-linked polyvinyl alcohol nanospheres and the synthesization of low-fouling modified membranes. Separation and Purification Technology. 2011;**77**:162.

[39] Fan LH, Harris JL, Roddick FA, Booker NA. Influence of the characteristics of natural organic matter on the fouling of microfiltration membranes. Water Research. 2001;**35**:4455-4463.

[40] Sun W, Liu J, Chu H, Dong B. Pretreatment and membrane hydrophilic modification to reduce membrane fouling. Membranes. 2013;**3**:226-241. DOI: 10.3390/membranes3030226

[41] Wan X, Ghosh R, Cui Z. Fractionation of proteins using ultrafiltration: Developments and challenges. Developments in Chemical Engineering & Mineral Processing. 2005;**13**:121-136

[42] Wang H, Yang L, Zhao X, Yu TQ, Du. Improvement of hydrophilicity and blood compatibility on polyethersulfone membrane by blending sulfonated polyethersulfone. Chinese Journal of Chemical Engineering. 2009;**17**:324-329.

[43] Wang X, Husson SM, Qian XS, Wickramasinghe R. Inverse colloidal crystal microfiltration membranes. Journal of Membrane Science. 2010;**365**:302-310

[44] Zhao X, Liu C. Irreversible fouling control of PVDF ultrafiltration membrane with "fouled surface" for mimetic sewage treatment. Royal Society of Chemistry Advances. 2016;**6**:94184

[45] Arnal JM, García-Fayos B, Sancho M. Membrane Cleaning, Chapter 3, Expanding Issues in Desalination. InTechOpen; 2011, DOI: 10.5772/19760

[46] Futselaar H, Blankert B, Spanjer T, Spenkelink F. Ultrafiltration used as pre-treatment for SWRO desalination: Dynamic coagulant control under extreme conditions. Desalination and Water Treatment. 2013;**51**:407-415. DOI: 10.1080/19443994.2012.699259

[47] Yu H, Cao Y, Kang G, Liu Z, Kuang W, Liu J, Zhou M. Enhancing the antifouling properties of polysulfone ultrafiltration membranes by the grafting of poly(ethylene glycol) derivatives via surface amidation reactions. Journal of Applied Polymer Science. 2015;**132**:41870. DOI: 10.1002/APP.41870

[48] Ahmed I, Ani I, Rajput R. Performance of microwave-synthesized dual solvent dope solution and lithium bromide additives on poly(ethersulfone) membranes. Journal of Chemical Technology & Biotechnology. 2011; **87**:177-188. DOI: 10.1002/jctb.2676

[49] Munari S, Bottino A, Camera Roda G, Annell GC. Preparation of ultrafiltration membranes. State of the art. Desalination. 1990;**77**:85-100

[50] Ahmad AL, Abdulkarim AA, Ooi BS, Ismail S. Recent development in additives modifications of polyethersulfone membrane for flux enhancement. Chemical Engineering Journal. 2013;**223**:246-267. DOI: 10.1016/j.cej.2013.02.130

[51] Reid CE, Breto EJ. Water and ion flow across cellulosic membranes. Journal of Applied Polymer Science. 1959;**1**:133-143.

[52] Mohanty K, Purkait MK. Membrane Technologies and Applications (Edited), CRC Press, Taylor & Francis Group, New York, USA. Chapter 2, 2012.

[53] Kesting, R, Irvine. Synthetic Polymeric Membranes: A Structural Perspective. Wiley-Interscience; 2nd edition, 1985. pp. 237-261

[54] AWWA. Microfiltration and Ultrafiltration Membranes for Drinking Water, Manual of Water Supply Practices-M53. 1st ed. 2005.

[55] Matsuura, T. Synthetic Membranes, and Membrane Separation Process, Chapter 5. CRC Press, Taylor & Francis Group, Florida, USA; 1994

[56] Sigma-Aldrich, Cellulose acetate (CAS Number: 9012-09-3), 2012. Available from: http://www.sigmaaldrich.com/catalog/product/sial/22199?lang=en&region=SA

[57] Sigma-Aldrich. Cellulose acetate (CAS Number: 9004-355-7), 2004. Available from: http://www.sigmaaldrich.com/catalog/search?term=Cellulose+acetate&interface=Product%20Name&N=0+&mode=mode%20matchpartialmax&lang=en&region=SA&focus=productN=0%20220003048%20219853286%20219853231

[58] Kenichi I, Hiroki I, Shouichi Y, C/O Nitto Electric Ind. Co. Ltd. Sulfonated polysulfone composite semipermeable membranes. EP 0165077 B2. 1998; US4818387. 1989.

[59] Li L, Zhang M, Rong M, Ruan W. Studies on the transformation process of PVDF from α to β phase by stretching. Royal Society of Chemistry Advances. 2014;**4**:3938-3943. DOI: 10.1039/C3RA45134H

[60] Kim IC, Choi JG, Tak TM. Sulfonated polyethersulfone by heterogeneous method and its membrane performances. Journal of Applied Polymer Science. 1999;**74**:2046-2055

[61] Huang L, Liao M, Yang X, Gong H, Ma L, Zhaoa Y, Huang K. Bisphenol analogs differently affect human islet polypeptide amyloid formation. Royal Society of Chemistry Advances. 2016;**6**:7239. DOI: 10.1039/c5ra21792j

[62] Li HB, Shi WY, Zhang YF, Liu DQ, Liu XF. Effects of additives on the morphology and performance of PPTA/PVDF in situ blend UF membrane. Polymers. 2014;**6**:1846-1861. DOI: 10.3390/polym6061846

[63] Pearce G. Introduction to membranes: Manufacturers' comparison: Part 1. Filtration & Separation. 2007;**44**:36-38.

[64] Khulbe KC, Feng C, Matsuura T. The art of surface modification of synthetic polymeric membranes. Journal of Applied Polymer Science. 2010;**115**:855-895

[65] Pinnau I, Freeman BD. Advanced Materials for Membrane Separations. Vol. 876, ACS Symposium Series. Washington, DC: American Chemical Society; 2004.

[66] Zeman L, Wales M. Steric rejection of polymeric solutes by membranes with uniform pore size distribution. Separation Science and Technology. 1981;**16**:275-290.

[67] Benham JL, Kinstle JF. Chemical Reactions on Polymers. ACS Symposium Series. American Chemical Society: Washington, DC; 1988.

[68] Xu Z, Huang X, Wan L. Surface Engineering of Polymer Membranes. Springer, Zhejiang University Press, Hangzhou, China; 2007

[69] Morao A, Escobar IC, Pessoa de Amorim MT, Lopes A, Gonc IC. Postsynthesis modification of a cellulose acetate ultrafiltration membrane for applications in water and wastewater treatment. Environmental Progress. 2005;**24**:367-382

[70] Good K, Escobar I, Xu X, Coleman M., Ponting M. Effects of ion beam irradiation on the transport properties and fouling control of commercial membranes. Desalination. 2002;**146**:259-264.

[71] Ma H, Bowman CN, Davis RH. Membrane fouling reduction by back pulsing and surface modification. Journal of Membrane Science. 2000;**173**:191-200.

[72] Pozniak G, Krakewska B, Trochimczuk W. Urease immobilized on modified polysulfone membrane—Preparation and properties. Biomaterials. 1995;**16**:129-134.

[73] Bottino A, Capannelli G, Monticelli O, Piaggio P. Poly(vinylidene fluoride) with improved functionalization for membrane production. Journal of Membrane Science. 2000;**166**:23-29.

[74] Bordawekar M, Escobar I, Lipscomb G. Application of a temperature-sensitive polymer brush to reduce fouling on membranes. Separation Science and Technology. 2009;**44**: 3369-3391

[75] Hegazy E, Elrehim HAA, Khalifa NA, Atwa SM, Shawky HA. Anionic/cationic membranes obtained by a radiation grafting method for use in waste water treatment. Polymer International. 1997;**43**:321-332.

[76] Jonsson C, Jonsson A. Influence of the membrane material on the adsorptive fouling of ultrafiltration membranes. Journal of Membrane Science. 1995;**10**:79-88

[77] Kim KJ, Fane AG, Fell CJD. The effect of Langmuir–Blodgett layer pretreatment on the performance of ultrafiltration membranes. Journal of Membrane Science. 1989;**43**:187-204.

[78] Stengaard, FF. Characteristics, and performance of new types of ultrafiltration membranes with chemically modified surfaces. Desalination. 1988;**70**:207-224.

[79] Jang H, Song DH, Kim IC, Kwon YN. Fouling control through the hydrophilic surface modification of poly(vinylidene fluoride) membranes. Journal of Applied Polymer Science. 2015;**132**:41712. DOI: 10.1002/APP.41712

[80] Fang Y, Xu ZK, Wu J. Surface Modification of Membranes. "Encyclopedia of Membrane Science and Technology". John Wiley & Sons Singapore Pte. Ltd.; 2013. DOI: 10.1002/9781118522318.emst061

[81] Strathmann H, Koch K. The formation mechanism of phase inversion membranes. Desalination. 1977;**21**:241-249.

[82] Pham VA, Santerre P, Matsuura T, Narbaitz RM. Application of surface modifying macromolecules in polyethersulfone membranes: Influence on PES surface chemistry and physical properties. Journal of Applied Polymer Science. 1999;**73**:1366-1378. DOI: 10.1002/(SICI)1097-4628(19990822)73:8<1363::AID-APP3>3.0.CO;2-P

[83] Gao W, Liang H, Ma J, Han M, Chen Z, Han Z. Membrane fouling control in ultrafiltration technology for drinking water production: A review. Desalination. 2011;**72**:1-8.

[84] Mehta A, Zydney AL. Permeability and selectivity analysis for ultrafiltration membranes. Journal of Membrane Science. 2005;**249**:245-249.

[85] Goddard JM, Hotchkiss JH. Polymer surface modification for the attachment of bioactive compounds. Progress in Polymer Science. 2007;**32**:698-725.

[86] O'Grady BR. Sulfonated polyethersulfone as a new platform for thin film composite membranes. Master Thesis. University of Connecticut (UConn), United States. 2013

[87] Atae-Allah C, Cabrerizo-Vílchez M, Gomez-Lopera JF, Holgado-Terriza1 JA, Roman-Rold R, Luque-Escamilla PL. Measurement of surface tension and contact angle using entropic edge detection. Measurement Science & Technology. 2001;**12**:288-298

[88] Bai H, Wang X, Zhou Y, Zhan L. Preparation and characterization of poly(vinylidene fluoride) composite membranes blended with nano-crystalline cellulose. Progress in Natural Science: Materials International. 2012;**22**:250-257

[89] Battaglia S. Quantitative X-ray diffraction analysis on thin samples deposited on cellulose membrane filters. X-Ray Spectrometry. 1985;**14**:16-19. DOI: 10.1002/xrs.1300140105

[90] Mohammad AW, Hilal N, Pei LY. Atomic force microscopy as a tool for asymmetric polymeric membrane characterization. Sains Malaysiana. 2011;**40**:237-244

[91] Kim KJ, Dickson MR, Chen V, Fane AG. Applications of field emission scanning electron microscopy to polymer membrane research. Micron and Microscopica Acta. 1992;**23**:259-271

[92] Bowen WR, Hilal N, Lovitt RW, Wright CJ. Atomic force microscopy as a tool for the membrane technologist. Microscopy and Analysis. 1998;**68**:13-16.

[93] Manzak A, Yıldız Y, Tutkun O. Characterization of Polymer Inclusion Membrane Containing Aliquat 336 as a Carrier. Membrane Water Treatment. 2015;**6**:5-102. DOI: http://dx.doi.org/10.12989/mwt.2015.6.2.095

[94] Fritzsche AK, Arerado AR, Moore MD, Hara CO. The surface structure and morphology of polyacrylonitrile membranes by atomic force microscope. Journal of Membrane Science. 1993;**81**:109-120

[95] Fritzsche AK, Arevalo AR, Connolly AF, Moore MD, Elings V, Wu CM. The structure and morphology of skin of polyethersulfone ultrafiltration membranes: A comparative atomic force microscope and scanning electron microscope study. Journal of Applied Polymer Science. 1992;**45**:1945-1956.

[96] Bottino A, Capannelli G, Grosso A, Monticelli O, Cavalleri O, Rolandi R, Soria R. Surface characterization of ceramic membranes by atomic force microscopy. Journal of Membrane Science. 1994;**95**:289-296.

[97] Chahboun A, Coratger R, Ajustron F, Beauvillain J, Aimar A, Sanchez V. Comparative study of micro- and ultrafiltration membranes using STM, AFM, and SEM techniques. Ultramicroscopy. 1992;**41**:235-244.

[98] Dietz P, Hansma PK, Inacker O, Lehmann HD, Herrmann KH. Surface pore structures of micro and ultrafiltration membranes imaged with the atomic force microscope. Journal of Membrane Science. 1992;**65**:101-111.

[99] Miwa T, Yamaki M, Yoshimura H, Ebina S, Nagayama K. Artifacts in atomic force microscopy images of fine particle and protein two-dimensional crystals as evaluated with scanning electron microscopy and simulations. Japanese Journal of Applied Physics. 1992;**31**:L1495.

[100] Kim YW, Kim JJ, Kim YH. Surface characterization of biocompatible polysulfone membranes modified with poly(ethylene glycol) derivatives. Korean Journal of Chemical Engineering. 2003;**20**:1158-1165.

[101] Abaticchio P, Boitino A, Roda GC, Capannell G, Munari S. Characterization of ultrafiltration polymeric. Desalination. 1990;**78**:235-255

[102] Mahendran R, Malaisamy R, Mohan DR. Cellulose acetate and polyethersulfone blend ultrafiltration membranes. Part I: Preparation and characterizations. Polym. Adv. Technol. 2004;**15**:149-157. DOI: 0.1002/pat.417

[103] Chakrabarty B, Ghoshal AK, Purkait MK. Effect of molecular weight of PEG on membrane morphology and transport properties. Journal of Membrane Science. 2008;**309**:209-221

[104] Calvo JI, Bottino A, Prádanos P, Palacio L, Hernández A. Porosity, Membrane Materials, Characterization, and Module Design. Encyclopedia of Membrane Science and Technology. John Wiley & Sons, Inc; 2013. DOI: 10.1002/9781118522318.emst065

[105] Yasuda H, Tsai JT. Pore size of microporous polymer membranes. Journal of Applied Polymer Science. 1974;**18**:805-819

[106] Kaneko K. Determination of pore size and pore size distribution 1. Adsorbents and catalysts. Journal of Membrane Science. 1994;**96**:59-89

[107] Youm KH, Kim WS. Prediction of intrinsic pore properties of ultrafiltration membrane by solute rejection curves: Effect of operating conditions on pore properties. Journal of Chemical Engineering of Japan. 1981;**24**:1-7.

[108] García NP, Rodriguez J, Vigo F, Chester S. UF Membranes Autopsies: An Approach to Hollow Fibre Membranes Surface The International Desalination Association World Congress on Desalination and Water Reuse 2015/San Diego, CA, USA, REF: IDAWC15- PEÑA

[109] Ebrahim S, Bou-Hamed S, Abdel-Jawad M, Burney N. Microfiltration system as a pre-treatment for RO units: Technical and economic assessment. Desalination. 1997:**109**: 165-175

[110] Zeman LJ, Zydney AL. Microfiltration and ultrafiltration: Principles and applications. New York: Marcel Dekker; 1996.

[111] Cheryan M. Ultrafiltration Handbook. Lancaster (PA): Technomic; 1986

[112] Lazarova V, Gallego S, García Molina V, Rougé P. Problems of operation and main reasons for failure of membranes in tertiary treatment systems. Water Science and Technology. 2008;**57**:1777-1784. DOI: 10.2166/wst.2008.241.

[113] Rautenbach, R, Linn T, A1-Gobaisi DMK. Present and future pretreatment concepts–Strategies for reliable and low-maintenance reverse osmosis seawater desalination. Desalination. 1997;**110**;97-106

[114] Strathmann H. Electrodialytic Membrane Processes and their Practical Application, Studies in Environmental Science. 1994;**59**:495-533

[115] BQ Magazine. Available from: http://www.bq-magazine.com/industries/2015/02/water-scarcity-in-the-middle-east

[116] Bcc Research, 2016. Available from: http://www.bccresearch.com

[117] Porcelli N, Judd S. Chemical cleaning of potable water membranes: A review. Separation and Purification Technology. 2010;**71**:137-143.

[118] Straub AP, Deshmukh A, Elimelech M. Pressure-retarded osmosis for power generation from salinity gradients: Is it viable? Energy & Environmental Science. 2016;**9**:31-48

Permissions

All chapters in this book were first published in DESALINATION, by InTech Open; hereby published with permission under the Creative Commons Attribution License or equivalent. Every chapter published in this book has been scrutinized by our experts. Their significance has been extensively debated. The topics covered herein carry significant findings which will fuel the growth of the discipline. They may even be implemented as practical applications or may be referred to as a beginning point for another development.

The contributors of this book come from diverse backgrounds, making this book a truly international effort. This book will bring forth new frontiers with its revolutionizing research information and detailed analysis of the nascent developments around the world.

We would like to thank all the contributing authors for lending their expertise to make the book truly unique. They have played a crucial role in the development of this book. Without their invaluable contributions this book wouldn't have been possible. They have made vital efforts to compile up to date information on the varied aspects of this subject to make this book a valuable addition to the collection of many professionals and students.

This book was conceptualized with the vision of imparting up-to-date information and advanced data in this field. To ensure the same, a matchless editorial board was set up. Every individual on the board went through rigorous rounds of assessment to prove their worth. After which they invested a large part of their time researching and compiling the most relevant data for our readers.

The editorial board has been involved in producing this book since its inception. They have spent rigorous hours researching and exploring the diverse topics which have resulted in the successful publishing of this book. They have passed on their knowledge of decades through this book. To expedite this challenging task, the publisher supported the team at every step. A small team of assistant editors was also appointed to further simplify the editing procedure and attain best results for the readers.

Apart from the editorial board, the designing team has also invested a significant amount of their time in understanding the subject and creating the most relevant covers. They scrutinized every image to scout for the most suitable representation of the subject and create an appropriate cover for the book.

The publishing team has been an ardent support to the editorial, designing and production team. Their endless efforts to recruit the best for this project, has resulted in the accomplishment of this book. They are a veteran in the field of academics and their pool of knowledge is as vast as their experience in printing. Their expertise and guidance has proved useful at every step. Their uncompromising quality standards have made this book an exceptional effort. Their encouragement from time to time has been an inspiration for everyone.

The publisher and the editorial board hope that this book will prove to be a valuable piece of knowledge for researchers, students, practitioners and scholars across the globe.

List of Contributors

Long Li and Tianlong Deng
Tianjin Key Laboratory of Marine Resources and Chemistry, College of Chemical Engineering and Materials Science, Tianjin University of Science and Technology, Tianjin, PR China

Yafei Guo
Tianjin Key Laboratory of Marine Resources and Chemistry, College of Chemical Engineering and Materials Science, Tianjin University of Science and Technology, Tianjin, PR China
College of Chemistry and Materials Science, Northwest University, Xi'an, PR China

Albert S. Kim
Civil and Environmental Engineering, University of Hawaii at Manoa, Honolulu, United States of America

Heyon-Ju Kim
Seawater Utilization Plant Research Center, Korea Research Institute of Ships and Ocean Engineering, Gangwon-do, Republic of Korea

Syed Javaid Zaidi
Center for Advanced Materials, Qatar University, Qatar

Farid Fadhillah
Chemical Engineering Department, Al Imam Mohammad Ibn Saud Islamic University (IMSIU), Riyadh, Kingdom of Saudi Arabia

Javier Miguel Ochando-Pulido and Antonio Martinez-Ferez
Department of Chemical Engineering, University of Granada, Granada, Spain

OrevaOghene Aliku
Department of Agronomy, University of Ibadan, Ibadan, Nigeria

André Pedral S. de Sena
Universidade Federal da Bahia/UFBa, Salvador, Brazil

Ben Xu
Department of Mechanical Engineering, The University of Texas Rio Grande Valley, Edinburg, TX, USA

Peiwen Li
Department of Aerospace and Mechanical Engineering, The University of Arizona, Tucson, AZ, USA

Penghua Guo
School of Energy and Power Engineering, Xi'an Jiaotong University, Xi'an, Shanxi, China

George Arampatzis, Avraam Kartalidis and Dionysis Assimacopoulos
Environmental and Energy Management Research Unit, School of Chemical Engineering, National Technical University of Athens, Athens, Greece

Armando A. Soares and João Silva
UTAD, University of Trás-os-Montes and Alto Douro, Vila Real, Portugal
CIENER-INEGI/University of Porto, Porto, Portugal

Eliseu Monteiro
CIENER-INEGI/University of Porto, Porto, Portugal
C3i - Interdisciplinary Center for Research and Innovation, Polytechnic Institute of Portalegre, Portalegre, Portugal

Abel Rouboa
UTAD, University of Trás-os-Montes and Alto Douro, Vila Real, Portugal
CIENER-INEGI/University of Porto, Porto, Portugal
Mechanical Engineering and Mechanics Department, University of Pennsylvania, Philadelphia, Pennsylvania, USA

Coskun Aydiner, Salim Oncel, Serif Cakmak, Tugba Nur Yilmaz, Emin Ender Celebi and Yasemin Melek Tilki
Department of Environmental Engineering, Engineering Faculty, Gebze Technical University, Kocaeli, Turkey

Derya Y. Koseoglu Imer
Department of Environmental Engineering, Engineering Faculty, Istanbul Technical University, Istanbul, Turkey
National Research Center on Membrane Technologies, Istanbul Technical University, Istanbul, Turkey

Esra Can Dogan and Ali Oguzhan Narci
Department of Environmental Engineering, Engineering Faculty, Kocaeli University, Kocaeli, Turkey

Pak Hunkyun
Electronics and Telecommunications Research Institute, Daejeon, Republic of Korea

Iqbal Ahmed
Center of Excellence in Desalination Technology, King Abdulaziz University, Jeddah, Saudi Arabia
Department of Mechanical Engineering, King Abdulaziz University, Jeddah, Saudi Arabia

Khaled S. Balkhair, Muhammad H. Albeiruttye and Amer Ahmed Jamil Shaiban
Center of Excellence in Desalination Technology, King Abdulaziz University, Jeddah, Saudi Arabia

Index

CPSIA information can be obtained
at www.ICGtesting.com
Printed in the USA
LVHW060854030122
707669LV00002B/148

9 781639 871551